D1004728

A Textbook of Entomology

A Textbook

3RD EDITION *Herbert H. Ross*

Assistant Chief, Illinois Natural History Survey

Professor of Entomology, University of Illinois

John Wiley & Sons, Inc.

New York · London · Sydney

# of Entomology

Library of Congress Catalog Card Number: 65-16424
Printed in the United States of America

With deep appreciation for his guidance, encouragement, and patience, this volume is gratefully inscribed to Professor George J. Spencer, who opened the door of entomology to me.

Preface

In entomology, as in other scientific disciplines, the upsurge of research and publications during recent years continues to increase our basic knowledge and perspectives about insects to a remarkable extent. As a result, there is a greater need now than when this book was first published for an introductory book that includes the fundamental aspects of entomology, organized so as to give students a general idea of the whole field.

In recent entomological investigations there has been renewed interest in relating insects to the living and nonliving components of the world in which they live. Only through a knowledge of these basic relationships can we achieve meaningful insights into the simplest questions about the "what, why, how, and when" of insects. This intimacy between insect and environment is emphasized especially in the chapters on physiology, geological history, ecology, and control.

It is increasingly evident that our information in such areas as comparative biochemistry, behavior, and anatomy will not be fully understood until the relevant facts are fitted into systems of evolutionary progressions. To provide the start of an evolutionary framework that might be useful for this purpose, preliminary family trees have been constructed for common families of the five large orders of insects, the Hemiptera, Hymenoptera, Coleoptera, Diptera, and Lepidoptera. These trees also group related families in a visual scheme that will be an aid in remembering them.

In the chapter on geological history an attempt has been made not to achieve completeness taxonomically, but rather to give a picture of the dynamic rise of insects in relation to the forces surrounding them. A word of caution is in order regarding the keys to orders and families. These are designed to accommodate only common members of common families, and hence they are far from complete. They are intended primarily to aid beginning students in realizing the differences used in delimiting orders and families and to give them practice in the manipulation of keys.

Again I wish to express my gratitude to many persons who were of assistance in planning and writing the initial edition. Of especial help were W. V. Balduf, B. D. Burks, G. C. Decker, D. M. DeLong, W. P. Hayes, Harlow B. Mills, C. O. Mohr, M. W. Sanderson, Kathryn

M. Sommerman, Roger C. Smith, and F. R. Steggerda, and the late T. H. Frison and H. J. VanCleave. I wish to express my thanks also to my wife, Jean, for her unstinting help in the preparation of all editions of this book.

I want to take this opportunity to express deep gratitude to the colleagues who have contributed suggestions, criticisms, and other material pertinent to revising this book. Their help has been of inestimable value. Limitations of time and space have made it impossible to follow all the good suggestions that were tendered, yet consideration of them frequently led to changed emphasis or a new approach to a problem.

I wish to thank many organizations and persons who loaned illustrations for this book. Individual acknowledgement is given under each illustration.

Herbert H. Ross

Illinois Natural History Survey,
Urbana, Illinois
December, 1964

Contents

Chapter One

Growth of North American Entomology

M<small>AN</small> has always had his troubles with insects. When he first emerged as man, he already had fleas and lice and was fed on by mosquitoes and pestered by flies. In those early days, when human populations were scattered and sparse, man's struggle was on a primitive plane—to find natural food from day to day and to escape the onslaughts of predatory animals. At this period it is doubtful that insects and insect-borne diseases were nearly so important deterrents to man as were other inimical factors of the environment. In fact, on the average, insects were probably of great help because termites, grasshoppers, grubs, and the like could be found and eaten when other foods were not obtainable.

From primeval conditions man's progress has been based essentially on changing various factors of his environment and making it better suited for his own survival and increase. But every change that benefited man also benefited a host of insects. Gradually, as the starker enemies of primeval life, such as the leopard and tiger, ceased to be a great threat to primitive man, insects became increasingly important as a challenge to his civilized status.

In the first place, increase in human populations allowed a great increase of such insect ectoparasites as lice and fleas. This was due to the ready accessibility of additional host individuals for the insects and, therefore, better opportunities for dissemination and chances for repro-

1

duction.   The same factors favored the increase and spread of pestilence, including insect-borne diseases.   When large cities arose, they were repeatedly swept by outbreaks of these maladies, in the same way that Imperial Rome was decimated by bubonic plague in the second century A.D.

Insects became a real factor with food as with health.   When man began to store food, it was attacked by a host of insects which before had been of no significance in the human environment.   In the tremendous food-storage organization of today, insects destroy thousands of tons of food annually in spite of widespread and expensive control programs.

When populations outstripped the food-producing capacity of natural surroundings, man domesticated animals.   The concentration of these allowed an increase of their ectoparasites and diseases, thus partially nullifying the effort to enlarge the food supply.   The cultivation of crops brought about the greatest change with regard to insects.   Agriculture congregated plant hosts so that their insect attackers could build up extensive populations on them.   The Egyptian writer in the time of Rameses II (1400 B.C.) commiserates with the peasant that "Worms have destroyed half of the wheat, and the hippopotami have eaten the rest; there are swarms of rats in the fields, and the grasshoppers alight there."   In the more recent period of crop improvement, new varieties of plants developed for increased yield have frequently been more attractive to certain insects than original wild hosts, with a resultant influx of destructive species to the cultivated crops.

This situation has been made more serious by man's development of transportation between all parts of the world.   Insects of many species have been carried to continents new to them, where they have found favorable climates, succulent acceptable cultivated hosts, and a freedom from the natural enemies which had kept their numbers in check in their original homes.   Sometimes the result has been disastrous, as, for example, the entry of the European corn borer and the Japanese beetle into North America.   These two species are of little economic importance in their native range, but in the United States they have caused losses to crops in the magnitude of millions of dollars per year.

North America has been especially hard-hit by losses due to insects. This is the result of the cultivation here in recent centuries of many crops not indigenous to the area, to introduction of many new pests, and to changes wrought by agriculture that have favored many endemic insect species.   Losses of crops, stored products, domestic products, and other commodities were estimated in 1951 at $10,200,000,000 for the United States alone.   In addition illness and deaths due to insect-borne diseases and secondary infections, sickness, and discomfort that result directly from insect bites were estimated as a cash loss of about $5,000,000,000. On the basis of these figures, insect damage in this country totals an estimated annual sum of $15,200,000,000.

A review of insect damage gives the entire group a sinister aspect. But the adage, "There is some good in everything," finds a real place even among this group of apparent despoilers. Many kinds of insects are definitely beneficial. The most conspicuous example is the honey bee, which not only produces a marketable crop of high cash value, but also pollinates many valuable plant species. Most of our fruits and legumes are dependent for pollination on a large number of insect groups such as bees, moths, flies, and beetles. Without these insects we would have no apples, pears, peas, beans, and seeds of other insect-pollinated plants.

Another group of economically beneficial insects embraces a large assemblage of predaceous and parasitic insects whose hosts are other insects. These include ichneumon flies, parasitic wasps, parasitic flies, and ladybird beetles. The adults or larvae of these species prey on or parasitize many important insect pests. In some cases they can be used as an efficient control method. The vedalia ladybird beetle, for example, is one of the chief means of combating the cottony cushion scale, an insect destructive to citrus orchards in California.

Man's efforts to combat destructive insects and encourage beneficial ones form a field of activity called applied entomology. It is comparable in many ways to the field of medicine that has arisen out of the challenge to combat sickness and disease. In North America, applied entomology involves a financial outlay of considerable proportions. Over five thousand people are employed primarily in the investigation of economic species and the development of control measures. Many firms make a specialty of manufacturing insecticides or apparatus for their application. In the United States alone $400,000,000 worth of insecticides are used annually. It must be remembered that the losses of $15,200,000,000 from insects were in addition to this control program, so that the total annual insect bill (exclusive of labor for applying insecticides) in the United States is in the neighborhood of $15,600,000,000.

It is difficult to visualize a loss in the magnitude of billions of dollars. Let us put it another way. In 1952 each person in the United States, on the average, paid out nearly $100 to insects. This means $400 for a family of four. Some of this sum was spent for insecticides and some for replacing damaged goods, but most of it was disguised as increased cost of commodities of plant or animal origin, such as lumber, clothing, and food.

Entomology, the study of insects, has developed into a very large division of the animal sciences owing to the proportions and importance of the applied field. For, although the primary objective of applied entomology is the reduction of insect damage, it has long been evident that a wide knowledge of fundamental information is necessary as a foundation for effective control. For this reason there has been an appreciation of basic entomological research in many directions. Some phases seemed of little importance when first started, yet later proved of inestimable value in control problems.

BEGINNINGS OF MODERN BIOLOGY

Because of their tremendous abundance, it might seem that insects would have been used a great deal in the early investigations in fundamental biology. The small size of the average insect, however, mitigated against this. Extremely delicate methods of dissection are required to study anatomy and physiology, and powerful microscopic equipment is necessary for taxonomic studies of almost any insect group. To a large extent, therefore, the fundamentals of biology were based on observations of larger animals. The early development of entomology was a process of transposing to insect studies the principles discovered in related fields. To gain a better appreciation of the growth of entomology in the New World, it is instructive to review the origin and evolution of its parent science, modern biology, which arose in the European theater of the Old World.

At the time of the discovery of America by the Spaniards the progress of world science was barricaded by the "age of authority." In the literature of the times were heated arguments over such matters as the number of teeth in a horse; learned authors were quoted, but apparently no one thought to examine a horse and actually count its teeth.

The subsequent sixteenth and seventeenth centuries, which saw the exploration and early colonization of the Americas, witnessed also the overthrow of authority and the return of observation and experiment in science. Both the exploration and scientific advance had their roots in the same fundamental causes, that in these centuries following the Renaissance men developed again the desire to look and think for themselves.

In the field of the biological sciences, Vesalius' work on human anatomy (1543) rejuvenated observation, and Harvey's proof of arterial and venous circulation of blood (1628) introduced experiment. Together, as Locy says, "they stand at the beginning of biological science after the Renaissance." Introduction of the microscope in the seventeenth century led to the microanatomical works of Malpighi and Swammerdam, and to discoveries of microorganisms with which Leeuwenhoek astonished the scientific world.

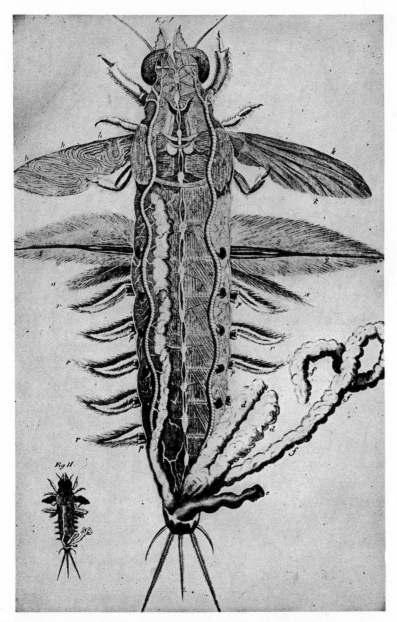

Fig. 1. Anatomy of a mayfly nymph, dissected and drawn by Swammerdam. One of the very early studies of insects, published about 1675. (From Essig, *College entomology,* by permission of The Macmillan Co.)

Fig. 2.   The famous museum of Olaus Worm, illustrated by its Swedish founder as it appeared in 1655.   (Reproduction loaned by Waldo Shumway)

During the seventeenth century entomology really started to develop. In fact, 1667 and 1668 may be considered almost its birth date, for in 1667 Redi used insects in demonstrations to test the theory of spontaneous creation.   He exposed meat in jars, some covered by parchment, others by fine wire screen, and some not covered.   The meat spoiled and attracted flies, which laid eggs in the exposed meat, resulting in a crop of maggots.   Of the two covered jars, no eggs were laid on that covered by parchment, but the flies were attracted to the screen-covered one and laid eggs on the screen, since they could not reach the meat itself.   Redi observed in this instance that, when the eggs hatched, maggots appeared on the screen instead of on the meat.   He concluded, therefore, that maggots in meat resulted from the eggs of insects, and not from spontaneous generation, as was previously supposed.

In 1668 Malpighi published anatomical studies of the silkworm, and Swammerdam published his first insect studies.   These men produced the first accurate studies of insect anatomy, preparing skilled illustrations showing details of minute structures and organs, fig. 1.   These model works were the inspiration for later work in insect anatomy.

Another important phase of the biological sciences paralleled these advances. As people began to observe nature, interest awakened in natural history, and books on the subject appeared. Early treatises by Wotton (1552) and Gesner (1551–1556) were elemental and general, and were characterized by a lack of discrimination between different kinds of related animals. Imaginary animals of folklore were even given consideration as if they actually existed. Later works by Ray at the end of the seventeenth century were on a sounder basis and introduced a clear species concept of living organisms essentially similar to that understood today. Natural history museums, fig. 2, came into being, stimulated by the many bizarre and unfamilar objects brought back to Europe by travelers and mariners. Many of these museums were operated as hobbies by wealthy persons and were the forerunners of the extensive private collections which later played an important role in the development of taxonomy.

PROGRESS IN THE EIGHTEENTH CENTURY

Historians picture in the eighteenth century the climax of the revolt against authority, in which despotism and ecclesiasticism were to some extent replaced by the individual asserting his right to be an end in himself. Historical expression of this tide of feeling is found in the American Revolution and the French Revolution. In the United States further expression is found in the rise of universality and state control of education.

In this same century, undoubtedly manifesting the same individualistic trend, the biological sciences in Europe progressed to a new peak. Entomology especially attracted a large number of talented workers. Lyonet, a Hollander, contributed anatomical work of the finest detail, his first and best publication describing the anatomy of the larva of the willow moth (1750). More important in this period from the standpoint of arousing widespread public interest were the voluminous works of the German, Roesel; the Frenchman, Réaumur; and the Swede, DeGeer. All three of these authors published detailed well-illustrated observations on many insects, their life histories, habits, and characteristics.

About the middle of the eighteenth century occurred a movement of extreme importance to the entire field of natural science. It has already been mentioned that John Ray introduced the first clear concept of species. But the names for these species were phrases or descriptions (in Latin), often several lines long, cumbersome, and inconsistently used. In most cases the first name was a noun and corresponded to our present-day usage of a generic name; the remainder of the phrase was adjectival and modified this "generic" name. In John Ray's time, for example, one of the butterflies had the name *Papilio media, alis pronis, praefertim interioribus, maculis oblongis argenteis perbelle depictis,* and another had the name *Papilio parva nigra duplici in alis exterioribus macula alba insignis.* The first word in each name was the noun *Papilio,* the remainder was an adjectival description of the species. Students in field classes, attempting to keep up with recording lists of plants and animals being found, undoubtedly were the first to shorten these names. Gradually they were shortened to such names as *Papilio media* and *Papilio parva,* each name composed of the original noun and only one adjective originally referring to some distinctive feature of the species. This is known as binomial nomenclature.

In this period the naturalist Linnaeus was coming into prominence as a systematist, organizing the known plants and animals into one of the first comprehensive classifications. In 1758 the tenth edition of his work "Systema Naturae" was published. In this, for the first time, the binomial method of names was employed uniformly throughout a large and comprehensive book. The method proved so successful that workers in all fields adopted it almost immediately. In fact, so profound was Linnaeus' influence on later workers that the tenth edition of his "Systema Naturae" has been designated as the official beginning point for zoological nomenclature. Although Latin is no longer the standard language of science, as it was until the end of the seventeenth century, the Latin names have been preserved and are used for scientific names throughout the world.

The stabilization of binomial nomenclature was of tremendous scientific advantage in two ways. First, it gave an easily designated and unambiguous "handle" to species, so that workers in different fields and different countries were better able to identify the species with which others were working. This was of prime importance for integrating advances in comparative anatomy, physiology, and other fields of biology. Second, it provided a system of names which could be expanded indefinitely in a simple manner to accommodate additional genera and species. How necessary to future progress such a simple method was, may be seen at once by this tabulation: For the entire world Linnaeus recognized about 4500 species of animals, including 2000 species of insects; today over 1,250,000 species of animals are recognized, and of these roughly 900,000 are insects.

Thus taxonomists after Linnaeus were presented with an open invitation to describe and name the myriad species occurring in all parts of the world. Much of the early work was superficial and has been criticized by many, but it furnished the basis for analyses that led to the formation of the theory of evolution and to the organization of such fields as ecology and limnology.

The field of insect taxonomy in particular had been handicapped under the old system. After Linnaeus' work it began to emerge as a specialized subject. The first outstanding insect taxonomist was Fabricius, a Danish student of Linnaeus. Fabricius' first work, "Systema Entomologica," appeared in 1775; others followed from 1782 to 1804. Fabricius treated the entire insect fauna of the world. By the end of his career it became apparent that this was too large a unit for intensive study by one person. As a result, many workers of the early nineteenth century following Fabricius studied either only one of the larger insect groups or the fauna of only one country.

Fig. 3. Carolus Linnaeus (1707–1778) at the age of forty. (After Shull)

The works of Réaumur, Linnaeus, DeGeer, and Fabricius stimulated a tremendous development of taxonomic study of insects among European entomologists. They also served as the most important basis for the beginning of entomology in North America.

DEVELOPMENT OF NORTH AMERICAN ENTOMOLOGY

American entomology came into existence about the beginning of the nineteenth century. For the first two thirds of the century development was slow, witnessing the appearance of scattered pioneer works such as form the backbone of further progress in any scientific movement. But after the Civil War many factors contributed to a hastening of the growth of entomology in the United States. The resultant demand for entomological investigation found eager and able enthusiasts available, with the result that by the end of the century, American entomology had blossomed into a well-balanced science of wide practical and theoretical scope.

Pre-Nineteenth Century Work

Prior to the nineteenth century only small fragments were known about North American insects. Naturalist Mark Catesby (1679–1749) was possibly the first to illustrate North American insects in his book, "A Natural History of Carolina, Florida and the Bahama Islands, containing figures of the Birds, Beasts, Fishes, Insects and Plants." Fabricius named some species, relying on specimens sent to him by various collectors or specimens which had been acquired by private collections in Europe. John Abbott, an Englishman who settled in Georgia, collected much material for European collectors in the period about 1780 and prepared many drawings of insects. The economic losses occasioned by insects were noticed with grave concern by Thomas Jefferson in 1782. He was particularly aware of the damage caused in stored grain and gave a few remarks on the problems of control, pointing out a need for further study. But until a few years before 1800 no concerted effort was evident by residents of the United States to investigate the native insect fauna.

Pioneering Period, Roughly 1800–1866

Work on American insects by American workers began about the turn of the nineteenth century. One of the first workers was W. D. Peck, who published many articles on the injurious insects of the New England states. These articles appeared from 1795 to 1819 in various agricultural journals. The pioneer work on North American entomology was "A Catalogue of Insects of Pennsylvania," published in 1806 by F. V. Melsheimer. The chief value of this little 60-page book was its stimulating effect. Its author, his collection, and his association with later workers were a real aid in opening up the subject. His insect collection, incidentally, was the first comprehensive one to be built up in North America and was ultimately purchased many years later by the Harvard Museum of Comparative Zoology.

In 1812 a group of enthusiastic naturalists organized the Academy of Natural Sciences of Philadelphia. This nucleus of scientists was the cradle of serious descriptive work in many fields of American biology, and among them was entomology. Thomas Say was the outstanding entomologist of the group. He published the first useful classic in the field, three well-illustrated volumes (1817–1828), "American Entomology, or descriptions of the insects of North America." The excellence of this work, together with his other papers on insects, has earned for Say the well-deserved title "Father of American Entomology." Say died at New Harmony, Indiana, in 1834.

In 1823 Dr. T. W. Harris of Massachusetts published the first of a series of papers on the life history and economic importance of many insects. Harris was a student of Peck, who taught natural history at Harvard, from which Harris graduated in 1815. Harris collected and observed insects constantly, and the breadth of his published work increased. It culminated in 1841 in his monumental "Report on Insects Injurious to Vegetation"; this was twice reprinted and revised, the last time in 1852. Harris received $175 from the State of Massachusetts for this work; this was the first tax-supported entomological program in North America. It was also the first real textbook of economic entomology, and Harris is justly regarded as the founder of applied entomology in America.

Fig. 4. Thomas Say (1787–1834), the father of American entomology. (After Howard, courtesy of U.S.D.A., E.R.B.)

The influence of Say and Harris took root immediately. In a few years a dozen authors published papers on the life history, habits, predations, and control measures of insects.

It is interesting to look back at the remedies in vogue for insect control during that period. It must be remembered that the arsenicals, pyrethrum, DDT, and many other effective insecticides were undiscovered at that time. A few of the standard remedies, to quote from Harris, included "Hand picking; sweeping into pans; spray with whitewash and glue; sulphur and Scotch snuff; fumigation with tobacco under a movable tent; syringe with whale-oil soap solution; soap and tobacco water," and many recommendations for cultural control. We can see in these the fore-runners of many control measures recommended today.

Interest in agriculture led to the establishment in 1853 of a new Bureau of Agriculture in the Federal Government. This Bureau appointed Townend Glover as Entomologist and Special Agent. Glover's duties were varied, including the preparation of exhibits of agricultural seeds, plants, and fruits, as well as insects. But in addition he did considerable investigational work, especially on the insects attacking orange trees and cotton. Glover had the belief that a picture of an insect is of much greater value than the prepared insect specimen. His greatest entomological efforts were consequently devoted to making copper etchings illustrating the insects of North America.

Fig. 5. Thaddeus William Harris (1795–1856), the founder of applied entomology in America. (After Howard, courtesy U.S.D.A., E.R.B.)

The farmers' losses caused by insect damage were attracting more and more attention. In response to this, the State of New York in 1854 appropriated $1000 for investigations on insects, especially those injurious to vegetation. Dr. Asa Fitch was chosen for this work, which he continued from 1854 to 1872. He wrote 14 fine reports, the result of a great amount of original observation performed with great care, which made available information on the life history of many insects. After Harris' work, these reports were the next great stimulus for further development of entomological work in the United States. Although such was not his title, Fitch was usually called State Entomologist; in his activities he was this in a very real sense and was the first one in the United States.

While Fitch was at the height of his career, several other entomologists were coming into prominence, including B. D. Walsh and C. V. Riley in Illinois, and E. T. Cresson and A. R. Grote in Philadelphia. Walsh's principal non-economic work was done from 1860 to 1864, but his great contribution to economic entomology belongs to the account of the last third of the century. Riley, Cresson, and Grote also made their great contributions in this later period.

The Civil War, which concluded the first two thirds of the century, seems to have had only slight effect on entomological work, most of which was being done north of the actual battle area. Its effects, however, had far-reaching entomological consequences in the years immediately following.

Science in Europe for this Period. While the foundation works of American entomology were being written by Say, Harris, Fitch, and Walsh, two very important series of events were taking place in Europe. These had only slight contemporaneous effect in America but were a great contributing factor to entomological development in the next period.

Fig. 6. Asa Fitch (1809–1878), the first of the state entomologists. (After Howard, courtesy U.S.D.A., E.R.B.)

In the first place, European taxonomists were making needed strides in redefining taxonomic concepts, especially families and genera, to accommodate the huge tide of insect species being discovered in the world. Every large order received some attention, and creditably workable systems of classification were set up for them. European workers had access to libraries and collections far superior to any in America. They prepared keys and illustrated works, many of which were simply transposed by later American taxonomists to fit American species of insects.

The second circumstance is one which concerned all biological science. In this period (1800–1866) there developed ideas which revolutionized the outlook in the entire field. Up to this point work had been almost entirely descriptive, with scarcely any concept of fundamental laws. Now these came to light in rapid succession. Owen brought forward the idea of analogy and homology of parts; Cuvier and Lamarck founded comparative anatomy; Milne Edwards propounded the idea of division of physiological labor; Müller demonstrated the interrelationship of anatomy and physiology; Schwann and Schleiden demonstrated the cell theory; Bichat founded histology; Von Baer founded modern embryology; and Schultze defined protoplasm. To climax this galaxy of ideas, Darwin and Wallace established the practicality of the theory of organic evolution.

That these discoveries were made in such rapid succession is not strange. Scientists had been on the verge of seeing them for many years, and as soon as one fundamental was discovered it served as a key to unlock the next half-anticipated secret. Together with genetics and bacteriology (both discovered later) these discoveries gave a preliminary outline for the entire range of known biological laws. These, of course, were as fundamental to basic progress in the study of insects as in the study of any other group of living things.

First Entomology in American Colleges.   In the early part of the pioneering period courses in natural history began to appear in various colleges in North America. Until the middle of the nineteenth century they were meager, mostly theoretical and classificatory. They were given chiefly by lecture, sometimes with demonstrations but with little or no field or laboratory work. This applied in large measure to chemistry and physics also. Louis Agassiz at Harvard was the first teacher in zoology to break away from this and introduce laboratory methods in teaching. The greatest impulse to the laboratory method of teaching, however, was the great upsurge of inquiry following the publication in 1859 of Darwin's *Origin of Species*. At about this same period such new institutions as Cornell and Johns Hopkins Universities emphasized the teaching of science. In the United States this was coincidental with the establishment of the land-grant colleges in 1862 by the Morrill Act of Congress. This promoted education in agriculture, mechanical arts, and natural sciences. Entomology was included only as a part of biology courses, but the foundations were being laid for the later development of its teaching.

AMERICAN EXPANSION PERIOD, ROUGHLY 1867–1900

In the two or three decades after the Civil War, American entomology expanded at a prodigious rate. Many reasons supplemented each other to this end. Important were the following:

1. As a result of the westward migration of thousands of people following the Civil War, agriculture expanded in the Middle West and the states on the Pacific Coast. Devastating insect outbreaks occurred periodically. The farmers' demand for entomological assistance resulted in the rapid development of both state and federal organizations in economic entomology.

2. American insect collections and libraries had gradually improved and, with the help obtainable from European literature, opened the door for more extensive and better descriptive work. The fundamentals of biology recently discovered provided avenues for many lines of investigation with insects.

3. Demand for trained entomologists brought about teaching of entomology in colleges and universities.

Economic Entomology

Before the beginning of this period (1867–1900) New York was the only state actively sponsoring entomology, through Asa Fitch. In 1866 Illinois appointed a State Entomologist (although he did not become active until 1867), and in 1868 Missouri followed suit. In Illinois the appointment was given to B. D. Walsh, who had written many fine articles on taxonomic and economic entomology. Walsh died by accident in 1869 but wrote three reports as State Entomologist before that tragedy. After him William LeBaron occupied the post for five years, followed by Cyrus Thomas from 1875 to 1882, and then by S. A. Forbes. In Missouri the appointment was given to C. V. Riley, who held the post from 1868 to 1876. Riley's annual reports were outstanding, in both scientific content and illustration, and were a tremendous stimulus to other authors.

From 1874 to 1876 the migratory locust invaded a number of the important grain-growing states. This outbreak was studied by Riley, who

Fig. 7. Benjamin Dann Walsh (1808–1869), an early vigorous writer on various phases of entomology, later first State Entomologist of Illinois. (After Forbes)

Fig. 8. C. V. Riley (1843–1895), insect illustrator *par excellence,* who first built up the Federal Bureau of Entomology. (After Howard, courtesy U.S.D.A., E.R.B.)

saw in it the need for action on a national scale against injurious insects. His efforts to secure national legislation persuaded Congress to establish the United States Entomological Commission. This was the first recognition in a broad way that economic entomology was of national importance and dealt with many problems whose thorough investigation transcended state lines. This Commission had Riley as chief, and A. S. Packard, Jr., and Cyrus Thomas as the two other members. The Commission was active officially for only three years but did some excellent work and published several extremely useful reports and bulletins. These treated not only the migratory locust but also a wide variety of other economic insects.

For a short period in 1878 Riley succeeded Glover as Entomologist for the Federal Department of Agriculture. J. H. Comstock then held the office for two years, after which Riley again held it, for 15 years. On Riley's return, the entomological work received such support that it was reorganized as a separate Division of Entomology. Under Riley's leadership it rapidly developed into a large and useful organization, with field stations in many parts of the country.

During these years several able entomologists in Canada began writing notable contributions on the insects of that country. The good work of two pioneers stands out conspicuously, that of the Rev. C. J. S. Bethune, who began publishing in 1867, and Dr. William Saunders, whose earliest paper appeared the following year. The efforts of these and other enthusiasts culminated in the founding of the Ontario Entomological Society in 1870 and the establishment of the office of Honorary Entomologist by the Department of Agriculture of Canada in 1884. This post was given to James Fletcher, who in 1887 was transferred to the staff of the Central Agricultural Experiment Station as Entomologist and Botanist. Fletcher had little help and a tremendous territory to cover; that he served the entomological needs of Canadian agriculture so well and so long is proof of his ability and industry. He died in 1908, following an operation.

Until 1887 organized work in economic entomology in the United States was being done only by the federal government and by New York, Illinois, and Missouri. Work was also being done by individuals in many other states, from Maine to California. But in 1887 the demands of an ever-expanding agriculture resulted in the Hatch Act, establishing agricultural experiment stations in all states. Investigations of injurious insects were stressed from the beginning, and there arose a need for trained entomologists which far exceeded the meager supply.

By the last decade of the century economic entomology was an influential and producing concern. Outstanding work was being done by many workers in the Federal Division of Entomology, notably L. O. Howard, who in 1894 succeeded Riley as chief. The state organizations (many connected with the agricultural experiment stations) included several brilliant men in their roster. To mention only a few, S. A. Forbes in Illinois, John B. Smith in New Jersey, and E. P. Felt in New York contributed immense amounts of original research and wrote monumental reports, many of lasting value.

Two items of interest had a special effect on entomological thought and procedure in this last third of the century: (1) About 1869 Paris green was discovered to be an effective insecticide, and its success opened up the entire field of stomach poisons for insects. (2) The cottony cushion scale, introduced about 1870 into California, had become such an abundant and serious pest of citrus trees in the '80's that it threatened the extinction of the citrus growing industry in the West. Known insecticides failed to deter the pest. Finally natural insect enemies of the scale were imported from Australia. One of these, the vedalia ladybird beetle, destroyed the scale with such persistence that in a few years it ceased to be a problem. Such a wonder-working event established the importance of biological control as a possible means of combating injurious insects.

Insect Taxonomy, Morphology, etc.

During this latter third of the nineteenth century an almost complete foundation was laid for the classification of North American insects. A large number of workers contributed to this, including J. L. LeConte and G. H. Horn on Coleoptera; A. S. Packard, Henry Edwards, and A. R. Grote on Lepidoptera; E. T. Cresson, Edward Norton, and L. O. Howard on Hymenoptera; S. W. Williston, Osten Sacken, and D. W. Coquillett on Diptera; S. H. Scudder on Orthoptera; P. R. Uhler and O. Heidemann on Hemiptera; J. H. Comstock on Coccidae or scale insects; Herbert Osborn on ectoparasites and Homoptera. In Canada the Abbé L. Provancher was outstanding, especially for his work on Hymenoptera.

These are only a few of the "old masters" who described the first great bulk of the North American insect fauna and gave us our first working synopses. Many outstanding European entomologists also contributed to this literature. It is noteworthy that many of the most outstanding taxonomists of this era and the one that followed were amateur entomologists and made their great contributions as a hobby, without remuneration. To mention a few: LeConte and Horn were practicing physicians, Edwards was an actor, Norton and Cresson were businessmen, Williston a geologist, Provancher a clergyman.

During the decades following Linnaeus, different points of view arose regarding many phases of the application of scientific names. As these differences became acute inconsistent usage threatened to nullify the

Fig. 9. John Henry Comstock (1849–1931), one of the "old masters" in the teaching of entomology in America. (After Howard, courtesy U.S.D.A., E.R.B.)

benefits of the binomial system, and taxonomists of all groups sought measures to bring about uniformity of practice. Success finally crowned their efforts at the International Zoological Congress held at Berlin in 1901, with the adoption by the zoological world of the International Rules of Zoological Nomenclature and the organization of the International Commission of Zoological Nomenclature. It was at this historic meeting that Linnaeus' tenth edition of "Systema Naturae" was designated as the beginning point for zoological scientific names.

Complementary to the development of better literature was the origin and growth of large research collections. Until the end of the first two thirds of the nineteenth century, North American insect collections were relatively small, usually consisting of at most a few thousand specimens. As the complexity of insect identification became obvious, the need became apparent for extensive collections to aid both accurate identification of unknowns and further progress in taxonomy. It was early recognized that accurate identification was essential for sound fundamental research in all fields and for consistent control recommendations. In the United States the first serious effort in this direction was made by Louis Agassiz, who in 1867 appointed Hermann A. Hagen to build up a collection of insects in the Museum of Comparative Zoology at Harvard University. Since then many institutions, including various academies of sciences, universities, and other state and Federal organizations, have amassed extensive insect collections. Many workers maintain personal collections of considerable size. At the present time institution collections in the United States and Canada house a combined total of about 45 million insect specimens, and personal collections probably another three or four million.

Teaching of Entomology

Until about 1867 entomology was taught in American colleges only as a portion of courses in biology or natural history. But in 1866 B. F. Mudge gave a course entitled "Insects Injurious to Vegetation" at Kansas State Agricultural College; in 1867 A. J. Cook gave a course in entomology at the Michigan Agricultural College; in 1870 Hagen gave rather informal courses in entomology at Harvard; in 1872 C. H. Fernald began teaching at Maine State College; in 1873 Comstock began teaching at Cornell University; and in 1879 Herbert Osborn taught at Iowa State College of Agriculture. These men were the real founders of the teaching of entomology in the United States. They had little organized or general literature to use as a basis for teaching and, to quote from Osborn, "were feeling their way in the matter of both content and method for entomological instruction."

The task confronting these men was enormous—learning or discovering the multitude of details about insects, including life histories, morphology, development, and classification, and combining it with the then new concepts of general physiology, embryology, phylogeny, and evolution. The splendid and coherent courses which developed from the welding of all this material represent a triumph indeed for these pioneer teachers. Of especial importance in this connection were the early textbooks written by A. S. Packard, who was a trail blazer in this field.

Fig. 10. A. S. Packard (1839–1905), who wrote some of the early entomology textbooks. (After Howard, courtesy U.S.D.A., E.R.B.)

Creation of the agricultural experiment stations in 1888 led to a tremendous demand for better-trained entomologists for economic positions and stimulated teaching in this field. By the end of the century entomology courses had been organized in most of the leading universities and colleges stressing natural sciences. This was especially true of the land grant colleges. Outstanding men, such as S. A. Forbes at Illinois, G. A. Dean at Kansas, and M. V. Slingerland at Cornell, set an early example of combining the fundamental and practical aspect in what may be called the first modern courses in economic entomology.

TWENTIETH CENTURY DEVELOPMENTS

In the present century the investigation of all known phases of entomology progressed and expanded at a remarkable rate. To a large degree this paralleled public appreciation of the tremendous damage caused by insects and the savings to be gained by control of them. This appreciation was expressed in the form of larger and larger appropriations to support entomological work. Of tremendous aid to entomological studies were improvement of microscopes, laboratory and field equipment, electric measuring and recording devices of many kinds, gradual improvement of travel facilities, and the increase in the number of journals and books for the publication of research results.

Certain phases or definite events attracted widespread popular attention at various times. Each of these was a stimulus to further expansion in all fields of entomology. Some of the outstanding items are cited in the following brief remarks.

Medical Entomology

In 1879 Patrick Manson in southern China discovered that mosquitoes transmit the agent of filariasis. About 1889 Curtice and Kilborne in Texas discovered that a tick transmits the organism that causes Texas fever of cattle. In 1898 Ronald Ross in India proved the association of malaria and anopheline mosquitoes. In 1900 Walter Reed and co-workers proved that the mosquito *Aedes aegypti* carries yellow fever. This series of discoveries solved the transmission mystery of some of the world's worst diseases and established the importance of the role insects and other arthropods play in relation to human health. This was the birth of medical entomology. Continued investigation has shown an ever-increasing number of diseases to be primarily insect- or arachnoid-borne, adding bubonic plague, dengue, typhus fever, trench fever, Rocky Mountain spotted fever, African sleeping sickness, and others to the list. For lack of known immunization methods, the medical world has in many cases turned to a control of the arthropod carrier as a means of combating the disease.

Economic Entomology

During the last decade of the nineteenth century and up to the present, many insects of foreign origin became established in the United States and Canada and produced catastrophic damage to agriculture. The gypsy moth threatened to wipe out fruit and other trees in the New England States from 1889 to well into the 1900's; the San Jose scale became a country-wide fruit tree scourge before 1900; the destructive cotton boll weevil had invaded the entire cotton belt between about 1895 and 1920; the European corn borer loomed as a possible serious threat to the Midwestern corn crop in the early 1930's; and so on.

Each of these "battles" between the entomologist and a new insect enemy brought forth discoveries of new insecticides, equipment, or methods which frequently had a wide application far beyond treatment of the insect under intensive study. But these discoveries, of course, required more men and more field stations and necessitated coordinated research in insect habits, morphology, physiology, taxonomy, agriculture, bacteriology, and other sciences.

Basic Entomology

The research advances in fundamental fields followed very closely the demands of the economic entomologist for more and better information about economic insects, suspected species, or related forms. With such a stimulus the number of research workers increased steadily, and the growth and scope of college teaching increased with it.

Research in insect taxonomy followed three principal lines: (1) making keys and synopses to groups not previously treated, (2) restudying groups for which new sets of diagnostic characters have been discovered, and (3) applying to problems of classification more recent biologic or ecologic information and evidence furnished by immature stages. Widespread road systems and the automobile brought to the entomologist tremendously greater collecting opportunities and resulted in a rapid increase in knowledge of species distribution.

In this century progress in insect morphology, physiology, embryology, and other fundamental fields followed a fairly definite pattern. This pattern was to work out for insects the details of the great biological fundamentals discovered in the nineteenth century. The disconnected anatomical works of Malpighi and Lyonet were re-investigated and extended along the lines of comparative anatomy, including histology, morphology, and embryology. Certain insects were found to be ideal subjects for experimentation in genetics and heredity, and in this way much was learned about both the insects and genetics in general. Research in insect physiology followed much the same pattern, that is, finding out how the gross functions were carried on in insects in comparison with other animals. In more recent years, researches with insects have played a role of increasing importance in contributing information and ideas of general biological interest. This is especially true in biochemistry, physiology, behavior, ecology, biogeography, and evolution.

Changing Entomological Economics

In 1945 and 1946 the commercial introduction of the new synthetic insecticide DDT heralded a revolution in economic entomology. Up to that time effective control of insect pests by insecticides had been a relatively expensive undertaking, because the necessary dosages of insecticides were high and the control often only moderately satisfactory. As a result insecticides were thought of as something to use either on a crop with a high value per acre, such as fruit or truck crops, or in an emergency, such as a grasshopper outbreak.

DDT and later discovered synthetic insecticides such as benzene hexachloride, dieldrin, organic phosphates, and others proved to be so effective in such small dosages that the cost of actual control was reduced drastically. This opened vast new possibilities for the use of insecticides on field crops, range animals, forests, and grasslands—in short, on crops with a relatively low per acre or per year value. Difficulties being encountered in plant toxicity of formulations, insecticide-resistant strains of insects, and backlashes of various sorts from drastic ecological changes in treated areas, indicate that we are still in the exploratory stages of the new era. There have been many instances of over-optimism which have emphasized the continuing need for nonchemical control wherever possible. There is no doubt, however, that the synthetic insecticides have altered the entire perspective on economic problems.

Agricultural Entomology. The usefulness of these entomological advances has been accentuated by monetary changes in the farm economy. Especially in irrigated areas and areas of relatively abundant rainfall, the per acre value of crops has risen steadily, as has also the per acre investment which the farmer actually underwrites. This investment includes increased application of fertilizer, increased use of better and more expensive seed, and more and better mechanized farm equipment. As a result the farmer has a greater monetary stake in each acre and is able and willing to spend more for insect control in order to protect his investment. In contrast to their indifference of only a decade ago, farmers as a whole are now consciously embracing scientific methods including the most up-to-date information that the entomologist can give them.

Forest Entomology. Depletion of forests combined with the great need for forest products, particularly building materials and paper, have focused attention on the importance of forests as a vital part of the national economy. In this setting forest entomology has "come of age." The enormous depredations caused by insects are now well realized, with the result that this field is receiving increased support from government agencies and private industry.

REFERENCES

Decker, G. C., 1964. The past is prologue. *Ent. Soc. Amer. Bull.* 10(1):8–15.

Essig, E. O., 1931. *A history of entomology.* New York: The Macmillan Co. Illus.

Gill, Theodor, 1908. Systematic zoology: its progress and purpose. *Smithsonian Inst. Rept. for 1907*: 449–472, illus.

Howard, L. O., 1930. A history of applied entomology (somewhat anecdotal). *Smithsonian Misc. Collections,* **84:**564 pp., illus.

Locy, W. A., 1910. *Biology and its makers.* New York: Henry Holt and Co. Illus.

Osborn, Herbert, 1937. *Fragments of entomological history.* Columbus, Ohio, by the author. Illus.

Spencer, G. J., 1964. A century of entomology in Canada. *Can. Ent.,* **96:**33–59.

Weiss, H. B., 1936. *The pioneer century of American entomology.* New Brunswick, N. J., by the author. Illus.

Chapter Two

Arthropoda: Insects and Their Allies

INSECTS belong in the great phylum of jointed-legged animals, the Arthropoda. Of this phylum the insects are a highly specialized group comprising the class Insecta. In the adult stage insects are characterized primarily by having the body divided into three regions, the head, thorax, and abdomen, and by the thorax bearing three pairs of legs. Both the body regions and number of legs are functional groupings of parts, groupings that are very different from those of their original ancestors.

The Arthropoda undoubtedly arose from a wormlike creature very similar in general organization to the Annelida or segmented worms. The body of this ancestor, fig. 11A, consisted of a series of uniform segments, each a full ring of the body. The head was a simple structure, probably bearing sensory bristles. The mouth was situated on the ventral side between the head and the first ring or segment of the body. Because of its position in front of the mouth or stomodeal opening, the head region in this early stage is termed the prostomium. Hypothetical steps in the evolutionary progress beginning with this simple stage and leading through generalized arthropods to insects are pictured in figs. 11A to 11E.

The first great step was the development of a pair of ventral appendages or legs on each body segment, aiding in locomotion, fig. 11B. Apparently the last segment, the periproct, bearing the anus, never had appendages. Paralleling this, an improvement in the sense organs of the head occurred; eyes and antennae were the ultimate result of this. The phylum Oncopoda is thought to represent this stage in evolutionary development, especially living members of the class Onychophora (*Peripatus* and its allies, fig. 12). Two other classes, the Pentastomida and the Tardigrada, may also be members of the Oncopoda, but they are so

25

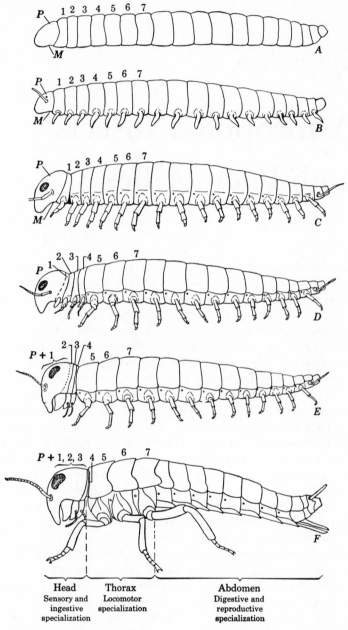

Head
Sensory and
ingestive
specialization

Thorax
Locomotor
specialization

Abdomen
Digestive and
reproductive
specialization

Fig. 11. Diagram showing hypothetical stages (*A* to *F*) in the development of body regions and appendages from a wormlike ancestor to an insect. *M*, mouth; *P*, prostomium. (Modified from Snodgrass)

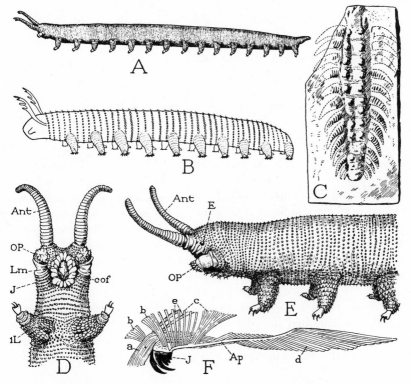

Fig. 12. Onychophora, ancient and modern. *A, Peripatoides novae-zealandiae; B*, restoration of the Middle Cambrian fossil *Aysheaia pedunculata; C*, a supposed onychophoran *Xenusion auerswaldae* from Pre-Cambrian or Early Cambrian quartzite; *D, E*, ventral and lateral views of anterior portion of *Peripatoides novae-zealandiae; F*, right jaw of same, showing muscles. *a–d*, jaw muscles; *Ant*, antenna; *Ap*, apodeme; *cof*, circumoral fold; *E*, eye; *J*, jaw; *L*, leg; *Lm*, labrum; *OP*, oral papilla. (After Snodgrass)

reduced in structure that their true relationships are open to question.

The Tardigrada or water bears, fig. 13, are minute animals that live in wet moss and in both fresh and salt water. The body, never more than a millimeter long, has four pairs of lobiform legs terminating in claws; the head has neither apparent mouthparts nor other appendages. The Pentastomid or linguatulids, fig. 14, are a small group in which the adult is wormlike and the earliest stages are minute four-legged creatures in general appearance resembling mites. The linguatulids are internal parasites of a variety of vertebrates.

The next step in the evolution of the arthropods was the development of joints or articulations in the legs, which resulted in great improvement

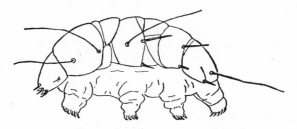

Fig. 13.   A tardigrade.   (After U.S.D.A., E.R.B.)

in locomotion, fig. 11 *C*.   At some point in their evolution, at least the first pair of legs was used in pushing food into the mouth opening, a function soon followed by the fusion of the first body segment with the prostomium. Judged from the condition of these two parts in the fossil group Trilobita, the fusion occurred at an early evolutionary stage.   Evidence from the same source indicates that eyes and antennae were well developed at this stage.   There are no living arthropods representing such a form as this, but the fossil group Trilobita had essentially this sort of body organization, fig. 16.

At some point near this stage, it appears that the evolving forms of arthropods separated into different paths.   One path led to the spider group, fig. 18, and the other led to the mandibulate arthropods which include the insects, centipedes, and crustaceans.

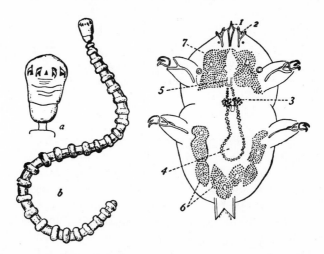

Fig. 14.   A linguatulid.   Left, *Porocephalus annulatus; a,* ventral view of head; *b,* ventral view of entire animal.   Right, larva of *Porocephalus proboscideus,* ventral view; 1, boring anterior end; 2, first pair of sclerotized processes seen between the forks of the second pair; 3, ventral nerve ganglion; 4, alimentary canal; 5, mouth; 6 and 7, gland cells.   (After Stiles and Shipley)

In the branch leading to the insects the next development was the utilization of appendages of segments 2, 3, and 4 as accessory feeding organs, fig. 11*D*. Not only did these appendages push food into the mouth, but also they acquired grinding surfaces to chew and shred the food preparatory to ingestion. Apparently the appendages of the first body segment never developed into strong mouthparts and actually atrophied in many groups. Appendages of the second body segment ultimately became the mandibles, those of the third became the maxillae, and those of the fourth became the second maxillae or labium. The three segments bearing the mouthparts are termed the *gnathal segments.*

The gnathal segments became consolidated with the prostomium, fig. 11*E*, resulting in a head structure of compound origin much like that in insects and their allies. This compound structure brought together in one functional unit all the organs intimately connected with feeding. In the primeval mandibulate ancestor, the rest of the body appendages formed a functional unit for locomotion. The classes Pauropoda and Chilopoda (centipedes) are present-day forms showing this type of organization.

Unlike the centipedes, however, the primeval mandibulate arthropod was undoubtedly aquatic. It gave rise to two lineages existing today. One lineage evolved into the Crustacea, fig. 21, characterized by having two pairs of antennae, and remaining aquatic. The other lineage evolved into the myriapod-insect groups, fig. 24. In this line the second antennae disappeared, the second pair of maxillae fused, forming the labium, the gnathal segments and prostomium formed a tightly consolidated structure, and the animals became terrestrial. It is likely that the preterrestrial progenitor of this line was a freshwater species crawling among leaves at the edges of ponds, much like some existing isopods. Possibly certain forms were able to survive in the damp leaf layer when their ponds dried up seasonally and thus became adapted to terrestrial existence in leaf litter or soil. This pathway to terrestrial existence seems to have been followed also by the isopods. The myriapods and most primitive insects still live primarily in this soil and duff habitat.

The terrestrial myriapod-insect ancestor gave rise to three lines that have persisted to the present, one represented by the class Chilopoda (centipedes); another by the three classes Symphyla, Pauropoda, and Diplopoda (millipeds); and the other by the class Insecta. In the first four classes, functional legs persisted on almost all segments of the trunk region.

In the insect branch a further body specialization evolved. The first three pairs of locomotor appendages enlarged; the remainder became reduced and finally disappeared or became modified into nonlocomotor structures, fig. 11*F*.   This centralized the locomotor function in the first three segments, behind the head, which then formed a well-marked body region, or thorax.   The posterior portion of the body containing most of the internal organs is called the abdomen.   The posterior appendages of the abdomen became modified as organs for mating or oviposition.   Some of the Crustacea have a distinct thorax and abdomen, but in these the thorax is usually composed of about eight segments.

Summarizing these developments from the primitive legless arthropod ancestor, it seems reasonable to suppose that: (1) Similar generalized appendages were developed on all postoral segments, and (2) these were continuously modified and became segregated into groups for specialized functions.   In the insects this has resulted in the present distinctive body form composed of three regions: head, with sensory appendages and mouthparts; thorax, bearing three pairs of legs; and abdomen, containing most of the vital organs and having terminal appendages adapted for reproductive functions.

A review of the major groups of the Arthropoda is of interest in visual-

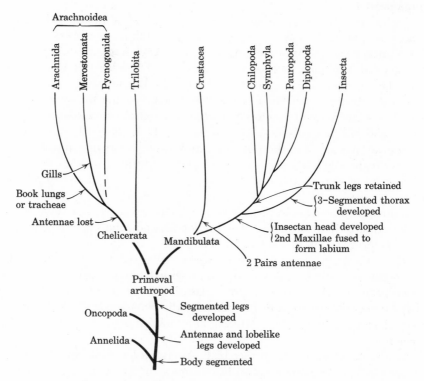

Fig. 15.   Suggested family tree of the classes of Arthropoda.

izing the place of insects among their relatives, fig. 15. In addition, it has a practical application, because the entomologist frequently encounters arthropods other than insects and is called on for information about them. This is especially true of the terrestrial forms that are of economic importance, such as mites.

CLASS TRILOBITA

The body was divided into head, thorax, and pygidium, the whole usually flattened and divided by two longitudinal furrows into three lobes, fig. 16. The head was a loosely organized region consisting of the primeval head (bearing a pair of long segmented antennae) and four body segments each bearing a pair of biramous appendages. Over this structure was a shell-like carapace. Many species had a pair of well-developed

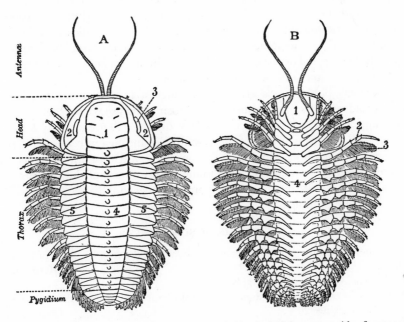

Fig. 16. Sketches of a complete trilobite *Triarthrus becki*. *A*, dorsal or upper side of carapace, showing three lobes, pleura (5), rachis or axis (4), glabella (1), and free cheeks (2) which bear the eyes (3). *B*, ventral or under side, showing biramous limbs (2, 3) attached to rachis, and upper lip or hypostoma (1) which covers mouth. The biramous legs were dual purpose: the upper feathered branch served for breathing gills and swimming paddles, the lower bare branch served for crawling. The short anterior appendages probably aided in feeding. (From Schuchert, after Beecher)

eyes. Each segment of the remainder of the body bore a pair of biramous appendages, except for the last segment, the telson. The various groups of trilobites exhibited a great variety of shapes and sizes. Most of them were two to three inches long, but some were quite tiny (10mm.) and some were giants, attaining a length of over two feet.

Trilobites were an abundant marine group in the early Paleozoic era but became extinct at the close of that portion of geologic time. In general actions most of them were probably similar to present-day isopods, swimming a little, running over the bottom, and feeding as scavengers. It is thought that some were carnivorous, others were pelagic and lived on plankton, and still others burrowed in the bottom and ingested mud and ooze.

CLASS ARACHNOIDEA

The body of a typical arachnoid is divided into two regions, the cephalothorax and the abdomen. The cephalothorax of the adult form usually bears six pairs of appendages, the anterior chelicerae (closely associated with the mouth), the pedipalps which are often chelicerate, and four pairs of walking legs. The abdomen bears no jointed appendages; its segmentation may or may not be visible externally.

Members of this class are of great interest to entomologists. Many arachnoids, such as mites and ticks, transmit disease organisms in much the same manner as do insect vectors of disease. A great many kinds of mites feed on crops, livestock, and man, causing heavy economic losses. These arachnoids usually occur in company with economic insects and are controlled in much the same manner and by use of the same equipment. As a result the arachnoid groups have gradually become a part of the field of entomology.

SYNOPSIS OF SUBCLASSES

1. Abdomen bearing pairs of external gills situated under platelike coverings, fig. 17
   Merostomata
   Abdomen without external gills 2
2. Abdomen consisting of a minute, fingerlike structure situated between bases of hind legs, fig. 20 .... **Pycnogonida**
   Abdomen large, usually larger than cephalothorax, fig. 18, or merged with it, fig. 19 .... **Arachnida**

Subclass Merostomata (Gigantostraca)

The abdomen bears pairs of appendages forming gills and platelike coverings; these are used both for respiration and, when the animal is off

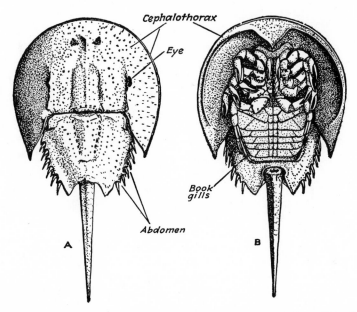

Fig. 17. King or horseshoe crab *Limulus*. *A*, dorsal view; *B*, ventral view. (From Wolcott, *Animal biology,* by permission of McGraw-Hill Book Co.)

the bottom, as swimming paddles. The extinct order Eurypterida lacked a carapace and looked somewhat like a scorpion. In the marine order Xiphosura the cephalothorax has a large horseshoe-shaped carapace; the horseshoe crab, *Limulus,* fig. 17, is the only living North American representative of the order.

Subclass Arachnida

The abdomen is large but has no external gills or locomotor organs. From front to back the cephalothorax bears a pair of chelicerae which are normally chelate, a pair of pedipalps which may be chelate, and four pairs of legs. In the ticks and mites the pedipalps and chelicerae are highly modified and closely associated with the mouth opening, this entire region having somewhat the appearance of a head.

This subclass is large and varied. Most of its groups are terrestrial animals but a few of the mites are aquatic.

KEY TO ORDERS (based on adults)

1. Pedipalps chelate, fig. 18*E*, or massive, long, and semichelate 2
   Pedipalps leglike or cylindrical, not at all chelate, fig. 18*B* 4

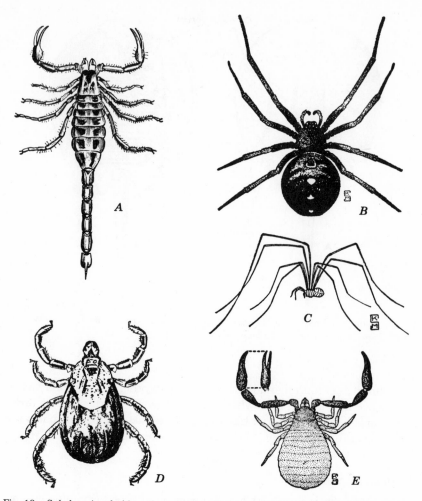

Fig. 18. Subclass Arachnida. *A*, a scorpion *Buthus carolinianus; B*, the black widow spider *Latrodectus mactans; C*, a harvestman or phalangid; *D*, the eastern dog tick *Dermacentor variabilis; E*, a pseudoscorpion *Larca granulata*. (*A* after Packard, *Guide to the study of insects*, Henry Holt & Co.; *D* from U.S.D.A.; *B C*, and *E* from Illinois Natural History Survey)

2. Tip of tail ending in a curved, sharp sting; tailpiece of abdomen stout and composed of 6 segments, fig. 18*A* .... **Scorpionida**
   Tip of tail without sting; tailpiece of abdomen either extremely thin and whiplike or no tail developed 3
3. First pair of legs having elongate, antennalike, segmented tarsus, quite unlike those of other legs; pedipalps massive but only semichelate; abdomen with either a very narrow tail-like apex or without tail .... **Pedipalpi**
   First pair of legs similar to others, tarsi short; pedipalps ending in well-developed, pincerlike chelae; abdomen oval, never with tail, fig. 18*E*. .**Pseudoscorpionida**

4. Apex of abdomen ending in a narrowed, tail-like portion composed of 15 segments
Palpigrada
Apex of abdomen not at all tail-like 5
5. Abdomen distinctly segmented (as in fig. 18*A*) 6
Abdomen unsegmented, fig. 18*B* 7
6. Abdomen as short as cephalothorax, fig. 18*C* .... **Phalangida**
Abdomen much longer than cephalothorax .... **Solpugida**
7. Junction of abdomen and cephalothorax constricted into a definite waist, best
seen from ventral aspect, fig. 18*B* .... **Araneae**
Junction of abdomen and cephalothorax not narrowed, fig. 18*D* ...... **Acarina**

SCORPIONIDA. This order is a small group containing only the true scorpions which range from about one half to 6 or 7 in. in length. In these arachnids the end of the tail bears a poisonous sting, fig. 18*A*, which the scorpion can use with great agility. The venoms of various species differ in the severity of their effects. In North America the group occurs chiefly in the Southwest, the range of one or two species extending as far northeastward as Illinois.

PEDIPALPI. A small order including the large whiptail scorpions (*Mastigoproctus*) and the smaller genus *Tarantula* which has no whip at the end of the abdomen. North American species occur in the South and Southwest.

PALPIGRADA. Another small order of minute species about 2.5 mm. long. They live under stones and in North America are restricted to the Southwest.

PSEUDOSCORPIONIDA. The pseudoscorpions, fig. 18*E*, are an abundant, widespread order. They are small, the largest not exceeding 7 or 8 mm. in length. Most of the species óccur under bark and in leaf mold but the genus *Chelifer* is found in houses also.

SOLPUGIDA. This is a small order containing hairy, agile arachnids of moderate size. They are often called vinegarones or sunspiders. The pedipalps and legs are short, and the chelicerae are massive. Vinegarones are tropicopolitan, in North America occurring in the Southwest and northeastward into Kansas.

PHALANGIDA. The harvestmen, or daddy longlegs, fig. 18*C*, have a broad abdomen, the entire body stout and oval or round in outline. In our larger species the legs are spindly, frequently five or more times as long as the body. The smallest species are minute, mitelike, and have short legs. These animals are common in damp shaded woods, where they move about on leaves, tree trunks, and in ground cover, seeking small insects and other food. In North America the group is widespread, represented by about two hundred species.

ARANEAE. This order comprises the spiders, fig. 18*B*, which in number of species are the dominant land group of the Arachnoidea. North

America alone has several thousand species. Neither legs nor pedipalps are chelicerate; the abdomen shows only traces of segmentation and is constricted at the base to form a thread-waisted joint with the cephalothorax. The pedipalps of the male are highly modified to transfer sperm to the female genital organs; these modifications assume varied shapes and are used extensively in the taxonomy of the group. Spiders are predaceous on insects and other small animals but aside from this are extremely varied in habits. Some species hunt their prey, running it down or jumping on it; the crab spiders wait in flowers or other places for the prey to come within reach; many groups spin webs in which prey is snared.

ACARINA. These are the mites and ticks, ranging from less than 1 to 15 mm. in length. The cephalothorax and abdomen are fused into a single continuous body region devoid of external segmentation, figs. 18D, 19. Mites are extremely varied in structure and habits. Although there are fewer species of mites than spiders, the mites are the most important group economically in the entire Arachnoidea. Several families, including ticks (Ixodidae), fig. 18D, chiggers (Trombiculidae), mange or itch mites (Sarcoptidae), and follicle mites (Demodicidae), are ectopara-

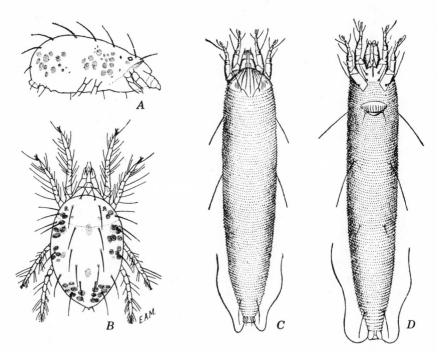

Fig. 19. Order Acarina. *A, B*, a red spider *Tetranychus pacificus*, lateral and dorsal aspects; *C, D*, a blister mite *Eriophyes pyri*, dorsal and ventral aspects. (From U.S.D.A., E.R.B.)

sites of birds and mammals. Species of "red spiders" (Tetranychidae), fig. 19*A*, *B*, attack leaves and cause defoliation of many crops. Various Tyroglyphidae, bulb mites, attack stored products and bulbs and roots of bulb crops. Members of the family Eriophyidae, fig. 19*C*, *D*, produce blisters and galls on several commercial crops, notably pears. In addition to economic species, there are many mites living in soil and ground cover and as ectoparasites on many native birds and mammals. One family, the Hydrachnidae, is aquatic, and is found in a wide variety of fresh-water habitats.

Subclass Pycnogonida

This subclass contains the sea spiders, fig. 20, a small marine group of spiderlike forms having a minute peglike abdomen. Sea spiders are found mostly on hydroids and sea anemones, but occasionally on jellyfish.

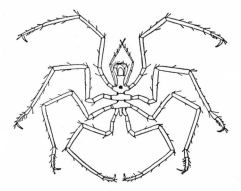

Fig. 20. A pycnogonid *Nymphon*. (From U.S.D.A., E.R.B.)

CLASS CRUSTACEA

To this class, illustrated in fig. 21, belongs such a varied assortment of forms that it is difficult to give a brief diagnosis that will apply to all. The majority have the following characteristics: body divided into head, thorax, and abdomen; head and thorax often closely joined and called the cephalothorax; head having two pairs of antennae, and, in a few groups, four pairs of accessory feeding appendages, including two pairs of maxillae, and a pair of maxillipeds; thorax usually having four to twenty distinct segments, each with a pair of segmented appendages; abdomen having one to many segments, with short appendages or none. A few parasitic or sedentary groups have extreme reduction in both body seg-

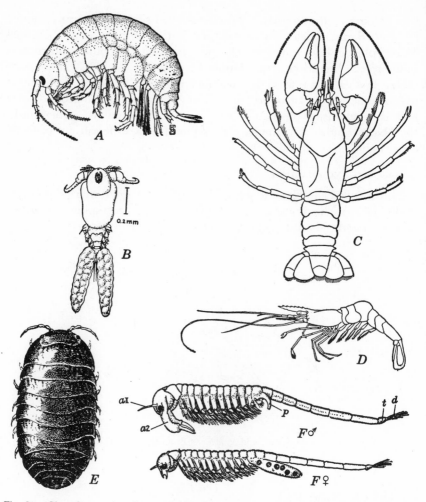

Fig. 21.   Class Crustacea.   *A*, an amphipod *Gammarus* sp.; *B*, a parasitic copepod *Ergasilus caerulus; C*, a crayfish *Cambarus bartoni; D*, a shrimp *Palaemonetes exilipes; E*, an isopod *Armadillidium vulgare; F*, a fairy shrimp *Branchinecta paludosa.*   All are fresh-water forms except *E*, which is terrestrial.   (*A* from Illinois Nat. Hist. Survey; *E* from U.S.D.A.; others from Ward and Whipple)

ments and appendages, as in parasitic Copepoda, fig. 21*B*, and the barnacles.   Several groups have a stout carapace covering much of the body, as in the crayfish, fig. 21*C*; some others have a shell, bivalve in appearance, that encloses most of the body and appendages.   In the sowbugs or pillbugs, fig. 21*E*, comprising the order Isopoda, the body is somewhat flattened.

Most groups of Crustacea are aquatic, either marine or fresh-water. A few are amphibious, such as certain species of crayfish. The only abundant terrestrial forms in North America are certain families of Isopoda. Species of these families occur chiefly in humid situations such as in and under rotting logs, under stones, or in the soil. At night some of these species leave their shelters and wander about freely.

Crustaceans rival insects in the variety of diverse forms developed in the class. They share another characteristic with the insects, that in many groups there is a succession of changes in form, or metamorphosis, in the life history of the individual. In the crustaceans this is well exemplified by the shrimp, fig. 22, which passes through four quite different immature stages before attaining the adult stage, giving a total of five distinctive body forms in the life cycle of the species.

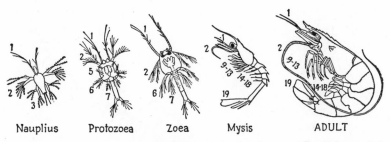

| Nauplius | Protozoea | Zoea | Mysis | ADULT |

Fig. 22. Metamorphic stages of the shrimp *Penaeus,* showing the succession of changes in body form and in the appendages (1–19). (From Storer, *General zoology,* McGraw-Hill Book Co., after Müller and Huxley)

Five subclasses of Crustacea are recognized. Four of these have terrestrial or fresh-water species; the fifth (barnacles) is exclusively marine.

MYRIAPOD GROUP

Four classes, the Diplopoda, Chilopoda, Symphyla, and Pauropoda, have centipede-like shapes and are often termed the myriapods. They all have a distinct head (composed of the original prostomium fused with several body segments whose appendages form the mouthparts) and an elongate trunk region bearing segmental ambulatory or walking legs. In each of these four classes the antennae are present, sometimes well developed. Although these classes share many superficial resemblances, they exhibit many differences in basic structure.

CLASS CHILOPODA

Here belong the centipedes, figs. 23, 24. They are elongate and many-segmented, with a pair of legs on each segment, and with the reproductive openings on the penultimate body segment. The head bears long antennae and has eyes which are either compound or composed of single facets. The mouthparts, fig. 25, consist of three pairs of appendages: the jawlike mandibles; the first maxillae, which are fused and resemble the insect labium; and the second maxillae or palpognaths, which are leglike,

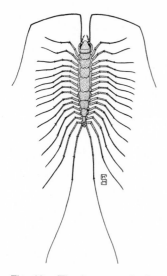

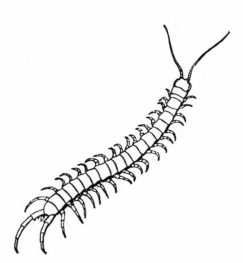

Fig. 23. The house centipede *Scutigera forceps*. (From Illinois Natural History Survey)

Fig. 24. A typical centipede. (After Snodgrass)

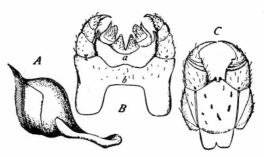

Fig. 25. Mouthparts and poison jaws of a centipede *Geophilus flavidus*. *A*, right mandible. *B*, the two pairs of maxillae; *a*, the united coxae of the first maxillae; *b*, the united coxae of the second pair or palpognaths. *C*, the poison claws or toxicognaths. (After Latzel)

sometimes with their bases fused to form a bridge below the first maxillae. An interesting feature of the Chilopoda is the poison claws, fig. 25*C*. These are appendages of the first trunk segment but are held beneath the head and superficially resemble mouthparts.

Chilopoda are predaceous in habits. About one hundred species are known from the United States. Most of them are nocturnal, moving about only at night in search of prey and hiding during the day in leaf mold, rotten logs, and galleries in soil. Species in temperate climates seldom exceed 1½ inches in length, but a few tropical species 8 or 10 inches long occur in the extreme southern United States. One species, fig. 23, having especially long legs is common in houses.

CLASS SYMPHYLA

Members of this class are about ¼ inch long and centipedelike in form, fig. 26, and have the trunk composed of about 15 segments (none fused in pairs) of which 11 or 12 bear legs. The reproductive openings are at the anterior end of the body. The head possesses long antennae, and the mouthparts consist of mandibles, maxillae, and labium. Members of the group are rare, usually occurring in humus, but certain species that eat the roots of plants are pests in gardens and greenhouses. In the United States the genus *Scolopendrella* is sometimes found in ground-cover samples. On cursory examination, small individuals may be confused with Pauropoda, which occur in the same type of situation.

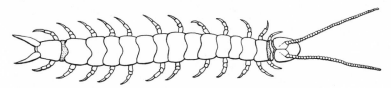

Fig. 26. A symphylid *Scutigerella immaculata.* (Adapted from Snodgrass)

CLASS DIPLOPODA

This includes the millipedes or thousand-legged worms, fig. 27. Except for a few segments at each end of the body, the body segments have fused into pairs so that each apparent segment has two pairs of legs. The mouthparts consist of a pair of mandibles and a platelike *gnathochilarium*, thought to be the fused maxillae. The labium is apparently atrophied. The reproductive organs open behind the second pair of legs.

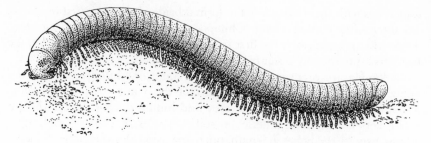

Fig. 27. A millipede *Parajulus* sp. (Original by C. O. Mohr)

Millipedes live in leaf mold, rotten logs, and other humid places. About one hundred species occur in the United States, especially in forested localities. A few species feed on living plants and become of local economic importance.

CLASS PAUROPODA

Several genera of minute animals comprise this class. The body consists of 11 or 12 segments, of which the dorsal portions are fused in pairs; eight or nine segments each bear a pair of legs, each pair evenly spaced from the next. The antennae are biramous, unlike any of their relatives. Each eye is represented by only a small spot. The mouthparts consist of a pair of mandibles and a curious complex lower lip thought to be the same as the gnathochilarium in the Diplopoda. As in the Diplopoda, the openings of the reproductive organs are located in the anterior part of the body.

The known North American species number about thirty-five and range in length from about a half to two millimeters. They occur on fallen, decayed twigs and logs, especially in situations where leaf mold has accumulated. Most of our species have slender white bodies, and look like small centipedes, fig. 28. Members of the family Eurypauropodidae

Fig. 28. A pauropod *Pauropus silvaticus*. (Adapted from Snodgrass)

are short and stout, resembling minute sowbugs (isopods). The class has been recorded from only scattered localities in eastern and western North America. Undoubtedly much remains to be discovered about these curious animals.

CLASS INSECTA

All the insects belong to this class. They are distinguished primarily by having a three-segmented thorax, each segment typically bearing a pair of legs. From this condition comes the term Hexapoda, meaning six-legged, which is sometimes used as the class name for insects.

Characteristics

A typical adult insect has three body regions, fig. 29. The anterior region is the head, which bears eyes, antennae, and three pairs of mouthparts. The next region is the thorax which is composed of three segments, each usually bearing a pair of legs; in many groups the second and third segments each bear a pair of wings. The posterior portion of the body is

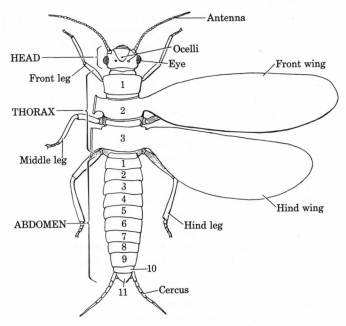

Fig. 29.   Diagram of a typical adult winged insect.   (Adapted from Snodgrass)

the abdomen. It consists of as many as eleven segments and has no legs. The eighth, ninth, and tenth segments usually have appendages modified for mating activities or egg laying. The exoskeleton in insects, as in other arthropods, provides both the protection for the vital organs and the support which maintains the body shape. The chief internal organs consist of the following parts: (1) a tubular digestive tract; (2) a long valvular heart for pumping the blood; (3) a system of pipelike tracheae for respiration; (4) paired reproductive organs opening at the posterior end of the body; (5) an intricate muscular system; and (6) a nervous system consisting of a brain, paired segmental ganglia, and connectives.

Insects lay eggs, except for a few forms which bear living young. The young insects molt from time to time in their development to the mature or adult stage, and generally at each molt an increase in size or the development of special parts takes place.

Immature insects do not have wings. The only known exception to this is found in the order Ephemeroptera, mayflies, in which the last brief immature instar has functional wings. Immature insects may be entirely unlike their adults in general appearance and may lack legs and many other structures typical not only of insects but also of arthropods.

Taxonomic Diversity

The insects have evolved into many strikingly different kinds of organisms. Living forms are classified in twenty-eight orders, listed below and described in Chapter 7.

Table 1. Classification outline of insect orders

Subclass Apterygota—wingless insects:
 Order Diplura—campodeans and japygids.
  Protura—proturans.
  Collembola—springtails.
  Microcoryphia—bristletails.
  Thysanura—silverfish.
Subclass Pterygota—winged insects:
 Series Paleoptera—ancient winged insects:
  Order Ephemeroptera—mayflies.
   Odonata—dragonflies.
 Series Neoptera—modern or folding-wing insects:
  Order Dictyoptera—cockroaches, walkingsticks.
   Isoptera—termites.
   Orthoptera—grasshoppers, crickets.
   Dermaptera—earwigs.
   Grylloblattodea—grylloblattids.
   Embioptera—embiids.

Table 1 (cont.)

Plecoptera—stoneflies.
Zoraptera—zorapterans.
Psocoptera—psocids, booklice.
Phthiraptera—chewing and sucking lice.
Thysanoptera—thrips.
Hemiptera—bugs (Heteroptera and Homoptera).
Hymenoptera—bees, ants, wasps.
Coleoptera—beetles.
Megaloptera—dobsonflies.
Neuroptera—lacewings.
Raphidiodea—snakeflies.
Mecoptera—scorpionflies.
Siphonaptera—fleas.
Diptera—two-winged flies.
Trichoptera—caddisflies.
Lepidoptera—moths, butterflies.

Success as a Group

Insects have attained the largest number of kinds of any animal group, with an estimated 900,000 described species; sometimes they occur in such numbers as to swarm in dark clouds or, attracted to lights, blanket the streets of a city a foot or more deep with their bodies.

Table 2. Estimated number of living animal species known at present for the world

Group	Number of Species
Chordata	60,000
Arthropoda exclusive of insects	73,000
Insecta	900,000
Mollusca	104,000
Echinodermata	5,000
Annelida	7,000
Molluscoidea	2,500
Platyhelminthes	6,500
Nemathelminthes	3,500
Trochelminthes	1,500
Coelenterata	9,000
Porifera	4,500
Protozoa	30,000
Total	1,206,500

In competition with other animals they have been able to fit into and populate almost every nook and cranny of the globe except the depths of the ocean. They abound throughout the tropics and also are one of the very few permanent animal inhabitants of the South Polar Region; aquatic insects may nearly pave the bottom of large rivers and lakes and also develop to maturity in the water in hoofprints; in one case grasshoppers may range over miles of prairie, in another a brood of maggots may feed and mature within a single rotting walnut husk, and in still another a wasp may mature within the tiny seed of a small plant.

These are but a few fragments of the evidence that insects are a remarkably successful group of organisms. With this in mind, it is interesting to speculate on some of the reasons why they have developed to such huge numbers of both species and individuals.

Adaptive Features

Among arthropods, insects represent the culmination of evolutionary development in terrestrial forms. They have exploited the mechanical advantages of an exoskeleton and used them as a basis on which to add specializations which give them still further advantages over their competitors. Chief advantages of an exoskeleton include (1) a large area for internal muscle attachment; (2) an excellent possibility of evaporation control, especially in small-bodied animals; and (3) almost complete protection of vital organs from external injury.

To this foundation other specializations have been added, some morphological and some physiological, which have been decided factors in assisting insects to attain their present development. The more outstanding of these specializations are enumerated in the following paragraphs.

Functional Wings. The power of flight greatly increased the statistical chances of survival and dispersal, except on wind-swept islands. It increased feeding and breeding range and provided a new means of eluding enemies. Increased feeding range undoubtedly opened the way for adoption of foods of more specific limitation, especially in those cases in which the host or breeding medium occurred in small quantities and scattered situations. For example, it would allow a species to adopt carrion as a food, since the individual with functional wings could seek out and reach carcasses which would be not only many miles away but also suitable as food for only a short period.

Small Size. In the main, insect evolution has followed the course of developing many small individuals rather than few large ones. This made available many new specific foods occurring in small quantity and has increased chances of hiding from and eluding enemies. Small size has the disadvantage that the total body surface increases tremendously in proportion to the body volume. This results in a high evaporation quotient which would make terrestrial life impossible for a thin-skinned animal. The exoskeleton of insects has provided a check to this evaporation quotient, and the possession of this exoskeleton is undoubtedly one of the principal factors which have allowed insects to develop small size.

Adaptability of Structures. The same structure has become adapted to perform different functions. For instance, the front legs of mantids and ambush bugs grasp and hold prey while it is being devoured, thus functioning as accessory mouthparts, rather than as ambulatory legs. In other instances essentially the same structure has become adapted to function under entirely different conditions, for example, the diverse modifications of the respiratory system, which has been adapted for many types of aquatic and terrestrial life.

Complete Metamorphosis. This is a specialization in which the life history is divided into four distinct parts, (1) the egg, (2) the larva or feeding stage; (3) the pupa, a quiescent transformation stage, and (4) the adult or reproductive stage. It occurs in the insect orders having the largest number of species, including the beetles and flies. In this type of life history all real growth is the result of larval feeding; the adult has only to maintain a more or less static metabolism and at the most provide sufficient food for maturation of sperms or eggs. This system has enabled the larva and adult to live in entirely different places and under different conditions, so that the larva has been able to take advantage of conditions most favorable for rapid growth and the adult to live in conditions best suited to fertilization, dispersal, and oviposition. Complete metamorphosis opened to the group an infinite variety of habitat and food possibilities. In connection with it there has often been a varied development of complex instinctive behavior. In addition, extremely short life cycles have frequently been developed, based on the extraordinary feeding and digestive ability of some of the larvae; a flesh-fly maggot, for instance, can develop from hatching to a full-grown larva in 3 days. In short, complete metamorphosis has enabled a species to combine the advantages of two entirely different ways of life and at the same time avoid many of the disadvantages of both.

Increase in Number of Species. Factors contributing to insect success would tend to assure the *persistence* of species, but would not cause an increase in the *number* of species. Other factors must be invoked to account for the extraordinary number of different insect species. According to evolutionary theory the following circumstances are especially important in this connection: (1) Many insect species can live only within narrow limits of certain ecological factors such as host, temperature, and humidity, and their species ranges become broken into isolated segments by relatively small long-range changes in climate such as those accompanying an Ice Age. (2) Because of their power of flight, winged insects may be prone to be carried over water gaps or other barriers by moving air masses; if this is true (and much evidence certainly indicates that it is), large numbers of new species would evolve by the process of colonization. (3) Changes in the genetic mechanism of many insect groups may produce genetic incompatibility between isolated populations with unusual rapidity.

None of these factors can be considered the most important reason why insects have achieved their present diversity and numbers. The process has been most complex, with various combinations of these factors and undoubtedly others working together and producing the end result. It should be borne in mind that not every insect lineage has every one of these specializations. For example, entire orders of insects, such as the chewing lice, sucking lice, and fleas, have lost all trace of wings, correlated with a limited sphere of activity on or near the host. Complete metamorphosis does not occur in about half the orders, but in these, other features come into play. It must be remembered, too, that these specializations are only a few of the most important of the very large number that have evolved in the class Insecta.

REFERENCES

Baker, E. W., and G. W. Wharton, 1952. *An introduction to acarology.* New York: The Macmillan Co. 465 pp.

Chamberlin, J. C., 1931. The arachnid order Chelonethida. *Stanford Univ. Publs., Univ. Ser. Biol. Sci.,* **7**:1–284.

Driver, Ernest C., 1950. *A guide to the identification of the common land and freshwater animals of the United States* (rev. ed.). Published by the author, 119 Prospect St., Northhampton, Mass. 558 pp.

Edmondson, W. T., 1959. *Freshwater biology.* 2nd ed. New York: John Wiley & Sons. 1248 pp.

Hoff, C. Clayton, 1949. The pseudoscorpions of Illinois. *Bull. Illinois Nat. Hist. Survey,* **24**:413–498.

Kaston, B. J., 1948. Spiders of Connecticut. *Connecticut Geol. Nat. Hist. Survey, Bull.* **70**:1–874.

MacSwain, J. W., and U. N. Lanham, 1948. New genera and species of Pauropoda from California. *Pan-Pacific Entomol.,* **24**:69–84.

McGregor, E. A., 1950. Mites of the family Tetranychidae. *Am. Midland Naturalist,* **44:**257–420.

Michelbacher, A. E., 1938. The biology of the garden centipede, *Scutigerella immaculata. Hilgardia,* **11:**55–148.

Pennak, R. W., 1953. *Freshwater invertebrates of the United States.* New York: Ronald Press Co. 769 pp.

Ross, H. H., 1962. *A synthesis of evolutionary theory.* Cliffside, New Jersey: Prentice-Hall, Inc. 387 pp.

Snodgrass, R. E., 1938. Evolution of the Annelida, Onychophora, and Arthropoda. *Smithsonian Misc. Collections,* **97,** No. 6:1–159.

1950. Comparative studies on the jaws of mandibulate arthropods. *Smithsonian Misc. Collections,* **116,** No. 1:1–85.

1952. *A textbook of arthropod anatomy.* Ithaca, N. Y.: Comstock (Cornell Univ.) Press. 363 pp.

Storer, Tracy I., 1951. *General zoology,* 2nd ed. New York: McGraw-Hill Book Co., Inc. 832 pp.

Wharton, G. W., 1952. A manual of the chiggers. *Entomol. Soc. Wash. Mem.,* **4:** 185 pp.

Williams, S. R., and R. A. Hefner, 1928. The millipedes and centipedes of Ohio. *Ohio Biol. Survey Bull.* **18,** 4(3): 91–146.

Chapter Three

External Anatomy

THIS chapter deals primarily with the external parts and divisions of the body and its appendages. Before proceeding to the description of these parts, it is necessary to understand a few generalities regarding the body wall, from which the parts are formed.

BODY WALL AND EXOSKELETON

The body wall of insects serves as an exoskeleton and is the only counterpart in insects to the internal skeleton of vertebrates. To the exoskeleton are attached the principal muscles which give the body cohesion. The body wall may have considerable spring or flexibility, but, except for a short time after a molt, *it will not stretch.*

Serving as both protection and a rigid attachment spot for muscles, various parts of the body wall are hardened or sclerotized. If all the body wall were uniformly hard, there would be no possibility of movement, whether for purposes of locomotion or for expansion to accommodate important activities such as food ingestion or development of eggs. Movement is possible because the hardened body areas form a series of plates, or *sclerites,* between which the body wall is soft and flexible, or membranous. This arrangement permits the development of hard exterior plates for protection and rigidity and at the same time allows many types of movement.

A simple example of how this works is found in the abdomen of the mosquito, fig. 30. When the abdomen is not engorged, a cross section of the body wall of the abdomen is narrow and elliptic, fig. 30*A*; the dorsal plate and ventral plate are connected at the sides by a strip of membrane which is very finely accordion-pleated. As the abdomen enlarges during a blood meal, these membranes simply unfold, allowing the dorsal and ventral plates to be pushed farther apart by the increasing volume of food

50

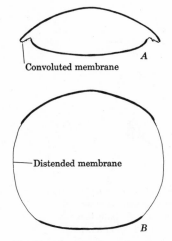

Fig. 30. Cross section of mosquito abdomen, diagrammatic. *A*, contracted; *B*, expanded.

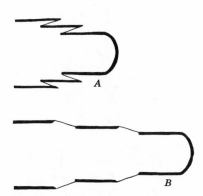

Fig. 31. Medial section of telescoping ring segments, diagrammatic. *A*, retracted; *B*, extended. The thin portion represents membrane.

being pumped into the abdomen. At its greatest expansion, the body cross section is nearly circular, fig. 30*B*.

Another common type of membranous connection, shown in fig. 31, works on the principle of telescoping rings. When the body is retracted, as in *A*, the rings overlap, and the membrane is drawn in with the telescoped section. When it is extended, as in *B*, the sections may be pushed out to the limit of the length of their connecting membranes.

Figure 32 illustrates how a flexible membranous strip affords articulation of a leg joint. To the right, a narrow strip of membrane forms a hinge between the leg and one plate; to the left a roll of membrane connects the leg and the other plate. In *A* the leg is held straight out; note that the left membrane forms a series of pleats. In *B* the appendage has been pulled forward; this movement stretched out the left membrane as the corner of the leg moved inward and back.

A common type of articulation found at leg joints of adult insects is illustrated in fig. 33. When the segment to the right is extended, as in *A*, the upper membrane *u* is folded, and the lower membrane *l* is stretched out. When the segment is pulled down, as in *B*, the upper membrane is pulled out, and the lower membrane is folded or pleated.

In all four of these examples movement has been made possible by flexibility, not elasticity. The illustrations do not show the muscles that actually *cause* the movements, but only the membranous connections that *permit* movement.

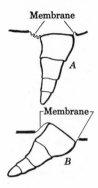

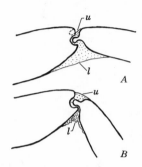

Fig. 32. Articulation of a membranous leg joint. *A*, held straight out; *B*, pulled forward.

Fig. 33. Membranous connections of a ball-and-socket leg joint. *A*, held straight out; *B*, pulled in. *u*, upper membrane; *l*, lower membrane.

Sclerites. The hardened or sclerotized areas of the body are called sclerites. The major sclerites are usually separated by areas or lines of membrane. Many major sclerites may be subdivided by furrows or new lines of membrane into additional sclerites. Sclerites may also unite, usually with an evident line, furrow, or seam along the line of fusion. In entomological usage these types of demarcation—membrane strips, furrows, and fusion lines—are called sutures. The term sclerite is applied loosely to any sclerotized surface area, bounded by sutures of any type.

External Processes. The surface of the integument usually bears many kinds of processes, including wrinkles, spurs, scales, spines, and hair. These are outgrowths of the body wall. They are of only incidental interest to the general subject of external anatomy but are extremely important in functions such as sound production and sensing various types of stimuli.

Internal Processes. There are many processes which are formed by the invagination of the body wall. These are called *apodemes*. Their point or line of invagination is almost always indicated by an external pit or groove. These pits and grooves provide some of the most reliable landmarks for identifying their parent sclerites. The apodemes provide internal areas for muscle attachment.

BODY REGIONS

The adult insect body is divided into three parts: the head, thorax, and abdomen. The phylogenetic origin of these regions is discussed on page 29. The head is usually a solidly constructed capsule no longer having obvious segmentation. The thorax and abdomen have both pre-served distinct segments of more or less ringlike form.

Legless immature instars of some insects have little differentiation be-tween body regions. The head is usually distinct, but the segments of both thorax and abdomen may be identical in appearance and form a single uniform body region, the trunk.

Orientation. In describing the relative position of various parts of an insect, several sets of terms are used to indicate direction or position. Certain body regions are used as a basis for orientation, chiefly the following:

1. ANTERIOR PORTION, the portion of the body bearing the head; or that portion of any part that is toward the head end.

2. POSTERIOR PORTION, the portion of the body bearing the cauda, or "tail end" of the abdomen; or that portion of any part that is toward the posterior end.

3. DORSUM, the top or upper side of the body or one of its parts.

4. VENTER, the underside or lower side of the body or one of its parts.

5. MESON, the longitudinal center line of the body, projected on either the dorsal or ventral aspect, or any point in between.

6. LATERAL PORTION, the side portion of the body or one of its parts.

7. BASE, APEX; in appendages or outgrowths of the body, such as anten-nae or legs, the point or area of attachment is called the base; the tip or furthermost point from the attachment is called the apex. In parts of appendages, such as a segment of a leg, the same orientation is used; the part articulated nearest the body is the base or proximal portion; the part away from the body is the apex or distal portion.

THE HEAD

The head, fig. 34, comprises the anterior body region of an insect. It is normally a capsule with a sclerotized upper portion, which contains the brain, and a membranous floor, in which is situated the oral opening or mouth.

Origin. As explained in Chapter 2, the insect head is a composite structure. It consists of a primeval area or prostomium anterior to the mouth, to which have fused the first four postoral segments. So complete is this fusion that little evidence remains to indicate the origin of the parts. Evidence from both phylogeny and embryology indicates clearly that in insects the first postoral segment (sometimes called the intercalary seg-ment) has no appendages except rudimentary ones in the embryo, and that the next three postoral segments bear the mandibles, maxillae, and labium, respectively.

There is considerable difference of opinion among various investigators concerning the origin and composition of the prostomium. Some believe that it consists only of a primeval head area, but others believe that it consists of a primeval head area to which one or more primeval segments fused very early in oncopodan or arthropod evolution. These differences of opinion are based on various interpretations of embryological evidence and innervation of head structures in Annelida and Arthropoda.

Position.   The head may assume various positions in relation to the long axis of the body. These positions are frequently used in classification. The two most important positions have been given definite names:

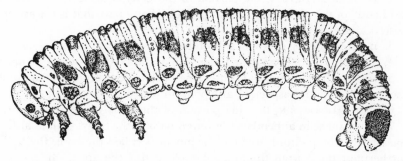

Fig. 34.   Sawfly larva *Neodiprion lecontei,* illustrating a hypognathous head.   (After Middleton)

HYPOGNATHOUS: the mouthparts are directed downward, and the head "segments" are in the same position as the trunk segments, fig. 34.   This is the generalized condition.

PROGNATHOUS: the head it tilted up at the neck so that the mouthparts project forward, fig. 35.

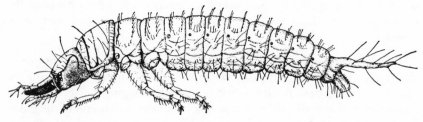

Fig. 35.   Ground beetle larva *Harpalus pennsylvanicus,* illustrating a prognathous head.   (From Illinois Nat. Hist. Survey)

Head Organization and Appendages. In a typical hypognathous head, fig. 36, the anterior region or face, the dorsal portion, and lateral portion form a continuous sclerotized capsule which is open beneath, like an inverted bowl. On this capsule are situated a pair of compound eyes, three ocelli, and a pair of antennae. The labrum hangs down from the lower front margin of the capsule to form a flap in front of the mouth. The ventral portion of the head forms a membranous floor posterior to the mouth; from this floor arises the hypopharynx, bearing the opening of the salivary duct. On each side of this floor hang down the three pairs of appendages forming the chewing organs or mouthparts, consisting of the mandibles, maxillae, and labium. These articulate with the ventral margin of the capsule. The posterior portion of the head is shaped like an inverted horseshoe, the capsule forming the dorsal and lateral portion, the labium closing the bottom of the shoe; the open center is called the occipital foramen, through which pass the oesophagus, nerve cord, salivary duct, aorta, tracheae, and free blood. Inside the head is a series of braces called the tentorium, fig. 40.

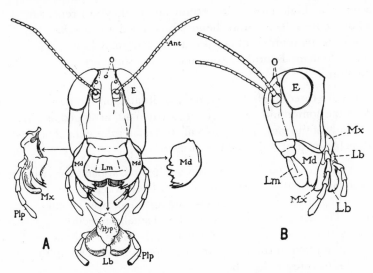

Fig. 36. Head of grasshopper, showing appendages and chief organs. *A*, anterior aspect; *B*, lateral aspect. *Ant,* antenna; *E,* compound eye; *Hyp,* hypopharynx; *Lb,* labium; *Lm,* labrum; *Md,* mandible; *Mx,* maxilla; *O,* ocelli; *Plp,* palpus. (After Snodgrass)

Special Structures of the Head Capsule. COMPOUND EYES are usually large, many-faceted structures situated on the dorsolateral portion of the capsule. Each eye is situated on or surrounded by a narrow ringlike or shelflike *ocular sclerite.* In many forms, especially larvae, the eyes are

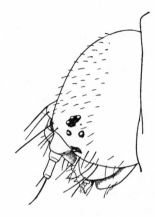

Fig. 37. Head of a caterpillar showing the ocularium.
(From Folsom and Wardle, *Entomology*, by permission of
The Blakiston Co.)

reduced to a single facet. In certain larvae they are represented by a group of separate facets, and the group is called an *ocularium*, fig. 37. In adult insects the number of facets may be extremely large. The housefly has about four thousand facets to an eye, and some beetles about twenty-five thousand.

OCELLI are three single-faceted organs situated on the face and usually between the compound eyes. The upper two are arranged as a pair, one on each side of the meson, and are called the lateral ocelli. The lower one is on the meson and is the median ocellus.

ANTENNAE are a pair of movable segmented appendages which arise from the face, usually between the eyes. They articulate in the antennal socket, which is sometimes surrounded by a narrow ringlike antennal sclerite. The periphery of the socket has a small projection on which the antenna articulates. Antennae are extremely varied in shape, and names have been applied to the more striking types. A few examples are listed here and illustrated in fig. 38:

Filiform or threadlike	*Clavate* or clubbed
Setaceous or tapering	*Capitate* or having a head
Moniliform or beadlike	*Lamellate* or leaflike
Serrate or sawlike	*Pectinate* or comblike

LABRUM is the movable flap attached to the ventral edge of the face. The inner side of the labrum forms the front of the preoral cavity and is called the *epipharynx*. The epipharynx frequently bears raised lobes and

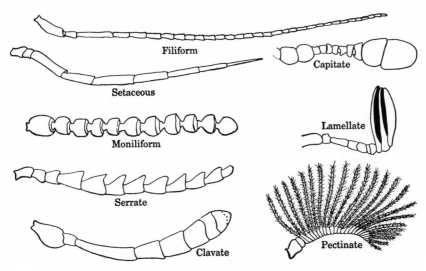

Fig. 38. Types of antennae.

complicated sets of sensory papillae and setae. These have proven very useful to the taxonomist as an aid in the identification of larval forms.

Principal Sutures and Areas. The head capsule is subdivided by several sutures. Most of these are considered secondary developments following the obliteration of the original segmental sutures. The principal head sutures and adjacent areas are as follows, fig. 39:

VERTEX is the entire dorsum of the head between and back of the eyes.

EPICRANIAL SUTURE is a Y-shaped suture whose stem begins on the back of the head, crosses the vertex, and forks on the face. The stem is called the epicranial stem; the two arms of the forked portion are the epicranial arms. These sutures are not fundamental divisions of the head but are lines of weakness associated with the bursting of the head capsule at molting. Because of their function, they are sometimes termed *ecdysial sutures.* In different groups of insects these sutures may cross quite different regions of the head. The sutures are most pronounced in the immature stages, as would be expected, but often occur in adults also. In spite of their lack of uniformity in different groups, these sutures are often of great use as landmarks or taxonomic characters within a group.

FRONS OR FRONT is the area on the anterior face which lies between or below the epicranial arms. The median ocellus occurs on this sclerite. It is bounded ventrally by the *frontoclypeal suture.*

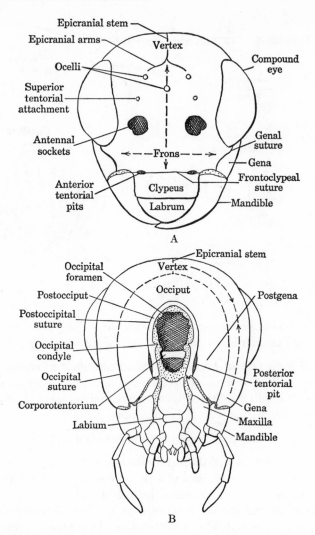

Fig. 39.   Diagrams illustrating the principal sutures and areas of the head.   *A*, anterior aspect (maxillae and labium omitted); *B*, posterior aspect.

CLYPEUS is the liplike area between the frontoclypeal suture and the labrum.   It never articulates with the frons but is joined solidly with it. The *labrum* hangs below it and articulates by means of the membranous connection between them.

GENA is the lower part of the head beneath the eyes and posterior to the frons.   There is sometimes a *genal suture* on the anterior portion of the face between the frons and gena; if this suture is absent, the division between frons and gena is indefinite.

OCCIPUT or OCCIPITAL ARCH is the area comprising most of the back of the head. It is divided from the vertex and genae by the *occipital suture;* in many groups this suture is either reduced to a crease or completely obliterated in which case the occiput can be defined only as a general area merging anteriorly with vertex and gena. The ventral portions of the entire occipital arch area are sometimes called the *postgenae.*

POSTOCCIPUT is the narrow ringlike sclerite which forms the margin of the occipital foramen. It is separated from the occiput by the *postoccipital*

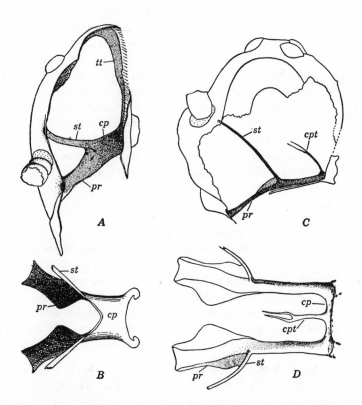

Fig. 40. Tentorium in Hymenoptera. *A*, head of *Macroxyela* cut away to show tentorium, lateral aspect; *B*, tentorium of same, dorsal aspect; *C*, head of *Aleiodes* cut away to show tentorium, lateral aspect; *D*, tentorium of same, dorsal aspect. *cp*, corporotentorium; *cpt*, corpotendon; *pr*, anterior arms; *st*, superior arms; *tt*, dorsal thickening extending from the tentorium along the side of the head.

suture, almost universally present in adult insects. The postocciput bears the *occipital condyle* on which the head articulates with the cervical sclerites of the neck region.

Tentorium, Fig. 40. The head is strengthened internally by a set of sclerotized apodemes or invaginations of the body wall which have evolved primarily as more rigid supports for the attachment of muscles connected with the mouthparts. In the apterygote insects and their allies the centipedes, the apodemes are more or less platelike or rodlike structures sometimes connected by ligamentous bridges. In the ancestors of the pterygote insects these structures enlarged, fused, and evolved into a strong internal skeleton of the head called the *tentorium.* Typically the tentorium is composed of four principal parts: the *anterior arms, posterior arms, corporotentorium* or central mass, and *dorsal arms.* The anterior arms are invaginated from the *anterior tentorial pits,* which usually are well defined externally as pits at each lower corner of the frons. The posterior arms are invaginated from the *posterior tentorial pits* which almost always persist as external slits on the postoccipital suture. The corporotentorium represents the inward extension, meeting, and fusing of the anterior and posterior arms. The dorsal arms are considered as secondary outgrowths of the anterior arms, because there is no large or persistent pit associated with their point of attachment with the head capsule, usually located between the antennal sockets and lateral ocelli. The shape and relative position of the tentorial parts are extremely different in various groups of insects.

The Mouthparts

The three most conspicuous elements of insect mouthparts are the mandibles, maxillae, and labium. These represent modifications of typical paired arthropod limbs. The shape of these parts in insects is so different from that in the original ancestral forms that evidence from other arthropod groups is necessary to demonstrate the relationship. A study of the appendages of fossil arthropods, together with an analysis of the comparative morphology of appendages of living forms, indicates that all present-day arthropod appendages arose from a simple generalized form.

Generalized Arthropod Appendage, Fig. 41. The basal segment or *coxopodite* is implanted in the side of the body wall. The apical segments form the *telopodite.* Each segment has potentialities for developing processes on both the lateral and mesal sides, the lateral processes called *exites,* and the mesal processes called *endites.* A primitive and early modification is illustrated by the leg of the trilobite, fig. 42. Note that the coxopodite has a gill-like exite and a spurlike endite; the telopodite is simple and without processes.

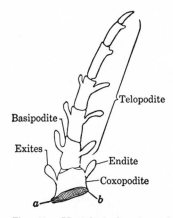

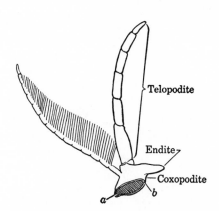

Fig. 41. Hypothetical arthropod appendage. (Modified from Snodgrass)

Fig. 42. Leg of a trilobite. (Redrawn from Snodgrass)

Second Antennae, Chelicerae. These structures, belonging, respectively, to the Crustacea and the Arachnida, are appendages of the first postoral segment, fig. 11*B–D*. There is no definite evidence of appendages on the heads of living Insecta which represent them. Presumably, they were lost in an early evolutionary stage.

Mandibles, Figs. 36, 44. These are the anterior or first pair of true insect mouthparts and lie directly behind the labrum. They are appendages of the second postoral segment. Typically they are hard and sclerotized and have various sets of teeth and brushes. They articulate with the head at the base of the lateral margin and (except in a few primitive insects) at the base of the mesal margin also. Near each of these articulations arise strong tendons which project into the head and serve as attachments for the strong muscles which operate the mandibles.

No insect mandibles preserve other characters that would assist in explaining how they were derived from a simple segmental appendage. In many Crustacea, however, the mandibles are of a much more primitive type, fig. 43. The muscular development indicates that this is a simple derivative of a form such as the trilibite leg, fig. 42. The chief modifications include (1) an enlargement and strengthening of the coxopodite, (2) development of its endite into a rasping toothed area, (3) loss of the exite, and (4) reduction of the telopodite. In all insect mandibles, fig. 44, the telopodite has been lost completely. The insectan mandible is therefore only a much-modified coxopodite and its endite.

Maxillae, Fig. 45. These lie directly behind the mandibles and are the appendages of the third postoral segment. The musculature indicates

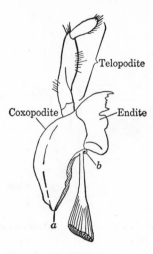

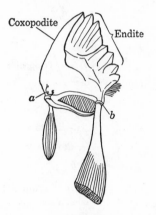

Fig. 43. Mandible of the crusta-
cean *Anaspides*. (Redrawn from
Snodgrass)

Fig. 44. Mandible of a grasshopper. (Redrawn from Snodgrass)

that their evolution follows very closely that of the mandible but with
these differences: (1) A mesal articulation has *not* been developed, (2) the
telopodite is retained as a tactile organ or palpus, (3) the coxopodite is
divided, and (4) the endite has developed into two distinct movable lobes.

In entomological literature, little reference is made to the terms coxopodite, telopodite, endite, etc., which refer to the fundamental divisions and
processes of arthropod appendages. In this account they have been used
until now to assist the student in correlating insect parts with those of
other Arthropoda. From this point on, however, it is pertinent to make
a change in terminology and employ those terms usually applied in entomological usage. These terms refer to parts which are in most instances
differentiated only in insects, and the terms are therefore necessary for
accurate identification of the part or area.

The generalized type of maxilla is a masticating structure which is
divided into several well-marked parts, fig. 45, as follows:

CARDO, the triangular basal sclerite which is attached to the head capsule, and which serves as a hinge for the movement of the remainder of the
maxilla.

STIPES, the central portion or body of the maxilla, usually somewhat
rectangular in shape. The stipes is situated above the cardo and is the
basis for the remaining parts of the maxilla.

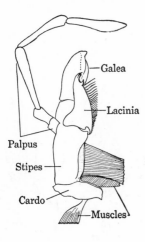

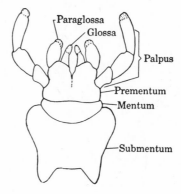

Fig. 45.   Maxilla of a cockroach,
illustrating   a   generalized   type.
(After Snodgrass)

Fig. 46.   Labium of a cockroach,
illustrating   a   generalized   type.
(Adapted from Imms)

GALEA, the outer (lateral) lobe articulating at the end of the stipes.   It is frequently developed as a sensory pad or bears a cap of sense organs.

LACINIA, the inner (mesal) lobe articulating at the apex of the stipes. It is usually mandible-like in general form with a series of spines or teeth along its mesal edge.

PALPUS, the antenna-like segmented appendage which arises from the lateral side of the stipes.   It is commonly five-segmented.   Presumably, it is entirely sensory in function.

Labium, Fig. 46.   This structure forms the lip posterior to the maxillae. It appears to be a single unit but really consists of a second pair of maxillae which have fused on the meson to form a single functional structure.   The parts of the labium correspond very closely to those of the maxillae, and their homologies have been established by studies of muscles and their point of attachment.

POSTLABIUM, the basal region of the labium, which hinges with the head membranes.   It is frequently divided into two parts: a basal *submentum* and an apical *mentum*.   The postlabium represents the fused cardines of the maxillae.

PRELABIUM, the apical region of the labium, including various lobes and processes.   The central portion or body is the *prementum* (sometimes also

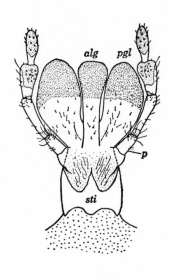

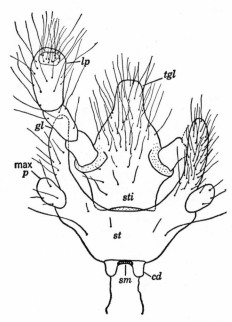

Fig. 47. Part of labium of hymen-opteron *Trichiosoma triangulum*, illustrating an alaglossa. *alg*, alaglossa; *p*, palpus; *pgl*, paraglossa; *sti*, stipulae.

Fig. 48. Labium and fused maxillae of hymen-opteron *Tremex columba*, illustrating a totoglossa. *cd*, cardo; *gl*, fused galea and lacinia; *lp*, labial palpus; *max p*, maxillary palpus; *sm*, submentum; *st*, fused stipites; *sti*, stipulae; *tgl*, totoglossa.

called *stipulae*), which bears a pair of *labial palpi,* one on each side of the prementum, and each usually three-segmented in generalized forms.

The apical portion of the prelabium frequently forms a sort of tongue and for this reason is called the *ligula.* It varies greatly in structure but usually is divided into two pairs of lobes: (1) the *glossae,* a pair of mesal lobes usually close together, and (2) the *paraglossae,* a pair of lateral lobes which usually parallel the glossae. In many groups such as the Hymenoptera, the glossae are fused to form an *alaglossa,* fig. 47. In other cases

	Maxilla		Labium	
Table *of* **Homologies**	cardo	corresponds to	postlabium { submentum / mentum	
	stipes	corresponds to	prementum or stipulae	
	palpi	corresponds to	palpi	
	lacinia	corresponds to	glossae	
	galea	corresponds to	paraglossae	

the glossae and paraglossae may be fused together into a single solid lobe called a *totoglossa,* fig. 48.

The accompanying table illustrates the homologies of the corresponding parts of the maxillae and labium. When consulting this table refer also to figs. 45 and 46.

Hypopharynx. From the ventral membranous floor of the head arises the hypopharynx, fig. 49. It usually forms a protruding lobe or mound. In generalized insects the hypopharynx is so closely associated with the base of the labium as to be considered a part of it. Unlike the other mouthparts, the hypopharynx is not an appendage but an unsegmented outgrowth of the body wall.

Superlinguae. In a few primitive insects a pair of simple lobes occur in close association with the hypopharynx. Embryologically they appear as lobes associated with the mandibles, but in advanced embryos they become more closely attached to the hypopharynx. These superlinguae occur in a few insect orders such as the Thysanura and Ephemeroptera, in some Crustacea, and in some members of the Symphyla. They apparently are a primitive arthropodan character which has atrophied in the more specialized insects.

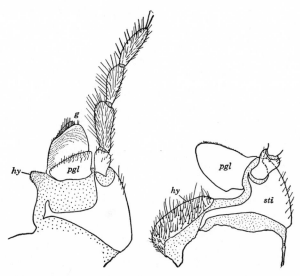

Fig. 49. Labium and hypopharynx of a sawfly *Arge pectoralis* (right), and a braconid *Aleiodes terminalis* (left). *g,* glossac; *hy,* hypopharynx; *pgl,* paraglossa; *sti,* stipulae.

Principal Types of Mouthparts

Insect mouthparts have become modified in various groups to perform the ingestion of different types of food and by different methods. The more diverse and interesting types are listed here, chosen to illustrate the varied shapes assumed by homologous parts, and the different uses to which they may be put. Many other types exist, many of them representing intermediate stages between some of the types treated here.

Chewing Type. In this type the various appendages are essentially as in the preceding figs. 44, 45, and 46. The mandibles cut off and grind solid food, and the maxillae and labium push it into the oesophagus. Grasshoppers and lepidopterous larvae are common examples. It seems certain that the chewing type of mouthparts is the generalized one from which the other types developed. This view is upheld by important evidence of two kinds. In the first place such mouthparts are most similar in structure to those of the centipedes and symphylids, which are the closest allies of the insects. In the second place, chewing mouthparts occur in almost all the generalized insect orders, such as the cockroaches, grasshoppers, and thysanurans; and they occur in the larvae of at least the primitive families of the neuropteroid orders. In many of the neuropteroid orders the adults frequently also have the chewing type of mouthparts which is little changed from the primitive type: for example, the Coleoptera and most of the Hymenoptera.

Cutting-Sponging Type, Fig. 50. In horse flies (Tabanidae) and certain other Diptera, the mandibles are produced into sharp blades, and the maxillae into long probing styles. The two cut and tear the integument of a mammal, causing blood to flow from the wound. This blood is collected by the spongelike development of the labium and conveyed to the end of the hypopharynx. The hypopharynx and epipharynx fit together to form a tube through which the blood is sucked into the oesophagus.

Sponging Type, Fig. 51. A large number of the nonbiting flies, including the house fly, have this type, fitted for using only foods which are either liquid or readily soluble in saliva. This type is most similar to the cutting-sponging type, but the mandibles and maxillae are nonfunctional, and the remaining parts form a proboscis with a spongelike apex, or labella. This is thrust into the liquid food, which is conveyed to the food channel by minute capillary channels on the surface of the labella. The food channel is formed in this type also by the interlocking elongate hypopharynx and epipharynx, which form a tube leading to the oesophagus. Certain solid foods, such as sugar, are eaten by flies with these mouthparts.

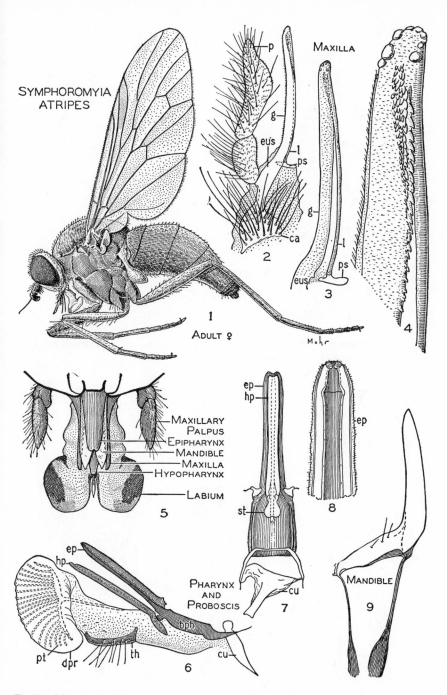

Fig. 50. Mouthparts of false black fly *Symphoromyia*, illustrating cutting-sponging type. (From Illinois Nat. Hist. Survey)

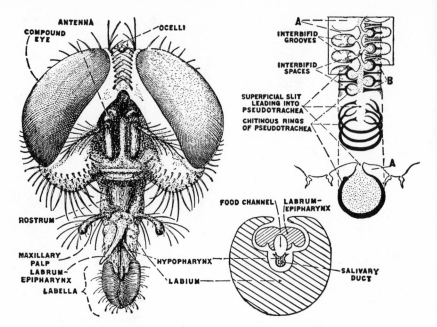

Fig. 51. Mouthparts of house fly, illustrating a sponging type. (From Metcalf and Flint, *Destructive and useful insects,* by permission of McGraw-Hill Book Co.)

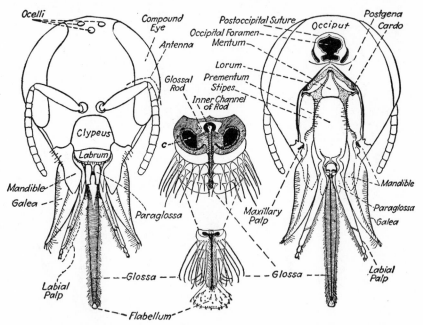

Fig. 52. Mouthparts of the honey bee, illustrating a chewing-lapping type. (From Metcalf and Flint, *Destructive and useful insects,* by permission of McGraw-Hill Book Co.)

This is accomplished as follows: First, the fly extrudes a droplet of saliva onto the food, which dissolves in the saliva; this solution is then drawn up into the mouth as a liquid.

Chewing-Lapping Type. Another type of mouthparts for taking up liquid food is found in the bees and wasps, exemplified by the honey bee, fig. 52. The mandibles and labrum are of the chewing type and are used for grasping prey or molding wax or nest materials. The maxillae and labium are developed into a series of flattened elongate structures, of which the alaglossa (usually called the glossa) forms an extensile channeled organ. This latter is used to probe deep into nectaries of blossoms. The other flaps of the maxillae and labium fit up against the glossa and form a series of channels down which the saliva is discharged and up which food is drawn. There is some difference of opinion among observers as to the exact mechanics by which passage of the liquids is attained.

Piercing-Sucking Type, Fig. 53. The mouthparts of many groups of insects are modified to pierce tissues and suck juices from them. This in-

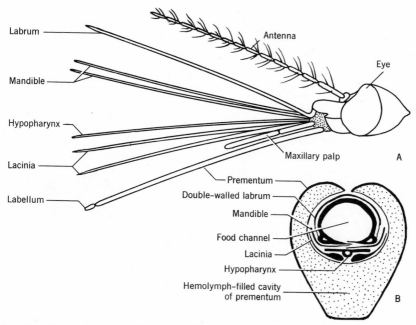

Fig. 53. Mouthparts of female mosquito, illustrating a piercing-sucking type. The double wall of the labrum permits this piece to be guided inside the host tissue. *A*, head and separated mouthparts; *B*, cross-section near middle of beak. (*A*, after Waldbauer; *B*, modified from Waldbauer.)

cludes aphids, cicadas, leafhoppers, scale insects, and others which suck juices from plants; assassin bugs, water striders, and predaceous forms of many sorts which suck juices from insects and other small animals; and mosquitoes, bedbugs, lice, and fleas, which suck blood from mammals and birds. In this group the labrum, mandibles, and maxillae (sometimes the hypopharynx also) are slender and long and fit together to form a delicate hollow needle. The labium forms a stout sheath that holds this needle rigid. The entire structure is called a beak. To feed, the insect presses the entire beak against the host, then inserts the needle into the host tissues, and sucks the host juices through the needle into the oesophagus.

An interesting and apparently generalized kind of this type of mouthparts is found in the thrips (see p. 261). Several components of the mouthparts are stylet-like but together form a rasping cone rather than a beak. On the extreme of complexity are the sucking lice or Anoplura, which have definite retractile beaks whose structure is so modified that only a few of the original parts remain (see p. 256).

Siphoning-Tube Type, Fig. 54. Adult Lepidoptera feed on nectar and other liquid food. These are sucked up by means of a long proboscis, composed only of the united galea of each maxilla. These form a tube which opens into the oesophagus.

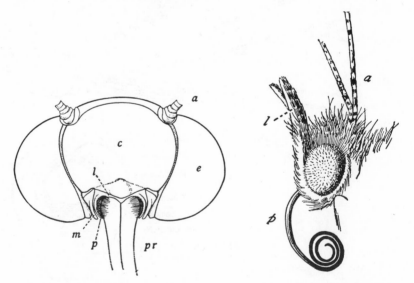

Fig. 54. Mouthparts of Lepidoptera, illustrating a siphoning type. Left, of a moth: *a*, antenna; *c*, clypeus; *e*, eye; *l*, labrum; *m*, mandible; *p*, pilifer; *pr*, proboscis. Right, of a butterfly; *a*, antennae; *l*, labial palpus; *p*, proboscis. (From Folsom and Wardle, *Entomology*, by permission of The Blakiston Co.)

CERVIX OR NECK

Between the head and trunk is a membranous region which forms a neck, or *cervix*. This has sometimes been regarded as a separate body segment, the microthorax, but little evidence has been found to support this view. It seems more likely that the cervix includes areas of both the labial head segment and the prothoracic segment, forming a flexible area between the two.

Embedded in the cervix are two pairs of *cervical sclerites,* fig. 55, which serve as points of articulation for the head with the trunk. The two sclerites on each side are hinged with each other to form a single unit, which articulates anteriorly with the occipital condyle on the postocciput of the head and posteriorly with the prothorax. Frequently the cervical sclerites are fused with the pleurae of the prothorax.

DEVELOPMENT OF THE GENERALIZED INSECT SEGMENT

The structure of existing Chilopoda and primitive insects suggests that the body segments of both groups evolved from a very simple type, fig. 56, composed of five elements:

1. The tergum or sclerotized dorsal plate, called the notum when referring to the thoracic tergum.

2. The sternum or sclerotized plate.

3. The pleural region connecting tergum and sternum; it is entirely membranous.

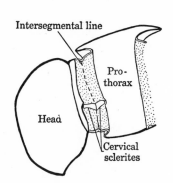

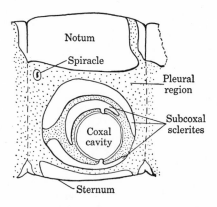

Fig. 55. Diagram of the cervical sclerites of an insect. (Modified from Snodgrass)

Fig. 56. Diagram of a simple insect body segment. (Redrawn from Snodgrass)

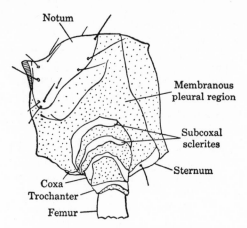

Fig. 57.   Mesothorax of a proturan *Acerentomon*.   (Redrawn from Snodgrass)

4. A pair of segmented legs; the basal segment or coxopodite of each leg is embedded in the membrane between the tergum and sternum. The coxopodite is divided into a basal portion (*subcoxa*) and an apical portion (*coxa*). In fig. 56 the subcoxa is divided into three sclerites.

5. A pair of spiracles, one in the membrane above each leg.

In a few archaic groups of insects and in the Chilopoda is found a type of segment, fig. 57, which represents the simple prototype. The tergum and sternum are unchanged. The subcoxa is represented by crescentic sclerites, or areas, one mesad of the coxa, and two laterad of it. The latter two appear to be units between the coxa of the leg and the tergum. These detached subcoxal sclerites in the pleural region are the forerunners of

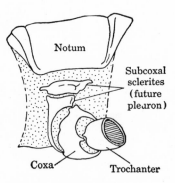

Fig. 58.   Prothorax of a stonefly nymph *Perla*. (Redrawn from Snodgrass)

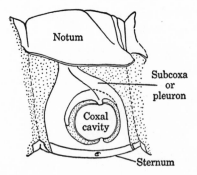

Fig. 59.   Diagram of a hypothetical step in the development of a winged segment.   (Redrawn from Snodgrass)

the pleural sclerites. The coxa forms the functional articulating base of the leg.

In the next evolutionary development, fig. 58, the subcoxal sclerites became immovably implanted in the segmental wall to form a solid base on which the functional leg articulates. The mesal subcoxal sclerites fused with the sternum, and the lateral subcoxal sclerites became flattened adjacent sidepieces together called the pleuron (pl., *pleura*). This condition is considered the generalized one from which developed both the specialized thoracic wing segments and the simplified abdominal segments. It is illustrated by many living forms, notably in the thorax of the immature stages of many such diverse groups as stoneflies, caddisflies, and lacewings, fig. 58.

THORAX

The thorax is the body region between the head and abdomen. It is composed of three segments, the prothorax, mesothorax, and metathorax, respectively.

In those orders which never developed wings, the three segments are nearly alike in general structure. The tergum and sternum are plate-like, and the pleural sclerites (the subcoxal arcs) are small or degenerate, fig. 57.

In the orders of winged insects, the three thoracic segments are extremely dissimilar. The prothorax has essentially the same parts as the basic condition, fig. 58, although the various sclerites may be consolidated or recombined to such an extent that their exact interpretation may be difficult. The mesothorax and metathorax have undergone a veritable morphological revolution correlated with the musculature necessary to combine both running and flying mechanisms in one segment. Many new sclerites have been added, and many of them regrouped.

Generalized Winged Segment. There are three principal areas in this segment as in the general wingless one: the tergum (called the *notum* when applied to segments of the thorax), the sternum, and the pleura. Their derivation from the primitive body segment is shown diagrammatically in Fig. 59. Each has many specializations, but those of the pleura are the most conspicuous external features accompanying the winged condition. In existing forms of winged insects the parts follow the generalized condition illustrated in Fig. 60.

PLEURON. This sclerite has become enlarged to form a conspicuous lateral plate. It has a ventral coxal process against which the leg articu-

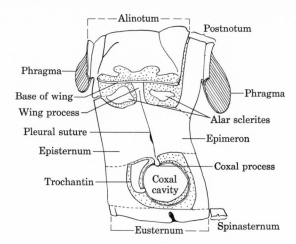

Fig. 60.   Diagram of a typical winged segment.   (Redrawn from Snodgrass)

lates and a dorsal wing process against which the wing articulates.   The pleuron is divided into an anterior portion, the *episternum,* and a posterior portion, the *epimeron,* by a *pleural suture* which extends from the coxal process to the wing process.   This suture marks the line of invagination of the internal pleural apodeme, the *pleurodema.*   Anteriorly and posteriorly the pleuron is fused with the sternum, the areas of fusion forming bridges before and behind the coxal cavities.

NOTUM.   This area is divided into two principal sclerites, the anterior *alinotum* and the posterior *postnotum.*   The alinotum has an anterior apodeme or *phragma* and also is the sclerite connecting directly with the wing. It is subdivided in various patterns in different groups.   The postnotum also bears a phragma and is connected laterally not with the wings but with the epimeron of the pleura to form a bridge behind the wings.   The postphragmal portion of the postnotum really is a part of the next posterior segment which has become a functional part of the one in front.   Transpositions of this type are frequent in insects.

STERNUM.   This plate is joined by anterior and posterior bands to the pleura, thus forming the socket in which the coxa is situated.   The central portion, called the *eusternum,* has a furrow which marks the line of invagination of the large internal apodeme called the *furca,* so named because it is forked or double at its apex, fig. 61.   Posterior to the eusternum is a small sclerite, the *spinasternum,* bearing internally a single apodeme, the *spina.*   This spinasternum had its origin in the membrane between the segments, but usually it is coalesced with the segment anterior to its place of origin.

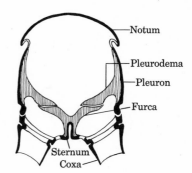

Fig. 61. Diagrammatic cross section of thoracic
segments illustrating the furca and pleurodemae.
(Redrawn from Snodgrass)

INTERNAL SKELETON. The various apodemes of the segments are frequently referred to collectively as the internal skeleton. They serve as
areas of attachment for many of the large leg and wing muscles. The
pleurodemae and furcae of a segment fit together closely as an almost
continuous band, fig. 61.

EXISTING FORMS. In a general discussion of the thorax it is impractical
to go into more detail than the preceding outline because of the almost
endless variety in the thoracic structure of existing insects. Many orders
exhibit a distinctive basic pattern, and in large orders such as the Coleoptera and Diptera there may be many extreme modifications within the
same order.

Often there is little apparent similarity between some existing form and
the generalized type. In such cases identification of the sclerites must be
preceded by orientation in relation to stabilized features or landmarks.
The apodemes and the sutures which are their external indications, plus
articulation points for legs and wings, are the most reliable.

LEGS. The typical thoracic leg consists of six parts, the *coxa, trochanter,
femur, tibia, tarsus,* and *pretarsus,* fig. 62A. The coxa is the segment which
articulates with the body; it may bear a posterior lobe called the *meron.*
The tarsus of adult insects is usually subdivided into two to five segments.
The pretarsus appears as a definite small end segment of the leg in larvae
of several orders and in the Collembola. In almost all other insects the
pretarsus is represented by a complex set of claws and minute sclerites set
in the end of the tarsus. The Collembola and Protura are also unique in
that the tibia and tarsus form a single tibiotarsus.

In general, insects have simple legs designed for walking or running,
fig. 62A. There have developed, however, a large number of modifications which fit the legs for other uses. These include jumping types with
greatly enlarged femora, as in the grasshoppers, fig. 62B; grasping types
armed with sharp opposing spurs and spines, fig. 62C, as in the praying

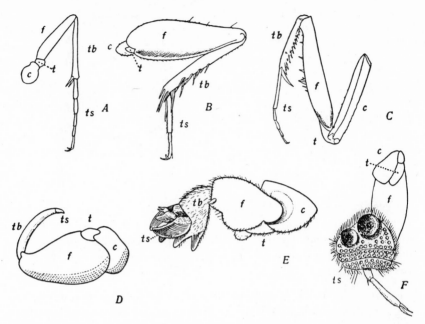

Fig. 62. Legs of insects adapted for: *A*, walking; *B*, jumping (hind leg of cricket); *C*, grasping (front leg of mantid); *D*, clasping (front leg of bug); *E*, digging (front leg of mole cricket); *F*, holding fast by suction (front leg of male diving beetle). *c*, coxa; *f*, femur; *s*, spur; *t*, trochanter; *tb*, tibia; *ts*, tarsus. (From Folsom and Wardle, *Entomology*, by permission of The Blakiston Co.)

mantis; swimming types, having long brushes of hair and flattened parts, as in the water boatmen; and digging types, with strong, scraperlike parts, fig. 62*E*, such as found in the mole crickets.

Wings

Insect wings are a unique evolutionary development found in no other organisms. In such flying animals as bats and birds, the wings are highly modified front legs. In insects, this is not the case. Their wings are outgrowths of the body wall along the lateral margins of the dorsal plate or notum. Unlike most other insect appendages, the wings have no muscles attached inside them.

Wings, giving the power of flight, have been one of the most important reasons for the success of the insect group as a whole. No other invertebrate group has ever developed wings. Because flight is so characteristic of the insect group, it is treated here in more detail than other activities such as walking which are achieved by many animal groups.

Typically, pterygote insects have two pairs of wings, the mesothorax and metathorax each bearing a pair. The prothorax is always wingless. In a few fossil forms the prothorax had lateral flaps, but there is no evidence to indicate that it ever had functional wings.

Origin. The manner in which wings and flight were developed by insects is not certain. Known fossils of the most ancient winged insects had wings which were as well developed as those of some present-day orders. No intermediate types have been discovered between these groups with functional wings and the primitive wingless forms such as the silverfish.

One of the more plausible theories regarding their origin is that wings began as flat lateral angulations of the notum that aided the insect in alighting right side up and ready to run when dropping from a higher to a lower level. Even so, it must have happened only under a particular set of circumstances. The insects must have frequented surfaces such as leaves several feet from the ground or vertical surfaces such as rocks or trees, otherwise the insects would not have attained enough speed in the course of dropping for the angular sides of the notum to "bite" into the air stream. It has been calculated that short-legged insects slightly under 1 cm. in length could have satisfied these requirements and have a stable, swooping glide at the end of their drop. The insects presumably were either predators chasing prey or, more likely, preyed on by swift predators such as spiders and centipedes. Under either circumstance, speed and maneuverability would have been an asset to survival. We can visualize something like a *Machilis,* fig. 165, with its long, rudder-like tails as filling all these requirements.

When stabilized dropping had been achieved, there would have come into play a selection pressure for the notal angles to enlarge into planing fins or flanges that would decrease the time required for the onset of gliding and increase the length of the glide. Some entomologists believe that these lateral flanges must have evolved a hinge at the base before they became very large, otherwise the flanges would have been a hindrance to the individual ducking under ground cover for shelter. The next step to flight would have been the flapping of these hinged extensions while the insect was airborne. This type of flight could have been controlled readily by primeval muscles still represented in some wingless segments of living insects. In the evolution from planing to flight, insect wings became longer and also narrower at the base where they connected with the body.

Structure. In basic design, insect wings are very simple. They are a flaplike extension of the body wall, with an upper and lower membrane, between which run supporting "fibers" called veins. The base of the

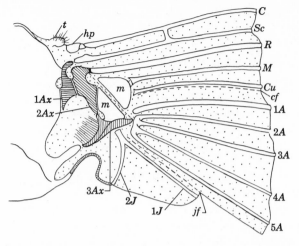

Fig. 63. Base of wing showing axillary sclerites connecting wing and thorax. *Ax*, principal axillary; *hp*, humeral plate; *m*, median plate; *t*, tegula. (Modified from Snodgrass)

wing connects with the body by a membranous hinge in which are set a group of small sclerites called *axillary sclerites*, fig. 63. These articulate with the edge of the notum. Closely associated with them are two alar sclerites, the *basalar* and *subalar sclerites*, which lie one on each side of the wing process of the pleuron.

Venation. In most wings, fig. 64, there are many thickened lines which strengthen the thin membrane. Some of these lines run from the base of

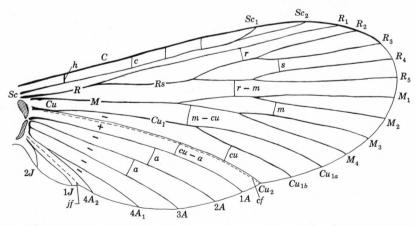

Fig. 64. Diagram of an insect wing showing typical veins and crossveins of a common type of venation.

the wing toward the apex; these are called *veins*. Others run more or less crosswise of the wing and connect veins; these are called *crossveins*. The pattern of veins and crossveins is termed *venation*.

Many investigators support the theory that veins and crossveins evolved from tracheae or air tubes which, in the adult stage, become detached at their bases and serve as wing supports. It is possible, however, that the venation arose as thickenings of the wing independently of tracheae.

The wings of insects exhibit innumerable differences in venation. These differences are of great importance in classification, because characters of venation identify many orders, families, and genera. All types of venation seem to have developed from the same basic pattern. This applies only to the main trunks of the veins.

The basic pattern of a common type of venation is shown in fig. 64. This is diagrammatic, since no such exact wing is known, but it combines evidence from the conditions of many. Each main vein has a definite name; these are listed here in order of occurrence from the anterior to the posterior margin of the wing. Certain veins have definite typical branches. Standard abbreviations are indicated for the veins.

COSTA (C) usually forms the thickened anterior margin of the wing. It is unbranched.

SUBCOSTA (Sc) runs immediately below costa, always in the bottom of a trough between costa and radius. Typically subcosta is divided into two branches.

RADIUS (R) is the next main vein. It is a stout one and connects at the base with the second axillary sclerite. It is divided into two main branches, R_1 and radial sector Rs. Radial sector is frequently divided into four main branches.

MEDIA (M) is one of two veins articulating with some of the small median axillary sclerites. The base is usually in a depression. In typical Paleoptera media is divided into two main branches, MA and MP. In the Neoptera MA is either atrophied or fused with, and appearing to be an integral part of Rs, and MP is called simply M. In these latter orders M is divided typically into four branches, fig. 64.

CUBITUS (Cu) also articulates with the median axillary sclerites, and has two main branches. Its basal portion and Cu_2 are in a depression; but Cu_1 runs along a ridge and is usually branched.

CUBITAL FURROW (cf) is a definite crease along which the wing folds. It is not a vein but is one of the most important landmarks for identifying the cubital and anal veins, which it separates.

ANAL VEINS ($1A$, $2A$, $3A$, etc.) form a set which are united or close together at the base and closely associated with the third axillary sclerites, $3Ax$.

JUGAL FURROW (jf) is a crease separating the anal region or fold from the jugal fold, which is the small area at the basal posterior corner of the wing. This fold also is one of the most stable wing landmarks.

JUGAL VEINS ($1J$, $2J$) are short veins in the jugal fold.

CROSSVEINS. Definite names are given to the kinds of crossveins, based on the veins which they connect. These crossveins have standard abbreviations which are never written in capital letters; these are outlined in Table 3. Numbers are used to denote individual crossveins in a series, for example, fourth costal crossvein, third radio-medial crossvein, and so on. There is one notable exception: The crossvein between costa and subcosta at the base of the wing is called the humeral crossvein and indicated by h. In orders such as the Trichoptera and Lepidoptera in which the crossveins are greatly reduced, the crossvein between R_1 and R_2 is called the radial crossvein r, and that between R_3 and R_4 is called the sectorial crossvein s.

Table 3.   Terminology of crossveins

Veins Connected	Name of Crossveins	Abbreviation
Costa to subcosta or R_1	Costal	c
Branches of radius	Radial	r
Radius to media	Radio-medial	$r-m$
Branches of media	Medial	m
Media to cubitus	Medio-cubital	$m-cu$
Branches of cubitus	Cubital	cu
Cubitus to anal	Cubito-anal	$cu-a$
Various anals	Anal	a

Evolution of Wing Design. The earliest functional wing undoubtedly had lengthwise fanlike folds from the base to the apical portion of the wing, fig. 65. The veins evolved as supporting structures running along each crease of the fan, with crossveins forming a system of braces between the veins. This combination of fanlike design and trusslike supports enabled a tough but thin membrane to have sufficient rigidity for flight. Among present-day insects only the mayflies and dragonflies have retained this simple fanlike type of wing. In these insects the axillary

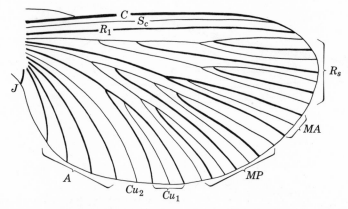

Fig. 65. Reconstruction of a possible prototype of the earliest functional insect wing (cross-veins omitted). The wing is fluted; in this drawing the veins on the crests of ridges are represented by thick lines, those at the bottom of furrows by thin lines. In nature this differential thickening does not occur. (Adapted from Edmunds)

sclerites are simple and provide no mechanism for folding the wings over the body.

The next step in the evolution of wings was the development of more complicated divisions of axillary sclerites, fig. 63, which aid in folding the wings compactly over the back. Along with this step, much of the wing surface lost its fanlike character and became almost flat. The vein *MA* either atrophied or fused with the radial sector. The vein still readily identifiable as media is usually labeled simply *M*, although it is homologous to *MP* of the archaic wing. This advanced type of wing is that illustrated in fig. 64, and is the basic pattern for the modern or neopteran orders of insects.

In the primitive wings there were probably a very large number of crossveins which acted as a bracing for the wing. Such a condition is found in such forms as the dragonflies, fig. 172, and the cockroaches.

There has been a steady evolution in almost every order of insects toward a stronger supporting system for the wing. In many cases this has involved a reduction in the number of both crossveins and vein branches, usually accompanied by either a union of various main veins or a realignment of veins or both. The wing in fig. 64 illustrates reduction of crossveins with no reduction of main veins. In the honey bee wing, fig. 66, the crossveins are not greatly reduced but the veins have become reduced in number, fused, and realigned. In the thrips wing, fig. 67, the venation is so reduced that identification of the few remaining veins is almost impossible.

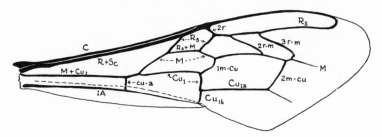

Fig. 66.   Front wing of a honey bee.

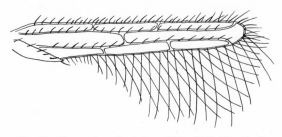

Fig. 67.   Front wing of a thrips.   (After D. L. Crawford)

ABDOMEN

The abdomen is the third and posterior region of the insect body.   It is relatively simple in structure, compared with the thorax, and in adults has no walking legs.   Primitively it is 12-segmented but this condition is distinct only in Protura, fig. 163, and certain embryonic stages.   Usually the abdomen consists of 10 or 11 segments, fig. 68A.   In some forms much greater reduction occurs, as in the Collembola which have only six abdominal segments, fig. 164.   Many groups, such as the housefly group, have the last several segments developed into a copulatory structure or an ovipositor which is normally retracted within the preceding segments.

Segmental Structure.   In adult insects, a typical segment consists of (1) a tergum or dorsal plate, (2) a sternum or ventral plate, (3) lateral areas of membrane connecting tergum and sternum, and (4) a spiracle on each side, usually situated in the lateral membrane.   In some larval forms, fig. 68B, and a few adults, there are sclerites in the lateral membrane; some of these undoubtedly represent vestiges of the subcoxal sclerites of the primitive appendages (see fig. 57, p. 72).

Appendages.   These may be divided roughly into two groups: those not associated with reproduction, and those developed for reproductive activities such as mating or oviposition.

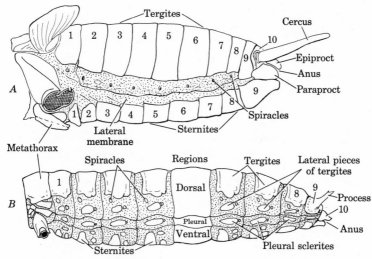

Fig. 68. Metathorax and abdomen of insects. *A*, adult cricket *Gryllus; B*, larva of ground beetle *Calosoma.* (Redrawn from Snodgrass)

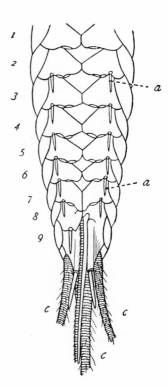

Fig. 69. Ventral aspect of the abdomen of a female *Machilis maritima,* to show rudimentary limbs (*a*) of segments 2 to 9. (The left appendage of the eighth segment is omitted.) *c, c, c,* lateral cerci and median pseudocercus. (From Folsom and Wardle after Oudemans)

NON-REPRODUCTIVE TYPES.   In most adult insects abdominal appendages are absent except on the terminal segments.   A few primitive forms
have retained degenerate legs represented by styli, as in the silverfish,
fig. 69.   The appendages of the eleventh segment, the cerci, are present in
most insects, fig. 68.   They are usually tactile organs and in such groups
as caddisflies become part of the male genitalia.   The cerci may appear
to belong to the tenth or ninth segment if the eleventh or tenth segment is
reduced.   In larvae and nymphs a great variety of abdominal appendages
are developed.   Well-known examples are the larvapods of caterpillars,
fig. 34, and the segmental gills of mayfly nymphs, fig. 168.

REPRODUCTIVE TYPES.   These generally include appendages of the
eighth and ninth segments.   Many morphologists believe that these appendages are homologous with true segmental appendages such as the
mouthparts and legs.   Others believe that the primeval appendages of the
eighth and ninth segments have been lost and that the appendages associated with reproduction are special structures arising from more mesal
outgrowths of the same segments.

FEMALE.   The ovipositor, fig. 70, is composed chiefly of three pairs of
blades, the first, second, and third valvulae, respectively.   The *first valvulae*
arise from a pair of plates, the *first valvifers* of the eighth segment.   The
valvifer and valvula probably correspond to the coxopodite and telopodite
of the generalized arthropod segment, fig. 41.   The *second valvifers* bear a
ventral pair of blades, the *second valvulae,* and a dorsal pair, the *third valvulae.*
In most insects with a well-developed ovipositor, such as the sawflies,
fig. 71, the first and second valvulae form a cutting or piercing organ with
an inner channel down which the eggs pass.   The third valvulae form a
scabbard or sheath into which the ovipositor folds when retracted.   In
Orthoptera, either all three pairs fit together to form the functional ovipositor, or the second valvulae form a short egg guide.
   In many insects the valvulae are only poorly or not at all developed, in
which case the apical segments of the abdomen generally form an extensile
tube that functions as an ovipositor.   This is exemplified by many of the
Lepidoptera and Diptera, fig. 72.

MALE.   In this sex the appendages of the ninth segment are usually
combined with parts of the ninth segment proper and sometimes parts of
the tenth to form a copulatory organ.   In each order this organ usually
displays fundamental peculiarities.   It is extremely difficult to homologize
the individual parts of these copulatory organs throughout the insect
orders or to be certain of their relation to what must have been the simple
parts and appendages from which they are derived.

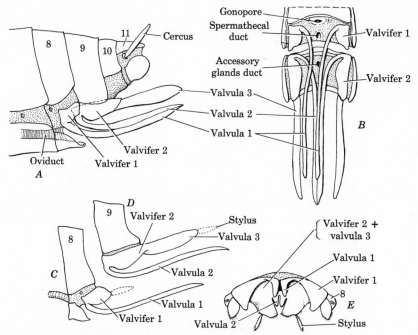

Fig. 70. Structure of the ovipositor of pterygote insects (*A–D*, diagrammatic). *A*, *B*, showing segmental relations of the parts of the ovipositor. *C*, *D*, lateral view of genital segments and parts of ovipositor dissociated. *E*, nymph of *Blatta orientalis*, ventral view of genital segments with lobes of ovipositor. (Redrawn from Snodgrass)

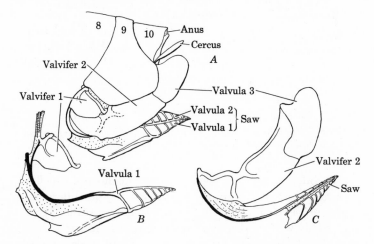

Fig. 71. The ovipositor of a sawfly. *A*, showing relation of basal parts to each other and to ninth tergum; *B*, first valvifer and valvula; *C*, second valvifer with second and third valvula. (Redrawn from Snodgrass)

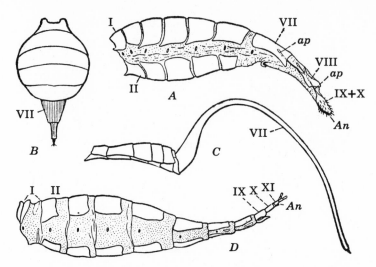

Fig. 72. Examples of an "ovipositor" formed of the terminal segments of the abdomen. *A*, a moth *Lymantria monacha* (from Eidmann, 1929). *B*, a fruit fly *Paracantha culta. C*, a fruit fly *Toxotrypania curvicauda. D*, a scorpionfly *Panorpa consuetudinis. ap*, apodeme; *An*, anus. (After Snodgrass)

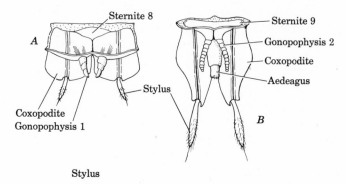

Fig. 73. Male genitalia of the microcoryphian *Machilis variabilis,* dorsal aspect. *A*, first gonopods, showing gonapophyses of eighth segment; *B*, second gonopods and median copulatory organ. (Modified from Snodgrass)

Structural differences in the copulatory organs furnish excellent taxonomic characters in many groups of insects for the differentiation of families, genera, or species. In any one group the constituent parts of the organ are usually well marked, and in each group there is a clear terminology for the designation of these parts. Until closer agreement is reached regarding the homologies of these structures in different orders, it is more practical to employ the terminology in established usage for any

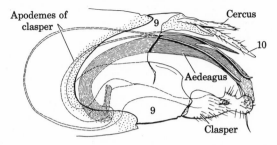

Fig. 74. Male genitalia of a caddisfly. (From Illinois Nat. Hist. Survey)

particular group. A simple type is illustrated by Microcoryphia, fig. 73, and a more complicated type occurs in the caddisflies, fig. 74. These two examples by no means give the full range of different types among the insect orders but will serve to indicate some of the variety in structure which may be found.

MUSICAL ORGANS

Insects make noises in various ways. In some cases the noise is produced by the insect's normal activities without aid of special noise-making structure. The most familiar example is the hum made by flying or hovering insects; the hum or note is produced by the extremely rapid vibrations of the wings and thoracic sclerites.

A few insect groups have special sound-producing structures. The sound waves are produced by the vibration of a wing membrane, a specialized portion of the body wall, or special membranes. Those areas are set in motion by structures specialized for the purpose. Grasshoppers which produce a crackling sound have a simple mechanism. The front margin of the hind wing scrapes over the thickened veins of the fore wing, causing the latter to vibrate. In other grasshoppers the inner face of the hind femur is provided with a file of minute teeth, fig. 75; this file is rubbed over the fore wing to make the latter vibrate. Various crickets have a file

Fig. 75. File on inner face of hind femur of a cricket. *A*, hind femur of *Stenobothrus;* *B*, file greatly enlarged. (After Comstock, *An introduction to entomology,* by permission of The Comstock Publishing Co.)

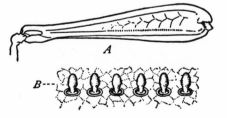

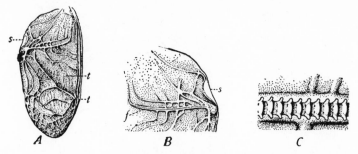

Fig. 76. Fore wing of *Gryllus,* showing file and scraper. *A,* as seen from above, that part of the wing which is bent down on the side of the abdomen is not shown; *s,* scraper; *t, t,* tympana. *B,* base of wing seen from below; *s,* scraper; *f,* file. *C,* file greatly enlarged. (After Comstock, *An introduction to entomology,* by permission of The Comstock Publishing Co.)

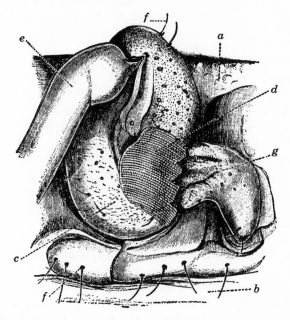

Fig. 77. Stridulating organ of a larva of *Passalus; a, b,* portions of the metathorax; *c,* coxa of the second leg; *d,* file; *e,* basal part of femur of middle leg; *f,* hairs with chitinous process at base of each; *g,* the diminutive third leg modified for scratching the file. (From Comstock, after Sharp)

on one or both wings, application of which causes a special area of the wing to vibrate, fig. 76. In other orders, such as the beetles, the scraper and file may be on the leg and body, respectively, fig. 77. It appears in these cases that the body wall itself serves as the vibrating surface. A

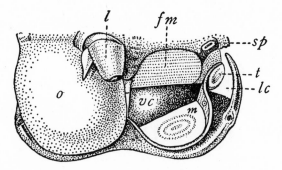

Fig. 78. The musical apparatus of a cicada; *fm*, folded membrane; *l*, base of leg; *lc*, lateral cavity; *m*, mirror; *o*, operculum, that of the opposite side removed; *sp*, spiracle; *t*, timbal; *vc*, ventral cavity. (After Carlet)

unique mechanism is developed in the cicadas, fig. 78. They possess a set of membranes situated in ventral pouches or cavities near the base of the abdomen. One of these membranes is connected internally with a muscle fiber. The contraction of this muscle pulls the membrane inward; the relaxation of the muscle allows the membrane to snap back to its original shape. These movements are alternated with great speed to produce sound waves. The other membranes act as sound reflectors.

REFERENCES

Edmunds, G. R., Jr. and Jay R. Traver, 1954. The flight mechanics and evolution of the wings of Ephemeroptera, with notes on the archetype insect wing. *J. Wash. Acad. Sci.,* **44**:390–399.

Flower, J. W., 1964. On the origin of flight in insects. *Jour. Ins. Physiol.,* **10**:81–88.

Imms, A. D., 1925. *A general textbook of entomology.* London: Methuen & Co. 698 pp.

Matsuda, Ryuichi, 1958. On the origin of the external genitalia of insects. *Ann. Ent. Soc. Amer.,* **51**:84–94.

Schmitt, J. B., 1938. The feeding mechanism of adult Lepidoptera. *Smithsonian Misc. Collections,* **97**, No. 4:1–28.

Scudder, G. G. E., 1961. The comparative morphology of the insect ovipositor. *Trans. R. Ent. Soc. London,* **113**:25–40.

Snodgrass, R. E., 1935. *Principles of insect morphology.* New York: McGraw-Hill Book Co. 667 pp.

1944. The feeding apparatus of biting and sucking insects affecting man and animals. *Smithsonian Misc. Collections,* **104**, No. 7:1–113.

1950. Comparative studies on the jaws of mandibulate arthropods. *Smithsonian Misc. Collections,* **116**, No. 1:1–85.

Weber, Hermann, 1952. Morphologie, Histologie und Entwicklungsgeschichte der Articulaten. *Fortschr. Zool.,* N. F., **9**:18–231.

1954. *Grundriss der Insektenkunde,* 3d rev. ed. Stuttgart, Fischer. 428 pp.

Chapter Four

Internal Anatomy

THE internal anatomy of insects involves primarily the organs which carry on the vital functions of life. These organs are protected from outside forces by the body wall. If parts of certain organs project as flaps or lobes beyond the body outline, they are incased by a thin mantle of body wall and are thus within the confines of the exoskeleton.

DIGESTIVE SYSTEM

The digestive system is the food tract and its accessory parts. It is composed of the alimentary canal and various glands connected with it either directly or indirectly. Typically these include the salivary glands, gastric caeca, and Malpighian tubules.

Alimentary Canal, Fig. 79. This organ is a tube passing through the central part of the body. Its anterior opening, the mouth, is situated at the base of the *preoral cavity* (the space enclosed by the mouthparts); its posterior opening, the *anus,* is on the posterior body segment. The alimentary canal is divided into three distinct parts: an anterior *stomodeum,* a middle *mesenteron,* and a posterior *proctodeum.* Usually between the stomodeum and mesenteron is the *stomodeal* or *cardiac valve,* and between

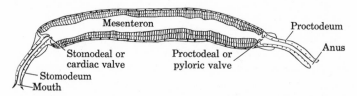

Fig. 79. The alimentary canal of a collembolan *Tomocerus niger,* showing the primary components of the food tract without secondary specializations. (Redrawn from Snodgrass)

90

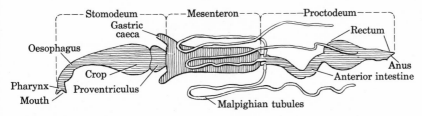

Fig. 80. Diagram showing the usual subdivisions and outgrowths of the alimentary canal. (Redrawn from Snodgrass)

the mesenteron and the proctodeum is the *proctodeal* or *pyloric valve.* The stomodeum and proctodeum result from embryonic infoldings of the ectoderm; the mesenteron is formed from the endoderm; this is discussed further in the portion on embryology.

In a few primitive insects the three parts of the alimentary canal are simple and tubular in shape, fig. 79. In most insects, however, each of these parts has become differentiated into functional subdivisions. The typical structure of these is as follows, fig. 80:

STOMODEUM. This portion is usually divided into three main portions: (1) an anterior, more or less tubular portion, the *oesophagus,* followed by (2) an enlarged portion, the *crop,* which narrows to (3) a valvelike *proventriculus* at the junction with the mesenteron. An indefinite portion of the oesophagus at the mouth opening is frequently called the *pharynx* but is difficult to identify without a knowledge of the musculature. The boundary between oesophagus and crop is frequently arbitrary, as in fig. 80; in some insects such as certain moths, fig. 81C, the crop is developed into a spherical chamber; this modification is carried still further by many flies, fig. 82, and the crop forms a sack connected to the oesophagus by a long lateral tube. The proventriculus may be a simple valve opening into the mesenteron; in insects which eat solid food, it bears a series of hooks for food shredding and is called the *gastric mill.*

MESENTERON. This middle portion of the alimentary canal is the place where most digestion takes place. It may be called the *ventriculus* or stomach. Usually it is tubular, but occasionally it is subdivided into definite parts. This subdivided condition is most pronounced in the Hemiptera, in which the mesenteron may have three or four sections. The mesenteron typically bears several fingerlike outgrowths, the *gastric caeca.* These usually occur at the anterior end of the stomach, fig. 80, but may be situated on more posterior portions.

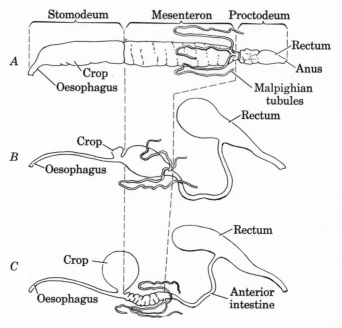

Fig. 81. Transformation of the alimentary canal of a moth *Malacosoma americana,* from the larva, *A*, through the pupa, *B*, to the adult, *C*. (Redrawn from Snodgrass)

PROCTODEUM. This posterior portion of the alimentary canal varies greatly in different insects but is usually divided into a tubular *anterior intestine* and an enlarged *posterior intestine*. This latter is termed the *rectum* and is connected directly with the anus.

Malpighian Tubules. With few exceptions insects possess a group of long slender tubules branching from the alimentary canal near the junction of the mesenteron with the proctodeum, fig. 80. These are the Malpighian tubules, which are excretory in function. The number of these tubules varies from 1 to 150. When a large number of these are present, they are often grouped into bundles of equal size.

Labial or Salivary Glands. Most insects possess a pair of glands lying below the mesenteron, fig. 82, and associated with the labium. Each gland has a duct running anteriorly. These unite, usually within the head, to form a single duct that opens into the preoral cavity between the labium and the hypopharynx. The function of these glands differs in various insects and in some has not been determined definitely. In most insects the labial glands secrete saliva, as in cockroaches. In lepidopterous

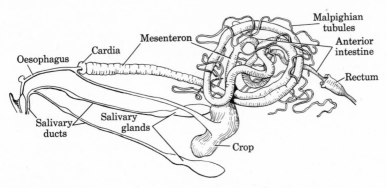

Fig. 82. Alimentary canal and salivary glands of a fruit fly *Rhagoletis pomonella,* showing the diverticular crop and the cardia of the mesenteron, characteristic of many Diptera. (Redrawn from Snodgrass)

and hymenopterous larvae these glands secrete silk, used in making larval nests and pupal cells. In blood-sucking insects the labial glands secrete an anticoagulin which keeps ingested blood in liquid form.

CIRCULATORY SYSTEM

The circulatory system comprises chiefly the blood and tissues and organs which cause its circulation through the body. In many animals, such as the vertebrates, the blood travels only through special vessels (arteries, capillaries, and veins) developed for this purpose. This condition is called a closed system. In insects this is not the case. For most of its course the blood simply flows through the body cavity, irrigating the various tissues and organs. There is a special pumping organ or heart situated dorsally in the insect body which pumps the blood from the posterior portion of the body and empties it into the internal cavity of the head. From this cavity the blood again flows back through the body, is drawn into the heart, and is again pumped forward, and so on. This kind of arrangement is called an *open system,* and the body cavity through which the blood flows is called a *hemocoel.*

Blood. The fluid that circulates through the body cavity is called the blood or *hemolymph.* It consists of a liquid part, the *plasma;* and an assortment of free floating cells, called blood corpuscles or *hemocytes.* Study of the blood involves histology and physiology and is discussed in the next chapter.

Dorsal Vessel. As its name implies, the dorsal vessel, fig. 83*A*, lies directly beneath the dorsum, or dorsal wall. It extends the length of the body, from the posterior end of the abdomen into the head. It is the principal pulsating organ which causes the flow of the blood.

The dorsal vessel is divided into two parts: a posterior portion called the *heart,* and an anterior portion called the *aorta.* In general, the heart is the pulsating portion, and the aorta is the tube which carries the blood forward and discharges it into the head.

The *heart* is usually more or less swollen in each segment to form *chambers* separated by constrictions. This chambered portion typically consists of nine parts, occurring in the first nine segments of the abdomen. Each chamber has a pair of lateral openings or *ostia,* through which blood enters the chamber. In certain insects the heart may depart radically from this typical condition. In cockroaches and japygids, for example, the first two chambers occur in the meso- and metathorax. In the bug *Nezara* the heart consists of a single large chamber having three pairs of ostia.

The *aorta* is typically a simple tubular extension of the heart. In some forms, such as cranefly larvae, the aorta also pulsates and thus is an accessory to the heart in causing circulation.

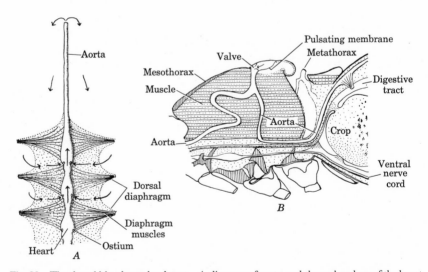

Fig. 83.   The dorsal blood vessel or heart. *A,* diagram of aorta and three chambers of the heart with corresponding part of the dorsal diaphragm, dorsal view, arrows indicating the course of blood circulation. *B,* vertical section of thorax and base of abdomen of *Sphinx convolvuli* showing pulsating membrane in mesothorax. (Redrawn from Snodgrass)

Dorsal Diaphragm and Sinus. Connected to the underside of the heart are pairs of muscle bands known as *wing muscles* or *alary muscles.* This name is applied because the muscles form flat fans or wings which connect the heart and the lateral portions of the tergites. These aliform muscles, when well developed, form a fairly complete partition between the main body cavity and the region around the heart. In such cases the partition is called the *dorsal diaphragm,* and the segregated heart region is termed the *dorsal sinus.* The diaphragm and sinus extend only as far as the heart and are not continued forward in the region of the aorta.

Accessory Pulsating Organs. In addition to the heart there may occur other pulsating organs for assisting in blood circulation. The two of most frequent occurrence are the thoracic pulsating organs and the ventral diaphragm. Other accessory structures are found only rarely.

THORACIC PULSATING ORGANS, FIG. 83*B*. In many insects, especially rapid fliers such as hawk moths, there is a pulsating organ which draws blood through the wings and discharges it into the aorta. The pulsating organ itself is a cavity in the scutellum provided with a flexible or pulsating membrane. The outlet of the structure is a tube called the aortic diverticulum which connects directly with the aorta.

VENTRAL DIAPHRAGM. Many Orthoptera, Hymenoptera, and Lepidoptera have muscle bands developed *over* the ventral nerve cord in much the same manner as the aliform muscles form a diaphragm *under* the heart, which is dorsal in position. When such a muscle band is formed over the nerve cord, it is known as the ventral diaphragm. By expansion and contraction it produces a flow of blood posteriorly and laterally.

TRACHEAL SYSTEM

Most insects possess a system of internal tubes or *tracheae,* for conducting free air to the cells of the body. This system of tubes is the *tracheal system,* and it performs the function of *respiration.* In almost all other animals respiration is a function of the blood stream, in conjunction with aerating surfaces such as skin or lungs. In addition to insects, however, a few groups of the Arthropoda possess a well-developed tracheal system. These groups include some of the Arachnida, a few Crustacea, and most of the Chilopoda. Rudimentary tracheal tubules are found in Onychophora and Diplopoda.

This tracheal system is of necessity highly complex, because it must branch into a myriad of fine tubules, each of which reaches intimately only a small group of cells. This intricate branching of tracheae in insects is analogous to that of the blood vessels and capillaries in the vertebrates.

Principal Components of Tracheal System

A common type of tracheal system is shown in fig. 84. The tracheae form definite groups in each segment and receive air from the exterior by means of segmentally arranged pairs of openings called *spiracles* (*s*). The spiracles join more or less directly with a main tracheal trunk (*t*), a pair of which usually run the full length of the body. In each segment there arise from these trunks various branches (always paired, since one comes from each trunk) which aerate the tissues of the organs. The number and position of these branches vary greatly in different insects, but generally there are three large branches given off on each side in any one segment:

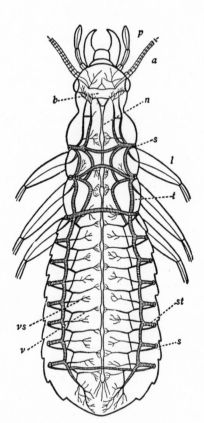

Fig. 84. Tracheal system of an insect. *a*, antenna; *b*, brain; *l*, leg; *n*, nerve cord; *p*, palpus; *s*, spiracle; *st*, spiracular branch; *t*, main tracheal trunk; *v*, ventral branch; *vs*, visceral branch. (From Folsom, after Kolbe)

(1) a dorsal branch aerating the dorsal vessel and dorsal muscles, (2) a ventral or visceral branch aerating the digestive and reproductive organs, and (3) a ventral branch aerating the ventral muscles and nerve cord.

The fine tips of the tracheae divide into minute capillary tubes, or tracheoles, usually one micron or less in diameter. These tracheoles ramify between and around cells of other tissues and are the functional part of the system through which oxygen diffuses into the body cells.

Tracheal Trunks. The segmental arrangement of clusters of tracheal branches indicates that originally insects had an independent tracheal system in each postoral segment, with no connection between tracheae of different segments. With only few exceptions, however, insects of the present have connections between the tracheae of adjoining segments if the tracheal system is developed. These connecting tubes form trunks. In many insects the main tracheal trunks are lateral in position and are called the *lateral tracheal trunks.* Frequently a second pair of *dorsal tracheal trunks* are found, one on each side of the heart. These are usually small in diameter and secondary to the lateral trunks. In most fly larvae the opposite is the case. There the dorsal trunks are greatly developed and are the chief respiratory passages, fig. 85.

Tracheal Air Sacs. An important feature in many groups is the development of air sacs that serve as air-storage pockets to aid respiration. These are often enlargements of the tracheal trunks, as in fig. 86. In many fast-flying insects, such as the housefly and the bees, the sacs fill a large part of the body cavity. These sacs can be squeezed and released by muscular contraction of the body to act like bellows and increase intake and expulsion of air.

Spiracles. When functional, spiracles are an important control over respiration. They are extremely varied in size, shape, and structure. If functional, they all have some sort of closing device. This device may be *external* (usually in the form of two opposed lips), or it may be *internal* (usually in the form of a clamp which pinches the trachea shut).

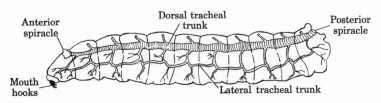

Fig. 85.   Fly larva illustrating large dorsal tracheal trunks.   (Redrawn from Snodgrass)

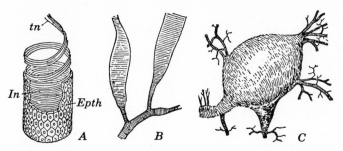

Fig. 86. Structure of a tracheal tube, *A*, and examples of tracheal air sacs, *B*, *C*. *Epth*, epithelium; *In*, intima; *tn*, taenidium in spiral band of cuticular intima artificially separated. (After Snodgrass)

Open Tracheal Systems

Systems in which spiracles are open and functional are called open systems. The more generalized type has ten pairs of spiracles, a pair on the mesothorax, metathorax, and each of the first eight abdominal segments. Many modifications of this type occur, including such examples as mosquito larvae, having spiracles on only the eighth abdominal segment; most of the maggots, which have only prothoracic spiracles and the terminal pair on the eighth abdominal segment; fly pupae, which have only the prothoracic spiracles; several aquatic forms, such as the rat-tailed maggots, having the posterior spiracles on a long extensile tube which is exserted through the breeding medium into the air.

Closed Tracheal Systems

In many forms of insects the spiracles are either functionless or entirely absent. In these cases the tracheal system is termed *closed.* It is usually well developed otherwise, as concerns the tracheal trunks and their interior branches. In most closed systems the spiracles are replaced by a network of fine tracheoles which run under the skin or into gills. This is illustrated by nymphs and larvae of many aquatic insects, such as mayflies, stoneflies, damselflies, and midges.

An interesting modification among the aquatic insects occurs in dragonfly nymphs. In these the rectum contains internal gill-like folds. Fine tracheae extend throughout these folds. The nymph periodically draws water into the rectum, and expels it, bathing these *rectal gills* and thus aerating the tracheae in them.

NERVOUS SYSTEM

The nervous system in insects is highly developed and consists of a central system and a stomodeal system. As in other animals, the nervous system serves to coordinate the activity of the insects with conditions both inside and outside the body.

Central Nervous System. The basic units of the central nervous system, fig. 87, are essentially: (1) the brain, situated in the head, and (2) paired nerve centers or *ganglia*, one ganglion for each segment. The ganglia are connected by double fibers into a cord, and the anterior ganglion is connected with the brain. Concurrently with the fusion of body segments which occurred in the evolution of the insect group, fig. 11, there occurred also a fusion of the ganglia belonging to each segment. For this reason the nerve centers in the head bear little apparent resemblance to the primitive condition, for the head is in reality composed of the archaic head, or prostomium, plus four body segments which join with it to make a solid mass.

BRAIN. The brain, fig. 88, is situated in the head above the oesophagus and for this reason is frequently referred to as the *supraoesophageal ganglion*. It has three principal divisions: (1) the *protocerebrum* which innervates the compound eyes and ocelli, (2) the *deutocerebrum* which innervates the antennae, and (3) the *tritocerebrum* which controls the major sympathetic nervous system. All three of these parts are definitely paired.

In its long evolutionary development, the various parts of the insect head have shifted somewhat in general orientation. Because of this shift-

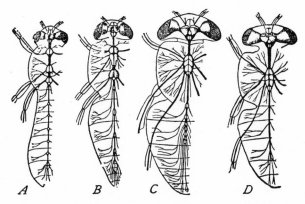

Fig. 87. Successive stages in the concentration of the central nervous system of Diptera. A, *Chironomus;* B, *Empis;* C, *Tabanus;* D, *Sarcophaga.* (From Folsom and Wardle, after Brandt)

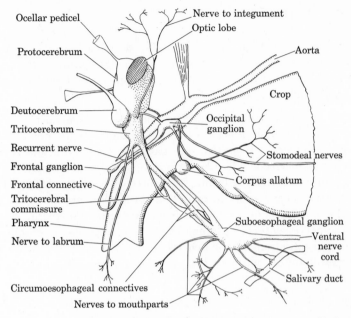

Ocellar pedicel
Nerve to integument
Optic lobe
Protocerebrum
Aorta
Crop
Deutocerebrum
Tritocerebrum
Occipital
ganglion
Recurrent nerve
Stomodeal nerves
Frontal ganglion
Frontal connective
Corpus allatum
Tritocerebral commissure
Pharynx
Suboesophageal ganglion
Nerve to labrum
Ventral nerve cord
Salivary duct
Circumoesophageal connectives
Nerves to mouthparts

Fig. 88.   Brain and associated structures of a grasshopper, lateral aspect.   (Redrawn from Snodgrass)

ing, the brain, which was originally *in front of* the mouth, is now *above* the mouth or oesophagus.   The protocerebrum and deutocerebrum are situated above the oesophagus and for this reason are considered to be the outgrowth of the primitive prostomial brain such as is found in the annelids.   The tritocerebrum is intimately joined with the deutocerebrum, but its two halves are connected by a commissure or connective fiber which passes *underneath* the oesophagus.   For this reason it is thought to be the ganglion of the first true body segment, now fused with the head.

SUBOESOPHAGEAL GANGLION.   Situated in the head, beneath the oesophagus and joined to the brain by a pair of large connectives, is a large nerve center, the suboesophageal ganglion.   It is in reality the fused ganglia of the original mandibular, maxillary, and labial segments.   This composite ganglion gives rise to the nerve trunks servicing the mouthparts.   From this nerve center a pair of connectives pass through the neck into the thorax.

VENTRAL NERVE CORD.   In the thorax and abdomen there is typically a nerve ganglion in the ventral portion of each segment.   The ganglia of adjoining segments are joined by paired connectives, the whole forming a chain of nerve centers stretching posteriorly from the prothorax, fig. 87. This chain is the ventral nerve cord. It is joined to the suboesophageal ganglion by the connective passing through the neck. The thoracic ganglia give rise to the nerves controlling the legs and wings, and the abdominal ganglia have branches and fibers to the abdominal muscles and abdominal appendages.

The generalized ventral nerve cord is composed of a chain of well-separated ganglia. In various groups of insects certain of these may fuse to form a smaller number of larger units. This type of modification is demonstrated strikingly in the order Diptera, fig. 87. Primitive members of this order possess a fairly generalized nerve cord; in more specialized families the thoracic ganglia fuse into a single large mass, and the abdominal ganglia become smaller and finally are scarcely discernible. The stages in this series of modifications are shown in fig. 87, *A* to *D*.

Stomodeal Nervous System. To control some of the "involuntary" motions of the anterior portions of the alimentary tract and dorsal blood vessel, insects possess a so-called sympathetic nervous system, fig. 89. There is considerable doubt, however, as to the exact function of many branches. It is more appropriate to term it the stomodeal system, because most of the parts are situated on the top or sides of the stomodeum. The central structure of this stomodeal system appears to be the *frontal ganglion,* which is situated in front of the brain and connected with the tritocerebrum by a pair of fibers. From the frontal ganglion a median recurrent nerve runs back beneath the brain and along the top of the oesophagus, where it connects with a system of small ganglia and nerves. This group innervates the stomodeum, salivary ducts, the aorta, and apparently certain muscles of the mouthparts.

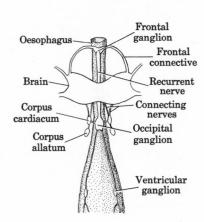

Fig. 89. Diagram of sympathetic nervous system of an insect. (Compiled from various sources)

MUSCULATURE

The insect body is provided with an extremely complex system of muscles. These are responsible for almost all the movements of the body and its appendages. Some insects may possess over two thousand muscle bands.

In a dissection, muscle tissue is one of the conspicuous features within the insect body. It does not form a continuous system but is distributed in different areas and enters into the composition of several organs. On the basis of distribution, muscle tissue may be grouped into three categories.

Visceral Muscles. The digestive tract and ducts of the reproductive system have an outer layer of muscle, which produces peristaltic movements. The muscle may be in circular, longitudinal, or oblique bands, or a combination of these. Special muscles occur in such places as the closing or opening mechanism of spiracles and in the mouth region. Muscles form pulsating bands which assist in the operation of the circulatory system.

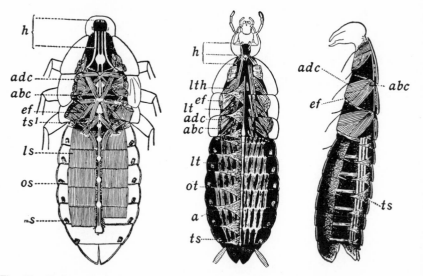

Fig. 90. Body musculature of a cockroach, showing ventral, dorsal, and lateral walls, respectively. *a*, alary muscle; *abc*, abductor of coxa; *adc*, adductor of coxa; *ef*, extensor of femur; *h*, head muscles; *ls*, longitudinal sternal; *lt*, longitudinal tergal; *lth*, lateral thoracic; *os*, oblique sternal; *ot*, oblique tergal; *ts*, tergosternal; *ts*, first tergosternal. (From Folsom and Wardle, after Miall and Denny)

Segmental Bands. The various segments of the body are connected by series of muscle bands which maintain body form, fig. 90. In the abdomen, the tergites are connected by longitudinal dorsal bands, and the sternites are connected by longitudinal ventral bands. The tergite and sternite of the same segment are connected by oblique or perpendicular tergosternal muscles. In the thorax the musculature appears entirely different. The most conspicuous muscles are large cordlike groups which operate the legs and wings; the other muscles are subordinate to these in size and prominence. In addition to these major muscle groups, there are many smaller bands which may be extremely complicated in pattern, fig. 91. In both thorax and abdomen the exact muscle pattern differs markedly in various kinds of insects.

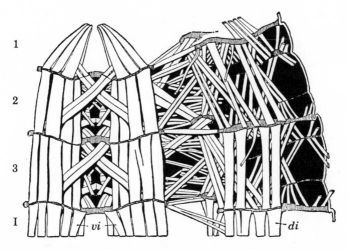

Fig. 91. Musculature of mesothorax and metathorax of a caterpillar. *di*, dorsal bands; *vi*, ventral bands. (After Snodgrass)

Muscles of the Appendages. The movable appendages have muscle bands of varying size and complexity. The mandibles of chewing insects have a few muscle groups which fill a large portion of the head capsule, but there are no muscles within the mandible itself, fig. 92. On the other hand, appendages that are divided into segments, such as the maxillae and legs, fig. 93, not only are activated by the large muscles inside the body, but in addition have muscles extending from segment to segment.

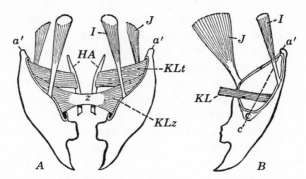

Fig. 92.   Diagram of the mandibular muscles of insects.   *A*, apterygote type with one articulation, *a'*; *B*, pterygote type with two articulations, *a'*, *c*.   Homologous muscles are lettered to match.   (After Snodgrass)

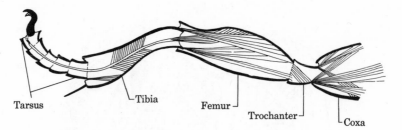

Fig. 93.   Diagram of an insect leg and its musculature.   (Adapted from Berlese)

REPRODUCTIVE SYSTEM

Insects are primarily dioecious, in that normally only one sex is represented in any one individual.   A few rare instances are known of hermaphroditic insects in which both sexes are represented in the same individual.   The most notable case is the cottony-cushion scale, *Icerya purchasi.*

In insects the reproductive system is a highly developed set of organs situated in the abdomen.   There is a close parallel between the parts of the male and female systems, and most parts of both are bilaterally symmetrical.

Female Reproductive System.   The female system consists essentially of a group of ovarioles in which the eggs are produced, a spermatheca in which sperms are stored, and a duct arrangement through which the eggs are discharged outside the body.   A typical system is illustrated in fig. 94*B*, *C*.   There are two ovaries, one on each side of the body.   An *ovary* consists of several to many *ovarioles* or tubules. Each ovariole ends in an attachment thread, called the terminal filament; the upper part of the ovariole contains the forming eggs and the lower larger portion contains the more matured eggs; the bottom of the ovariole forms a small duct or pedicel. The pedicels of each group unite to form a *calyx*. Each calyx opens into a lateral oviduct. The oviducts of the two sides join to form a common *oviduct*. This opens into an egg-holding chamber, or *vagina*, which opens directly into the external ovipositor, or egg-laying mechanism.

Two glands are connected with the dorsal wall of the oviduct. One is the *spermatheca*, a single bulbous organ with a gland attached to its duct. The other is a paired structure, the *accessory glands* or *colleterial glands,* which secrete adhesive material used in making a covering over egg masses or gluing eggs to a support.

In some of the more primitive groups such as the Orthoptera, the vagina may be only a pouchlike invagination of the eighth sternite.

Many deviations are found from the system just described with differences occurring in the number and shape of ovarioles and tubules, ducts, and glands. In many groups the spermatheca exhibits many shapes which are of considerable taxonomic value.

The primitive family Japygidae has a most interesting reproductive

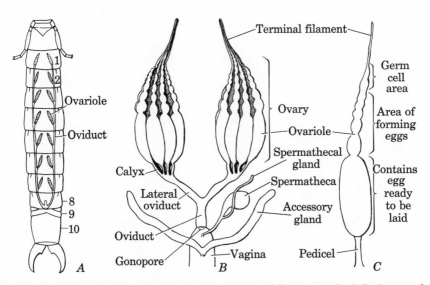

Fig. 94. Female reproductive system. *A*, primitive type of *Heterojapyx gallardi; B*, diagram of common type found in many insects; *C*, diagram of a single ovariole. (Redrawn from Snodgrass)

system. The ovarioles, fig. 94*A*, are arranged segmentally, linked together by a pair of long lateral oviducts which fuse to form a single oviduct near the egg-laying aperture. This condition suggests that the ancestral insect groups possessed independent ovaries in each segment and that there has occurred a constant migration and consolidation of these to the posterior end of the body, evolving finally the typical system shown in fig. 94*B, C*.

Male Reproductive System. In general organization the male system is similar to the female. It consists primarily of a pair of testes, associated ducts and sperm reservoirs, and outlets to the outide of the body. A common type is shown in fig. 95.

Each testis consists of a group of *sperm tubes,* in which the sperms are produced. The sperm tubes open into a common duct, the *vas deferens,* which in turn opens into a reservoir, the *seminal vesicle.* From each seminal vesicle proceeds a duct, the two ducts joining to form a common ejaculatory duct. This duct runs through the penis, at the end of which is the sperm escape opening. The penis is usually associated with structures of the external male genitalia; the structure called the *aedeagus* usually forms a rigid sheath around the true membranous penis. Associated with the internal part of the ejaculatory duct are *accessory glands,* which may be single or paired.

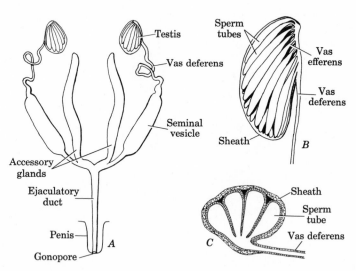

Fig. 95. Typical male reproductive system of an insect. *A*, entire system; *B*, structure of a testis; *C*, section of a testis and duct. (Redrawn from Snodgrass)

SPECIALIZED TISSUES

In addition to the extensive systems outlined in the preceding pages, the insect body contains some other smaller or less definitely organized tissues. The most important are the *fat body*, the *enocytes*, and the *corpora allata*.

The Fat Body. This is a loosely organized aggregation of cells which occurs throughout the body, especially in the later larval or nymphal instars. The cells of the fat body may be packed so tightly as to appear like an organized tissue. The function of this fat body is partly to store food and partly to aid in excretion.

Enocytes. These are clusters of cells or single large cells which occur at various points in the body cavity. Their function is not yet demonstrated satisfactorily.

Corpora Allata. These are a pair of ganglia-like bodies, fig. 88, closely associated with the stomodeal nervous system. They secrete hormones important in regulating metamorphosis and the development of some adult tissues.

REFERENCES

Bonhag, P. F., 1949. The thoracic mechanism of the adult horsefly (Diptera: Tabanidae). *Cornell U. Agr. Expt. Sta. Mem.*, **285**:1–39.

Daly, H. V., 1963. Close-packed and fibrillar muscles of the Hymenoptera. *Ann. Ent. Soc. Amer.*, **56**:295–306.

Imms, A. D., 1925. A general textbook of entomology. London: Methuen & Co. 698 pp., illus.

Richards, A. G., 1963. The ventral diaphragm of insects. *J. Morph.*, **113**:17–48.

Snodgrass, R. E., 1935. Principles of insect morphology. New York: McGraw-Hill Book Co. 667 pp., illus.

Chapter Five

Physiology

In insects as in all living organisms, persistence from generation to generation (in other words, survival of the species) depends upon successful *growth* and *reproduction*. These are two of the basic functions or activities of life. The third function is called *irritability,* and this is essentially the set of responses that coordinates the organism with its environment. From this viewpoint, an insect is a mechanical-chemical unit of life that extracts needed chemicals (nutrients) from its surroundings, converts these to substances needed for its growth, then builds these substances into the mature reproducing form, and finally reproduces.

Nearly all of the basic chemical and physical processes involved in insect physiology are the same as those occurring in other forms of animal life. These include the oxidation of foods in metabolism, oxygen and carbon dioxide exchange in respiration, fertilization in reproduction, and the transmission of an impulse over a nerve fiber.

The various processes by which these activities are performed constitute the function of the organs and tissues described in the previous two chapters. This chapter deals with the actual workings of the organism, its *physiology.* Because the processes of organ and tissue physiology rest on cell physiology, it has seemed practical to combine the discussion of physiology with histology.

GROWTH AND MAINTENANCE

Insects are complete, multicellular animals in which the body contains nonreproductive tissues and organs, the *soma,* and a group of cells, the *germ cells,* which ultimately produce the sperm or eggs. In these complex animals the function of growth includes not only growth itself but maintenance of the soma during the maturation and production of the sperm or eggs.

108

The processes of growth and maintenance include digestion and assimilation (nutrition), excretion, respiration, general metabolism and its regulation, and functions of the body wall, blood, and other structures.

Structure and Function of the Body Wall

The body wall or integument is the surface layer of ectoderm that surrounds the body and appendages. It is a complex organ containing many kinds of external hairs and sense receptors and internal processes of many types for attachment of muscles.

The body wall has three primary functions: (1) to protect the organism from outside forces, such as evaporation (insects' most important enemy), inimical organisms, and disease; (2) to receive external stimuli through specialized sensory hairs, processes, or areas; and (3) to act as the agent of the locomotor system, since the motivating muscles of the legs, wings, and movable sclerites are attached to the exoskeleton. In addition, the integument cannot stretch and in immature insects must be shed regularly to allow growth. These functions are accomplished by a surprisingly simple cellular structure.

Evaporation. Loss of water by evaporation is the greatest threat to terrestrial organisms, and all insects are terrestrial or aerial for at least some portion of their lives. Evaporation is a function of surface, not volume, and as size decreases, the ratio of surface to volume increases. Thus because insects are both small and terrestrial, they are faced with a major problem of protecting from excessive evaporation the small total amount of water contained in their bodies. The protection lies in the impermeable nature of the insect cuticle, which is remarkably resistant to the passage of water or water vapor. Without such efficient protection it is doubtful if an insect flying in the air, for even a short time, could escape desiccation to a fatal point.

Structure of Integument. The body wall, fig. 96, consists primarily of a layer of cells, the *epidermis,* and an outside covering, the *cuticle,* which lies on top of and is secreted by the epidermis. Formation of cuticle is the chief function of the epidermis. The cuticle forms a mechanical protective layer whose properties contain the key to much insect physiology.

EPIDERMIS. The cells comprising most of the epidermis are simple in type, with large nuclei, united by an indistinct basement membrane. Certain cells of this layer, however, are highly specialized and produce hairs and other surface structures of peculiar types.

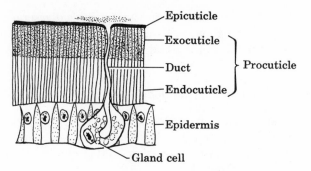

Fig. 96.   Diagram of body wall structure.   (Adapted from Wigglesworth)

CUTICLE.   The cuticle is made up of a relatively thick inner layer, the *procuticle,* and a very thin outside *epicuticle.*

The epicuticle is usually only a micron thick but seems to be the layer which gives the entire cuticle its property of impermeability.   The epicuticle may be composed of several layers.   The inner layer is composed chiefly of *cuticulin* (possibly a lipoprotein), and the outer layers usually contain waxes and other organic substances.

The procuticle is composed of *chitin,* proteins, and other compounds. Chitin, the distinctive component of the procuticle, is susceptible to some acids but is resistant to alkalis.   The procuticle may be differentiated into a more or less definite outer layer, the *exocuticle,* and an inner layer, the *endocuticle.*   The exocuticle is often impregnated with cuticulin and colored substances such as carotin and melanin.   These substances strengthen and color the soft chitin and give the impregnated areas hardness and much greater impermeability.   Such strengthened areas are called "sclerotized" and may contain as little as 20 per cent chitin.   Soft areas, which may consist of as much as 80 per cent chitin, are called "membranous."

The procuticle has the form of a fairly elastic jelly, traversed by extremely fine openings or pore canals.   The pore canals run from the epidermal cells to or into the epicuticle but not completely through it. Their function is dubious.   In very thick, hard cuticle, as on the elytra of beetles, the cuticle may be laid down as successive series of minute parallel rods, which give the structure additional strength.

SPECIALIZED CELLS.   Some epidermal cells have special functions, either the secretion of fluids or the formation of definite structures such as hairs.

DERMAL GLANDS.   Single epidermal cells or groups of cells develop into large cells which produce various secretions.   These cells, fig. 96, are connected to the exterior by a duct running through the cuticle.   Secretions of different types are produced by a variety of these dermal glands, including wax (often forming definite external patterns), many types of ill-smelling scent compounds, and irritating skin poisons.

Some dermal glands (formerly thought to produce molting fluid) are believed to secrete the outer waxy covering of the epicuticle.

SETAE, FIG. 97. Most of the flexible hairs or bristles of insects are formed by epidermal cells called *trichogen cells*. At the time of the actual formation of the hair, the trichogen cell is large and nucleated and has a duct which passes through the cuticle to the surface. From this point the products of the cell build up the hair. Closely associated with the trichogen cell is a *tormogen cell*, which forms a socket (usually flexible) around the base of the hair. A hair or bristle of this histological origin is called a *seta* (pl., *setae*). The parent cells may degenerate after the seta is formed.

Specialized setae originate in the same manner. These include scales, poison hairs, and sensory setae, discussed in later pages.

COLOR. The majority of insect colors are located in the epidermis or its vesture. Insect colors are of two types: pigments and structural colors.

Pigments such as carotin and melanin are deposited in the exocuticle and produce different colors by selective action on different wavelengths of light. These pigments are responsible for practically all nonmetallic insect colors and a few metallic ones.

Structural colors are produced by extremely delicate and minute vanes which break up light into various wavelengths by reflection and interference. These vanes may be produced by the epicuticle, as is the case with many beetles, especially those with metallic colors. The most com-

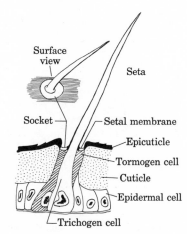

Fig. 97. A seta and its socket.
(Adapted from Snodgrass)

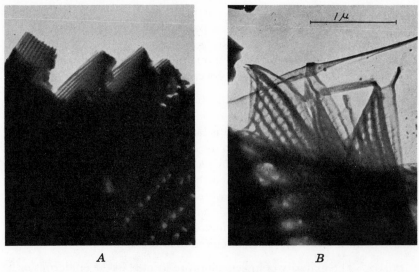

A   B

Fig. 98. An iridescent scale of a butterfly wing, photographed with the electron microscope.
The pictures are of fractured ends of broken scales to obtain different views of the surface vanes.
A, $\times 6000$; B, $\times 12,000$. (After Anderson and Richards)

mon example of this occurs in the moths and butterflies. In these the
wings are covered with scales (modified setae), and the scales bear ribs
running the length of the scale, fig. 98. Studies with the electron microscope have shown that each rib is composed of several parallel extremely

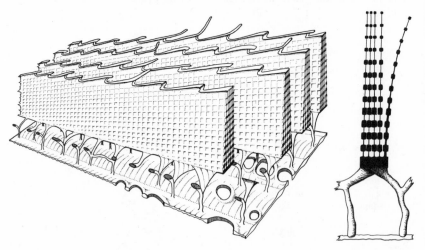

Fig. 99. Diagram of light-breaking structure of ribs of butterfly scale shown in fig. 98. $\times$ cir.
18,000. (After Anderson and Richards)

thin fenestrate vanes. Studies on the tropical *Morpho* butterflies indicate that ribs of more simple structure produce nonmetallic colors and that ribs of great complexity, fig. 99, produce the dazzling iridescent colors for which these butterflies are famous.

Epidermal Growth: Molting. In immature insects the cuticle normally does not stretch. The epidermal cells are bound to the cuticle and cannot grow until they are detached from the cuticle.

To increase in body size, therefore, an insect must produce periodically a larger new cuticle and shed the old. This phenomenon of shedding the old "skin" is termed *molting* or *ecdysis*. It is one of the most important physiological processes of insects. The actual act of molting is preceded by the formation of the new cuticle under the old. The following steps in this process have been observed.

1. First the old cuticle is loosened to form a small space between it and the epidermal cells, fig. 100*A*.

2. The epidermal cells enlarge and multiply. This results in a discontinuous pattern in the multiplication of these cells, with bursts of growth at molting and periods of no growth in between.

3. Enzymes are secreted into the space below the cuticle and begin to digest it, the digested material being absorbed by the epithelial cells.

4. The epidermal cells begin to secrete the new cuticle.

5. The epidermal cells apparently continue to absorb the digested old cuticle and use this material in adding to the new cuticle. Up to 85 per cent of the old cuticle may be digested.

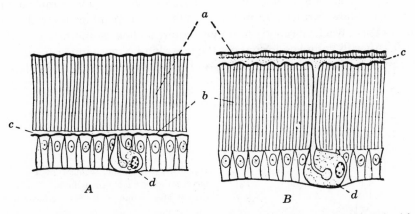

Fig. 100. Diagram of the production of new cuticle prior to molting. *A*, a new epicuticle formed; digestion of old endocuticle scarcely begun. *B*, digestion and absorption of old endocuticle almost complete. *a*, old cuticle; *b*, new cuticle; *c*, space between old cuticle and layers beneath it; *d*, enlarged dermal gland. (From Wigglesworth, *Principles of insect physiology*, by permission of E. P. Dutton and Co.)

6. When the new cuticle is otherwise completed, fig. 100*B*, certain enlarged dermal glands discharge their contents over the outside of the new cuticle. This secretion forms the final waxy layer of the epicuticle.

7. When the new cuticle is fully formed, the insect has to break out of the old one. The initial rupture is made along a mesal line of weak cuticle which typically extends along the dorsum of the thorax. This rupture is caused by the pressure of the blood. The insect contracts the abdomen, forcing the blood into the thorax and causing it to bulge until the cuticle breaks along the line of weakness. The insect may swallow air (or water, if aquatic) to aid in this process. The insect then wriggles and squirms free from the old skin. At or before this time, the molting fluid is usually reabsorbed by the body, so that at the time of molting the area between the old and new skins may be dry.

8. For a short period after molting the new cuticle can be stretched, at least in the nonsclerotized (membranous) portions. During this short period, the insect stretches the cuticle first by swallowing air or water, thus increasing its internal volume, and then by increasing the blood pressure in first one body region and then another. These actions "blow up" the regions and stretch the integument. When the blood pressure is reduced, the stretched integument *does not shrink again* but contracts into a series of small folds or minute accordionlike pleats. In larvae with no sclerotized body areas, these folds may occur over the entire body, fig. 101. In insects with definite sclerotized plates, the folds occur in the membrane between the sclerites. As the body increases in size with subsequent growth, the integument increases by a simple expansion of the folds, as in figs. 30 and 31. When this avenue for increase is exhausted, the insect must molt again to allow further size increase. The act of shedding the old skin may take only a few seconds, or it may require an hour or more.

9. After its complete formation, the new skin becomes impermeable to many substances, especially water, and is locally sclerotized and colored

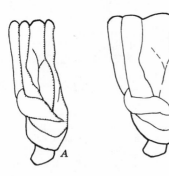

Fig. 101. Abdominal segment of a sawfly larva, showing membranous folds. *A*, immediately after molting; *B*, after growth and expansion.

to assume its normal condition. In many groups, such as the grass-hoppers, this occurs just after the stretching process which follows molting. In other cases, for instance in adult Trichoptera, Lepidoptera, and many Hymenoptera, this occurs before molting while the adult is still incased in the pupal skin. It was thought formerly that hardening and coloring of the skin were a result of its exposure to air after molting, but experiments on partial dissection have shown this not to be true.

Nutrition

In common with protozoans and mammals—in fact, almost all living things except the green plants—insects require water, a large selection of amino acids, some nucleic acid derivatives such as guanine or cytidine, a number of vitamins, and a large number of inorganic substances including phosphorus, potassium, iron, copper, zinc, cobalt, and calcium. These are called the *minimum nutritional requirements*. Although foods such as carbohydrates and fats are eaten and used as energy sources by many insects, they are not essential because some of the other organic compounds can be "burned" by the tissues for energy.

Even though watching a whole insect may obscure the fact, the ultimate basis of both growth and reproduction is cell division. Cell division itself requires duplication of the nuclear deoxyribonucleic acid (DNA) and its associated complex nuclear proteins. Certain cells, such as those producing setae, must be supplied with large quantities of nutrients that the cell can convert into extrusions that are highly elaborated chemical compounds. These circumstances explain the dietary needs for the amino acids and other organic compounds that are essential for the manufacture of proteins and nucleic acids (DNA, RNA).

Insects, however, like other animals, cannot synthesize the amino acids and other organic compounds they need. That is why these compounds are nutritional requirements. As a consequence, insects obtain them by eating other living or dead organisms that have either synthesized these compounds (green plants) or have themselves obtained these compounds directly or indirectly from green plants (other animals or parasitic plants).

These nutritional requirements are not daily but life-time requirements. Thus a caterpillar eats green plants and stores large quantities of required nutrients that suffice for maturation and egg-laying of the adult, which itself may feed primarily on carbohydrates, or may not feed at all. In other insects such as certain mosquitoes, however, the egg-laying capacity is dependent in large measure on the protein-rich blood meals of the adult female. In these same insects the male (which takes no blood meal) matures and produces sperm with an adult diet of only nectar and water.

The mosquitoes present another interesting phenomenon common in insects. The larva feeds chiefly on whole microorganisms, the adult male on nectar, the female (of most species) on mammalian blood. These are quite different foods, yet are only a small fraction of the entire insect dietary. Some species feed only on green plants (leaves, bark, roots, pollen, or heart-wood), some only on dung, rotten wood, or animal carcasses, others on various kinds of living animals. Whatever its food, in each kind of insect there has evolved a system of digestion by which the insect can normally procure its nutritional requirements from its food.

VITAMINS. Investigations to date indicate that insects need the vitamins of the B complex and cholesterol, a close relative of the sterols of the vitamin D group. Cholesterol may be of especial importance as a base compound for the formation of several hormones. Vitamin C is synthesized by the roach *Blattella*, and may be a requirement for insect growth. Insects apparently neither require nor synthesize the fat-soluble vitamins A, D, or E. It has been shown with experiments on blow fly larvae that some of the needed vitamins are obtained from symbionts or from microorganisms mixed with the normal diet which may itself be deficient in these substances.

WATER REQUIREMENTS. As in other organisms water is a fundamental basis of metabolic processes because practically all of them occur in aqueous solution. Hence water is a very important item of the insect diet. Insects have developed many structural and physiological specializations to conserve water. Most of them obtain an abundance for their needs in foodstuffs with a fairly high water content, such as foliage and blood. There are cases in which water is conserved to such a high degree that the insect is able to exist entirely on dry materials. In these instances the insect makes use of the water resulting from the oxidation of the foodstuffs. But even in most of these cases the food must contain a small percentage of water to supplement the metabolic water. An unusual property is found in the mealworm: in atmospheres of high humidity it can actually absorb water from the air.

Digestion

Digestion is the process of dissolving and chemically changing food so that it can be assimilated by the blood and furnish nutriment to the body. Because the food of different insects includes an array of diverse materials, many modifications are found in the digestive systems, each adapted to handling a particular type of food. The digestive system may be entirely different in larval and adult stages of the same insect, fig. 81, especially in such forms as the Diptera, in which the food of the various stages is entirely different. Within an order there may be strikingly different digestive systems. The Hymenoptera, for instance, include such diverse forms as sawfly larvae which are herbivorous and wasp larvae which are internal parasites, each having a different type of digestive system.

A generalized type of digestion system is found in herbivorous and omnivorous insects such as cockroaches, grasshoppers, and many beetle larvae and adults. This type is used as a basis for the following account. A few of the more conspicuous modifications from this type are discussed; the enormous number of other modifications must be relegated to a specialized study of the subject.

Salivation. In a great variety of insects saliva is mixed with the food before it is ingested. In chewing insects the saliva is secreted into the mouth and mixed with the food there. In sucking insects the saliva is ejected into the liquid food, and the mixture is then siphoned into the pharynx. The saliva is usually produced by the labial glands.

Typically each gland is like a long bunch of grapes, each "grape" a small cluster or *acinus* of secreting cells. Each acinus has its own duct; these join successively to form the large duct of the whole gland. The acinus may contain cells of different histological structure. The labial glands having a function connected with food may be segregated into two general groups, based on the principal substance they are known to secrete.

1. DIGESTIVE GROUP. In many insects the labial glands are the chief source of amylase. This is usually secreted into the food mass before it is swallowed, and the actual digestion takes place in the digestive tract. In adult Lepidoptera and bees the glands secrete invertase, which is exuded at the tip of the proboscis and drawn up into the stomach with the nectar. The enzymes are secreted in the acini, fig. 102.

Fig. 102.   Acini of salivary gland of a cockroach.
A, small acinus with duct; *B*, cross section of acinus.

2. ANTICOAGULIN GROUP. The labial glands of blood-sucking insects secrete no digestive enzymes but instead produce an anticoagulin. The anticoagulin prevents the ingested blood meal from clotting and plugging up the beak and digestive tract.

Extraintestinal Digestion.   In special cases, digestive enzymes are extruded from the body onto or into the food and effect at least partial digestion before the food is taken into the digestive tract.   This is called extraintestinal digestion.   Plant lice, for instance, extrude saliva containing amylase from the beak into the host tissues and in this way digest starch in the host plant cells.   Many predaceous beetles that lack salivary glands eject their intestinal enzymes through the mouth onto their prey.   When digestion has occurred, the fluids produced are reabsorbed.   Flesh-feeding maggots extrude proteolytic enzymes from the anus and effect extraintestinal digestion of the tissues in which they live and which form their food.

Ingestion.   Insects take their food into the alimentary canal by way of the mouth.   In insects with chewing mouthparts, the mandibles and maxillae cut off and shred the food.   The closing together of these opposing structures presses the food to the back of the mouth or cibarium, at the base of the hypopharynx, fig. 103.   The hypopharynx is then pulled upward and forward, forcing the food into the pharynx, which is the anterior end of the oesophagus.   From this point the food is moved along the digestive tract by peristaltic action.   In insects with sucking mouthparts, fig. 104, the pharynx forms a bulblike pump, which expands and contracts by action of head muscles.   The *pharyngeal pump,* as it is called, pulls the liquid food through the beak and into the region of peristaltic control.   Digestive enzymes or other secretions may be mixed with the food before it is swallowed.

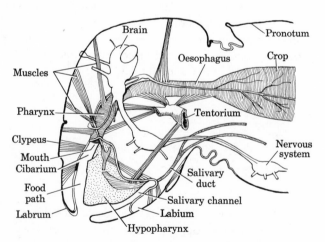

Fig. 103.   Sectional diagram of the head of a chewing insect, showing the generalized parts, areas, and musculature used in swallowing.   (Redrawn from Snodgrass)

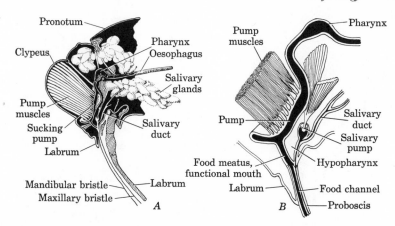

Fig. 104. The sucking pump and salivary syringe of a cicada. *A*, section of the head showing position of the sucking pump (cibarium) with dilator muscles arising on the clypeus. *B*, section through the mouth region, showing food meatus, suck pump, and salivary syringe. (Redrawn from Snodgrass)

The only known exception to oral ingestion occurs in the earliest larval stages of some internal parasites which absorb their nutriment through the general body surface from the tissues or blood of their host.

Stomodeum, or Fore-Intestine. The food is passed through the oesophagus into the stomodeum. There is considerable variety in the functions of the stomodeum; it may serve simply as a passage into the mesenteron, or may be enlarged to form a capacious crop in which the food can be stored and partial digestion may take place. In some cases, as in the Orthoptera, digestive juices are passed from the mesenteron to the stomodeum.

The stomodeum typically consists of a layer of simple epithelial cells, fig. 106*A*, which secrete a definite cuticle. It is believed that this cuticle is practically impermeable to enzymes and to the products of digestion and that little or no absorption takes place through it. The function of the cuticle is probably to prevent absorption of only partially digested compounds, because such premature absorption would interfere with complete digestion.

Proventriculus. Orthoptera and other groups that eat coarse food have a set of powerful shredding teeth in the proventriculus for dividing the food into smaller particles. A typical arrangement is found in the cockroach, fig. 105, where six stout teeth do the shredding. Fleas employ a mass of sharp, needle-like teeth directed backward. During digestion these are driven backward at the same time that the blood meal in the mesenteron

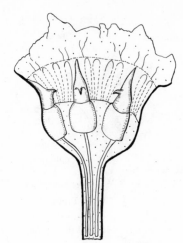

Fig. 105. Proventriculus of a cockroach laid open to show three of the six macerating teeth.

is thrust forward, and the fine teeth crush the blood corpuscles, causing them to disintegrate. These movements are caused by rhythmic and opposing muscle contractions. In other insects the proventriculus is simply the narrowed end of the stomodeum.

Mesenteron or Mid-Intestine. In this portion of the alimentary canal, the epithelial cells, fig. 106*B*, are exposed, since they do not secrete a

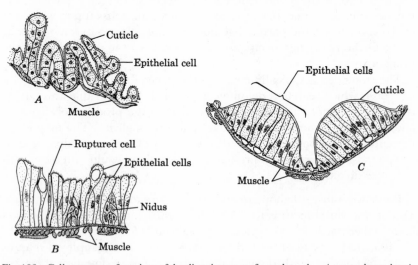

Fig. 106. Cell structure of portions of the digestive tract of a cockroach. *A*, stomodeum, longitudinal section; *B*, mesenteron, longitudinal section; *C*, proctodeum, cross section.

cuticle. Some of these exposed cells perform most of the actual food absorption, and other cells carry on enzyme secretion.

The actual secretion of enzymes by the epithelial cells is accomplished by two methods: (1) *holocrine secretion*, in which the cells disintegrate in the process, emptying their contents into the lumen of the intestine; and (2) *merocrine secretion*, in which the enzymes diffuse through the cell membrane into the lumen. The former is illustrated in fig. 106B. This shows the clusters of regenerative cells, or nidi, which replace the cells used up during holocrine secretion.

ENZYMES. In insects as a whole the mesenteron produces a wide variety of enzymes for digesting carbohydrates, fats, and proteins. In the main these are the same enzymes present in the mammalian system. Production of enzymes is usually correlated with diet. Omnivorous insects such as the cockroach produce the full complement of enzymes for digesting all types of food. Blood-sucking insects, however, produce chiefly proteolytic enzymes. Some insects secrete cellulase for digesting cellulose. Certain clothes moths are able to digest keratin with the aid of a common insect proteinase combined with peculiar pH conditions of the mid-intestine. The wax moths digest wax but the enzyme which enables them to do this has not been isolated.

PERITROPHIC MEMBRANE. The epithelial cells of the mesenteron are exposed and delicate. If the food bolus were to be pushed over the unprotected surface of these cells, it would undoubtedly injure them severely and interfere with their functions of secretion and absorption. In vertebrates, mucous glands coat and lubricate the food boluses and hard particles to avoid abrasive injury to stomach epithelium. Insects have no mucous glands and they obtain protection for the epithelium by the formation of a *peritrophic membrane*, fig. 107. This membrane forms a continuous tubular covering around the food mass. The membrane is composed of chitin; it is freely permeable to digestive enzymes and all the products of digestion. Its remarkable permeability has been demonstrated experimentally by the use of dyes.

The formation of the peritrophic membrane is a topic of considerable interest. In a great many insects, it originates from a secretion of the general surface of the mesenteron. This chitinous secretion is formed into a layer over the parent epithelial cells and then separated from them to form a sort of tube around the food mass. The tube usually remains attached at the anterior of the mid-intestine where the fore-intestine projects into it.

The peritrophic membrane is not formed in certain insect groups which take only liquid food, including Hemiptera, Anoplura, and adults of fleas,

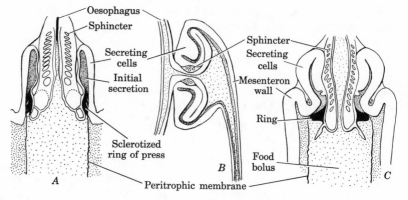

Fig. 107.   Annular molds producing a peritrophic membrane.   *A*, larva of mosquito *Anopheles; B*, tsetse fly *Glossina; C*, earwig *Forficula*.   (Modified from Wigglesworth)

mosquitoes, and horseflies.   It is absent also in a few other groups, notably the Carabidae, Dytiscidae, and Formicidae.

CARDIA.   In some groups of insects the peritrophic membrane is secreted by a specialized group of cells around the anterior end of the mesenteron. The secretion is pressed or molded into a membrane by the outward pressure of either the mesenteron entrance as it is distended by incoming food, or by a special organ called a *cardia*.   This structure is most highly developed in the Diptera and Dermaptera, fig. 107.   It consists principally of a sclerotized ring around the opening into the mesenteron, which presses against the walls of the intestine a flow of secretion from the group of cells just anterior to it.   As the membrane is formed, it is passed back through the intestine as a sheath around the food.

Proctodeum or Hind-Intestine.   The function of this part of the digestive tract is still not well understood in many insects, although it is thought that normally no absorption of food occurs here.   The epithelial cells secrete a definite cuticle, fig. 106*C*, as in the stomodeum, but this cuticle is readily permeable to water.   The posterior part, forming the rectum, is usually heavily muscled to compress the residue of the food after digestion and so form the excrement into pellets before defecation.   Two other functions are well established:

1. WATER ABSORPTION.   All insects must conserve water to the utmost and to do so they rely on the proctodeum to absorb water from the excrement and return it to the body.   In the mealworms, the epithelial cells in the rectum may extract almost all the water from the excrement, leaving

it a dry pellet. The water absorption plays an especially important role in excretion, under which it is discussed more fully.

2. SYMBIOTIC DIGESTION. Termites, certain wood cockroaches, and certain scarab-beetle larvae whose chief diet is wood fiber, have no enzyme for digesting the cellulose they eat. They rely instead on a rich symbiotic fauna of microorganisms in the hind intestine, which digest the cellulose to form acetic acid. The acetic acid is absorbed by the proctodeum. Investigations of symbionts in other insects have brought out many apparent contradictions and questions, showing the need for more research in this field.

Adaptations to Liquid Diet. Various insects that suck blood or plant juices have evolved methods for extracting much of the water from the food before it comes into contact with the digestive enzymes. This arrangement has two advantages: (1) Some of the assimilable sugars in the food may be absorbed rapidly, and (2) the enzymes do not suffer excessive dilution. The partial dehydration is accomplished by the following methods.

1. In adults of many Diptera the mesenteron is divided into several sections, each with a different type of epithelium. It is thought that the first section acts as an absorption area to take most of the water out of imbibed liquids.

2. In such blood-sucking Hemiptera as the bedbugs the first part of the mesenteron forms a large crop in which the blood meal is received. This "crop" absorbs most of the water and so concentrates the blood before it is passed to the region in which the enzymes are produced. Note that in the Diptera and the bedbug the water is absorbed from the mesenteron and passed into the insect's blood stream. From the blood stream it is excreted through the Malpighian tubules into the proctodeum.

3. In the scale insects, cicadas, and many other Homoptera the digestive tract has a curious structure called the *filter chamber*, fig. 108. The anterior part of the mid-intestine lies beside a part of the hind-intestine. In some forms the two parts may be bound together by a common sheath, and in some instances the mid-intestine may loop through an invagination of the hind-intestine. Because these animals live on plant juices, it has been thought that this filter chamber allowed water in newly ingested sap to pass rapidly from mid-intestine to hind-intestine, thus concentrating the sap prior to its digestion in the posterior region of the mid-intestine. This excess fluid is passed from the anus as honeydew. Honeydew analyses have cast considerable doubt on the correctness of these deductions, because the honeydew is found to contain large amounts of amino acids

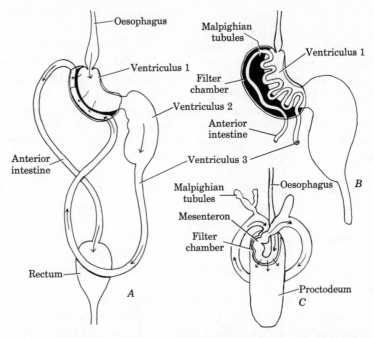

Fig. 108.   The filter chamber of Homoptera.   *A*, diagram of a simple type of filter chamber in which the two extremities of the ventriculus (mesenteron) and the anterior end of the hind intestine are bound together in a common sheath.   *B*, the ventriculus convoluted in the filter chamber and the anterior part of the hind intestine issuing from its posterior end.   *C*, the filter chamber of the scale insect *Lecanium,* diagrammatic.   (Redrawn from Snodgrass, after Weber)

and carbohydrates.   It is evident that more information is needed to obtain a better understanding of the function of the filter chamber.

Larval Adaptations.   A peculiar modification of the digestive tract is found in the larvae of the higher Hymenoptera and Neuroptera.   The end of the mesenteron is closed and does not connect with the proctodeum. The mesenteron, during larval development, becomes greatly distended with fecal matter.   Prior to pupation the two sections of intestine become joined, and the fecal pellet for the *entire larval life* is evacuated.

Stomach Reaction.   The average contents of the digestive tract in most insects are slightly acid, with a pH of 6–7.   The saliva is usually neutral. In plant-feeding insects the intestine averages more alkaline and has been recorded as high as pH 8.4 to 10.3 in the silkworm larva.   Carnivorous or flesh-feeding insects usually average on the more acid side.   Acidity of pH 4.8 to 5.2 has been observed in the crop of the cockroach after a carbohydrate meal; it has been suggested that this acidity is the result of fermentation by microorganisms, but it might be due to acid secretion of certain cells of the salivary glands. The greatest acidity recorded is pH 3.0 in a portion of blow fly larvae intestine.

Cardiac or Gastric Caeca. Little is known about the function of these processes. It has been suggested that they house the regenerative supply of the normal bacterial fauna of the intestine.

Assimilation

The products of digestion are absorbed by the epithelium of the digestive tract. The greater part of this absorption occurs in the mid-intestine, but various products may be absorbed in the fore-intestine or hind-intestine, depending on the kind of insect. It is believed that much intermediary metabolism occurs in the mid-intestine, including a certain amount of glycogen synthesis and protein breakdown. The ability to carry on these latter activities marks the cells of the insect mid-intestine as more primitive than intestinal cells of mammals.

Storage. The principal area of glycogen synthesis and storage is in the fat body. Parts of the mid-intestine and other areas may also be storage sites for food reserves. In insects with complete metamorphosis these reserves are usually highest just before pupation and may be nearly depleted by the end of metamorphosis.

Excretion

Many waste products of metabolism either are of no value to the organism or would be harmful if allowed to accumulate. The process of eliminating these waste products is excretion. The elimination of carbon dioxide and some water is technically excretion but for convenience it is treated under respiration. Excretion, as treated here, is restricted to the elimination of excess water, salts, nitrogenous wastes such as uric acid, and various undesirable organic compounds.

In insects the Malpighian tubules are the chief known organs of excretion. In addition, certain excretions may be deposited in the cuticle or setae as pigments. The use of the dye indocarmine as an indicator has demonstrated that in the Thysanura part of the salivary glands may also be excretory in function.

Several tissues, such as the fat bodies and molting glands, have been considered as excretory in function because of the deposition of uric acid crystals within their cells. Uric acid, however, is the end product of pro-

tein metabolism and precipitates out as crystals very readily.   It is believed that the observed uric acid crystals in many tissues are due simply
to rapid protein metabolism, resulting in the production of uric acid too
fast to be entirely taken up by the blood.   Under these conditions the
excess uric acid is precipitated as crystals in the cells in which it is formed,
to be dissolved and eliminated at a later time.

Malpighian Tubules.   These organs excrete chiefly uric acid.   In most
insects the excretion is accomplished by circulation of water, as is the case
in most of the vertebrates.

In its simplest form in insects this process is as follows: the uric acid
(probably in the form of a sodium or potassium salt) in the body cells is
diffused into the blood, which eventually circulates around the Malpighian tubules.   All or part of the cells of these tubules absorb the uric
acid from the blood and discharge it, in aqueous solution, into the lumen
of the tubule.   From this point uric acid solution, or urine, is forced into
the proctodeum and voided through the anus.

This method of excretion requires a continuous supply of water, and
with many insects water is at a high premium.   Very likely the basic solvents (sodium and potassium salts) are equally valuable.   Various methods have evolved which conserve these compounds by extracting the water

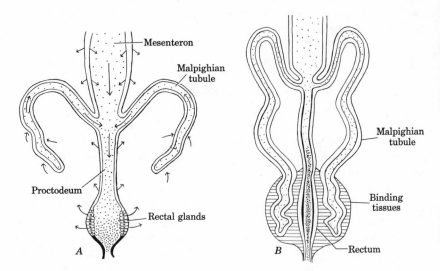

Fig. 109.   Diagram of water circulation in alimentary and excretory systems of an insect.
A, common type, arrows indicating direction of water movement; *B*, type in mealworm
Tenebrio, showing close relation of rectum and tips of Malpighian tubules.   (Adapted from
Wigglesworth)

and base from the urine and returning them to the blood or to the upper end of the tubules.

One method, exemplified by earwigs and grasshoppers, is the development of absorptive areas in the rectum, fig. 109*A*. These areas extract the water from the excrement and return it to the blood. In forms such as the mealworm, fig. 109*B*, the tips of the Malpighian tubules are bound to the rectum by a membrane. Apparently here the absorptive powers of the tubules are added to those of the rectum, for in these forms the excrement is dried to a powder. It is also probable that the absorbed water is returned directly into the tubules, and thus the same water is reabsorbed and used over and over again.

In a second modification the cells of the lower part of the Malpighian tubules extract water and base from the urine. In these cases the upper part of the tubules contains a clear liquid, and the lower part contains solid crystals of precipitated uric acid. These are pushed into the proctodeum for evacuation. The cells of the two areas have well-marked histological differences. It may be that here also some of the water is used continuously, much as arrows in fig. 109*A* indicate. Many insects, such as the larvae of Lepidoptera, combine all these methods.

In the Malpighian tubules of many insects, deposits of carbonates have been found. It is not known why these are present or what becomes of them. Their presence suggests that the Malpighian tubules may have other excretory functions, in addition to those relating to uric acid and water.

Pigments. Certain wastes of metabolism may be converted into pigments (some of them derivatives of uric acid) and are often deposited in the cuticle. In the butterfly family Pieridae these pigments are deposited in the scales of the wings, and in the Colorado potato beetle they are deposited in the sclerites; in both they form the color pattern of the insect. The conversion of metabolic wastes into pigments is likely a common phenomenon among insects.

Glandular Secretions. Certain glands which secrete wax, scents, and other substances may be able to utilize waste products as a basis for their secretions. There is little experimental evidence to support this conjecture.

Respiration

The process of respiration consists of the uptake and use of oxygen by the tissues and the liberation and disposal of the carbon dioxide. The exchange of gases between the cell and the environment is called *external* respiration, whereas the exchange and utilization of gases within the cell is called *internal* respiration or respiratory metabolism.

EXTERNAL RESPIRATION

This is accomplished in most insects by the tracheal system, in essence a system of open tubes, called *tracheae* and *tracheoles,* through which air is brought directly to the tissue cells.

Tracheae and Tracheoles.   The tracheae are invaginations of the ectoderm, and their general character is similar to that of the epidermis, fig. 110.   The foundation structure is a layer of flat epithelial cells, which secrete the lining of the tracheae, a cuticlelike substance called the *intima.*   The surface of the intima is thickened by spiral filaments or *taenidia,* which strengthen the trachea and ensure that it remains round and open even under conditions of bending and pressure.   The tracheae divide and redivide, becoming smaller and smaller; finally each ends in a cluster of minute branches, the *tracheoles.*   The tracheoles are less than 2μ (microns) wide; they possess taenidia but have no regular layer of epithelial cells.   The base of each cluster of tracheoles has a weblike cell (the tracheole cell) with extremely thin protoplasmic extensions.   These extensions appear to surround and follow the tracheoles.   The tips of the tracheoles lie alongside, between and actually within the tissue cells of the body.   It is believed that most of the respiratory gas exchange of the tissues occurs through these tracheole tips.

The properties of the tracheae and tracheoles are quite different from those of the epidermis.   Both tracheae and tracheoles are permeable to

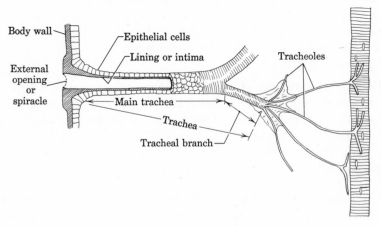

Fig. 110.   Diagram of an open trachea of an insect.

gases, presumably extremely so where the wall is as delicate as in a tracheole. The tracheae are impermeable to liquids; the spiracles at least are extremely hydrophobic, that is, the surface resists the entry of water. The tracheoles, especially their tips, are readily permeable to liquids.

TRACHEOLE LIQUOR. In many insects the tips of the tracheoles contain a certain amount of liquid that is either blood or a similar substance. When associated with relaxed muscle, fig. 111*A*, this liquor may rise a considerable distance in the tracheoles; when the muscle is fatigued, a large part of the liquor is withdrawn from the tracheole into the cells, fig. 111*B*. It is possible that this retraction is due to the increased osmotic

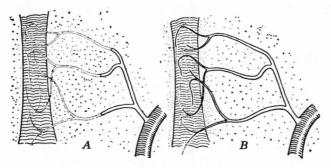

Fig. 111. Rise and fall of liquor in tracheoles. *A*, high level in rested state; *B*, low level in fatigued state. Tracheoles shown with dotted lines contain liquid, those with plain lines contain air. (From Wigglesworth, *Principles of insect physiology*, by permission of E. P. Dutton and Co.)

pressure of the muscles resulting from acid metabolites incurred during contraction. The withdrawal of liquor into the cells brings the air into contact with the fatigued cell and presumably aids in increasing the oxygen supply for working tissue in which the oxygen need is greatest.

Diffusion. The actual mechanics by which oxygen passes through the length of the tracheae and tracheoles, and finally into the tissues, and by which carbon dioxide is eliminated along the reverse path, has been the subject of many theories. It is now generally accepted that these gases are conveyed by diffusion, with the help of some mechanical ventilating in certain insects. Analyses have been made of dimensions of tracheae, oxygen consumption, and the diffusion coefficient of oxygen of various insects. They have shown that, even in the case of large caterpillars, diffusion alone will provide a sufficient stream of oxygen to the tracheole endings if the oxygen pressure in them is only 2 or 3 per cent below that of the atmosphere.

This same reasoning accounts also for the elimination of carbon dioxide, since it has a rate of diffusion only slightly less than that of oxygen. Analysis of carbon dioxide elimination, however, has shown that nearly a fourth of the amount produced in the body is eliminated over the general body surface. This is explained by the fact that carbon dioxide diffuses *through* *animal tissues* about 35 times as fast as oxygen. Consequently, any carbon dioxide formed in metabolism diffuses not only into the tracheoles but also into surrounding tissues in all directions and eventually to the exterior through the body wall.

Blood Respiration. Normally the blood plays no important part in transporting oxygen from the atmosphere to the tissues. But it should be borne in mind that the blood itself is an extensive living tissue which requires oxygen for its maintenance and functioning and carbon dioxide disposal to remain healthy. Because the blood passes over and among many trachea and tracheoles, it has a ready supply of oxygen throughout its course in the body cavity. Any excess carbon dioxide in the blood will ultimately escape through either the tracheal or body wall.

Ventilation of the Tracheal System. For many small or sluggish insects, gaseous diffusion alone is sufficient to satisfy the needs of respiration, but it is not adequate for active running and flying forms with a high metabolic rate and large energy consumption. In these forms, diffusion is supplemented with mechanical ventilation of the tracheal system. Two principal types of structures are used for this purpose.

1. The taenidia of the trachea prevent their being flattened but in some instances allow a longitudinal contraction and expansion like an accordion. The contraction may result in a reduction of as much as 30 per cent of the expanded volume.

2. Certain portions of the tracheae may be elliptic instead of round and have weak taenidia or none at all. These elliptic portions form sacs which can be flattened by an increase in blood pressure or by bending. In many instances these air sacs form distinct enlarged chambers, resembling the elliptic tracheal structures in having no taenidia and being readily compressible, fig. 112. The action of these is like that of a bellows.

Both these structures act as air sacs analogous to lungs. The respiratory movements of the insect body cause alternate filling and emptying of these sacs. When the body is contracted, the accordionlike sections are contracted, or the blood pressure is increased and results in a compression of the air sacs. Both actions cause an ejection of air from the sacs through the spiracles. When the body is relaxed, the air chambers expand, owing to their own elasticity, and fill up with air from the outside.

The effect of this ventilation, fig. 113, is to keep the air sacs and tracheal

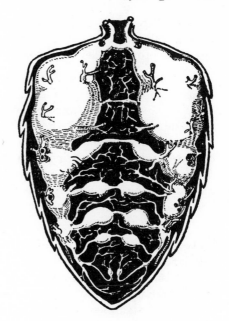

Fig. 112. Tracheal system in the abdomen of the honeybee worker, showing the air sacs. Dorsal tracheae and air sacs have been removed. (From Wigglesworth, after Snodgrass, *Principles of insect physiology*, by permission of E. P. Dutton and Co.)

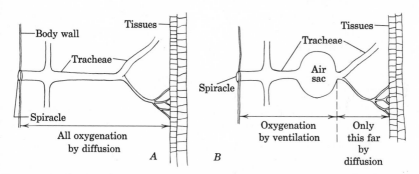

Fig. 113. Diagram to illustrate relation between diffusion and ventilation. *A*, system without ventilation, relying entirely on diffusion; *B*, system supplementing diffusion with ventilation.

trunks filled with air similar in composition to that of the atmosphere. Diffusion acts along the remaining short distance to the tissues, through tracheae branching from the sacs or tracheal trunks.

Respiratory Movements. The air sacs have no muscles of their own. They are operated by a combination of opening and closing of the spiracles and compression of the abdomen. The compression of the abdomen is usually accomplished by the tergosternal and longitudinal intersegmental muscles. These muscles pull the terga and sterna toward each other and telescope the segments. The two movements usually occur simultaneously and provide a device for squeezing the air sacs.

Spiracle Control and Evaporation. Oxygen and carbon dioxide diffuse readily through the tracheal system, and so does water, in the form of water vapor. If the spiracles remain open indefinitely, the insect loses water steadily, and water is normally a precious commodity to the insect. Because of this situation, natural selection has exerted constant pressure on terrestrial insects favoring changes that conserved water without hindering respiration. These pressures have resulted in the evolution of many mechanical devices that close the spiracles for long or short periods, and thus eliminate water loss when the insect does not need the spiracles open for oxygenation.

Respiration Control. The opening and closing of the spiracles and the regulation of the respiratory movements are controlled by some sensitive mechanism whose exact nature is unknown but which appears in general to be correlated with oxygen and carbon dioxide levels in the blood. From the mass of experimental data there have emerged a number of interesting generalities, the more important being listed here:

1. The immediate sensory control of respiration is in the segmental ganglia of the ventral nerve cord. Each ganglion controls only its own segment, so that each segment normally acts as an isolated unit as far as respiration is concerned.

2. Experimental evidence indicates that there is a modulating or coordinating center that can produce rhythmic action of all or several segments. The mode of action and identity of this center are not established, although the thoracic ganglia and muscle tissues have been suggested.

3. During times of rest the respiratory movements may cease altogether, and the spiracles close. In some insects an excess of oxygen will effect the same reaction.

4. Practically any external nervous stimulation (visual, tactile, etc.) will initiate or increase respiratory activities.

5. Various internal chemical stimuli will increase respiration. It is believed that in most species the respiratory nerve center is stimulated by increased acidity of its receptive tissues, caused equally well by either high carbon dioxide tension or acid metabolites produced because of oxygen want. Thus, in cockroaches a high tension of carbon dioxide causes respiratory activity. In mosquito larvae, on the other hand, the carbon dioxide diffuses from the body rapidly and seldom builds up in excess amounts; it is oxygen want that drives these forms to the surface for more air.

6. In many sluggish or inactive insects, the carbon dioxide is released in cyclic bursts. These range from bursts of a minute each every two or three minutes to thirty-minute bursts once in 24 hours. The mechanism and significance of these bursts are not known.

Adaptations for Aquatic Life. The foregoing discussion deals with the type of respiration found in terrestrial insects. But many forms either live in water or spend a great deal of time submerged in water. There are several types of adaptations for obtaining respiratory needs in aquatic situations.

1. DIVING AIR STORES. When they dive beneath the surface, certain insects carry with them a film or bubble of air attached to some part of the body. Both adults and nymphs of water boatmen (Corixidae) and backswimmers (Notonectidae) carry a film of air in the pile on the ventral surface of the body; this film is kept in place by hydrophobe hairs which resist penetration of the air film by the water. Adults of the diving beetles (Hydrophilidae, Dytiscidae) have an air space under the wing covers or elytra, into which space the spiracles open. This air store serves not only as a supply of oxygen for the insect but also as a sort of lung and gill, obtaining added oxygen from the water and discharging carbon dioxide into it by diffusion. It cannot provide for respiratory needs indefinitely in this way, but it enables the insect to remain under water a considerable period before having to come to the surface for more air.

2. AIR TUBES. Many insects that live submerged all the time, breathe through a tube or pair of tubes which can break through the surface of the water. Only the pair of spiracles connected with these tubes is functional; the others are either closed or not developed. The mosquito larva, fig. 303, has a rigid tube; when in need of oxygen, the larva swims to the surface and thrusts the end of the tube through the surface-tension membrane and into contact with the air. The rat-tailed maggot, fig. 308, a fly larva which lives in a viscous or liquid medium, does not swim to the surface but has a respiratory tube which can be extended 3 to 4 inches to the surface. Several other kinds of tubes occur in aquatic, semi-aquatic and some parasitic groups.

3. CUTANEOUS RESPIRATION. Large numbers of aquatic insect larvae make no contact with the atmosphere and have no external devices or special structures for respiration. The same is true of many parasitic insect larvae that live within the tissues of their hosts. In these the gas exchange is made by diffusion through the body wall. The insect utilizes the oxygen dissolved in the water, and excess carbon dioxide diffuses into the water. There are two distinct types of cutaneous respiration. In the first (including very small or first-instar larvae) there is no tracheal system present; in these the gas exchange within the body is by diffusion through

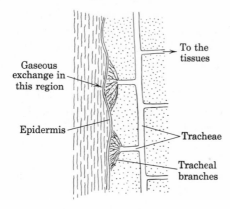

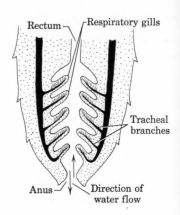

Fig. 114. Diagram of cutaneous respiration in aquatic insects.

Fig. 115. Diagram of rectal gills of dragonfly nymphs. (Adapted from Wigglesworth)

the tissues, including the blood. In the second type (including most of the larger gill-less forms such as late-stage midge larvae and many caddisfly larvae) the tracheal system is developed, but instead of spiracles there are clusters of fine tracheae in the epidermis, fig. 114. Here the gas exchange takes place first through the epidermis and then into the fine peripheral tracheae. From this point the diffusion pattern is the same as in a spiracular system.

4. GILL RESPIRATION. Among the most conspicuous adaptations for aquatic life are the frondlike gills of damselfly nymphs, fig. 171, and mayfly nymphs, fig. 168. These typify many aquatic nymphs and larvae, which have developed gills for their respiratory exchange. The tracheae extend into these gills, and the diffusion of gases takes place through the epidermis between the tracheal threads and the water. An unusual structure occurs in dragonfly nymphs, fig. 115. The rectum is enlarged, and gills provided with abundant fine tracheae extend into the pouch so formed. Into this rectal chamber the insect draws water and then expels it; the respiratory exchange occurs through the thin walls of the gills.

INTERNAL RESPIRATION

The utilization of oxygen in metabolic processes in the cell, with the liberation of CO_2, is accomplished through the action of a set of complex respiratory aids and enzymes called the cytochrome system. It has been found that peculiar changes in the components of this system are associated with the extremely low respiratory rates prevailing during diapause. The cytochrome system is localized in the mitochondria of the cell. The

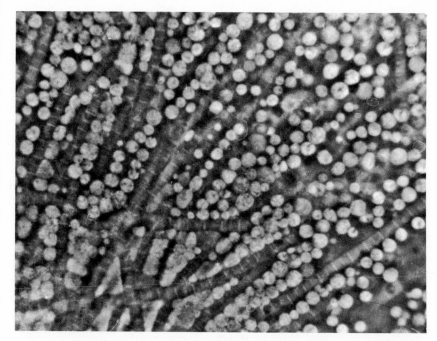

Fig. 116.   Fresh fibrillar muscle of *Drosophila* after minimal teasing, showing striated muscle fibrils and the giant, round mitochondria associated with them.   ×1600.   (After Watanabe and Williams)

flight muscles of Diptera and Hymenoptera contain giant mitochondria (called sarcosomes), fig. 116, which are more than twice the size of ordinary types.   It is thought that these giant mitochondria assist in achieving the high respiratory rates needed for rapid flight.

The Blood and Circulation

The blood or hemolymph of insects is the tissue liquid comprising the body fluid that bathes the internal organs.   The blood effects the chief distribution of food products to the tissues and carries waste products from them.   Normally it has only a secondary although important role in respiration.   It flows through closed ducts for only a short part of its course, its progress through the tissues being by percolation.   Thus, in both distribution characteristics and circulation method, insect blood resembles mammalian lymph more than mammalian blood.   In insects the blood, in addition to the aforementioned functions, constitutes a hydraulic pressure system with its own peculiar functions.

Blood Properties. Insect blood is usually a greenish or yellowish liquid, but it may be clear and colorless. Its specific gravity is close to that of water, varying from about 1.03 to 1.05. It is usually slightly acid, but the pH varies with the species, instar, age, and sex. The dissolved substances in insect blood include about the same array of salts, proteins, carbohydrates, urea, and fats as mammalian blood, but the proportions are often quite different. The most striking features include a very low chloride content and an extraordinary amount of amino acids, which may be 20 to 30 times as abundant as in human blood.

Insect blood differs greatly in its clotting properties. In many kinds of insects it does not clot at all, and wounds are simply stopped by a plug of cells. In other species the blood clots readily.

Almost without exception the blood contains no hemoglobin and apparently has no mechanism for absorbing oxygen in chemical combination. It will take up oxygen and carbon dioxide in physical solution.

Blood Cells. The blood cells or hemocytes are found freely suspended

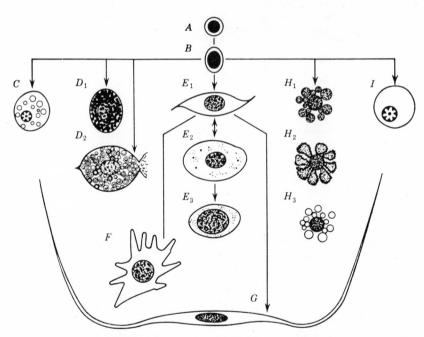

Fig. 117. Principal classes of hemocytes and their possible ontogeny, based upon those found in *Prodenia* and *Tenebrio*. *A*, proleucocyte; *B*, smooth-contour chromophile cell; *C*, spheroidocyte; *D*, cystocytes; *E*, plasmatocytes; *F*, polypodocyte; *G*, vermiform cell; *H*, eruptive cells; *I*, oenocyte-like cell. (From Munson in Roeder, *Insect physiology,* after Jones)

in the blood. In insects as a whole the hemocytes exhibit great diversity; in one species 10 types subdivided into 32 subtypes have been found. Figure 117 illustrates the major types thus far recognized. The hemocytes multiply and grow throughout the insect's life. They appear first as small, dark-staining bodies incapable of phagocytosis. As they mature they assume various shapes. Hemocytes may adhere to tissues, in which case they spread out in a starlike form. In some insects all the hemocytes circulate with the blood fluid; in others all adhere to tissues, forming clusters of phagocytic "tissue"; in many others both circulating and attached phases occur.

Hemocytes are as varied in function as in appearance. They ingest some living and all dead bacteria, collect at wounds and form a plug to close such breaks in the body wall, and form a partition to exclude certain parasites from the body cavity. In addition, the blood cells frequently play an important part in histolysis during advanced metamorphosis.

Functions. The blood of insects has four known functions. The first three listed here are functions of the blood as a living tissue, whereas the fourth function is purely mechanical.

1. TRANSPORTATION. Digested food materials are absorbed from the digestive system and conveyed to the tissues, and waste products are carried from the tissues to the excretory organs. In addition, certain hormones are transported from their source to the tissues.

2. RESPIRATION. Presumably in all insects at least some of the cells are not provided with tracheoles for direct respiratory exchange. These cells undoubtedly obtain their oxygen from the dissolved store in the blood. We have seen that much carbon dioxide diffuses through the tissues and finally through the cuticle; the blood aids in this process. In larvae of certain species of *Chironomus* the blood contains dissolved hemoglobin. This is not nearly so effective in absorbing oxygen as mammalian hemoglobin. It does take up considerable oxygen, however, which the larvae may use when hiding in the oxygen-deficient ooze on the bottom of a pond.

3. PROTECTION. The hemocytes dispose of certain bacteria and parasites. The healing of wounds is effected by the blood or its hemocytes.

4. HYDRAULIC FUNCTION. The entire volume of blood inclosed within the body wall forms a closed hydraulic system capable of transmitting pressure from one part of the body to another. In this purely mechanical sense it is put to many uses by the body. The pressure of the blood is regulated by contractions of the thorax or abdomen or both. Alternate increase and decrease of blood pressure, brought about by respiratory movements, causes the emptying and filling of the tracheal air sacs and pouches. Localized blood pressure is responsible for stretching of the

exoskeleton after molting, inflation of the wings, and frequently operation of the egg-breaking device at time of hatching.

Circulation. In general, fig. 118, insect blood circulation follows this path: It is pumped forward by the heart from the abdomen, through the aorta, and emptied into the head; from the head it percolates back between the tissues until it reaches the abdomen, where the circle starts forward again through the heart.

Blood is sucked into the heart through the ostia and then driven forward by peristaltic movements which flow along the entire length of the heart. The negative pressure of the heart chambers which aspirates the blood and the systolic pressure which causes the forward flow of blood are due to the elasticity and muscular manipulation of the heart, the aliform muscles, and other muscles which may be associated with them. At times the flow is reversed, and blood pours from the heart back into the visceral cavity. In most insects the heart is unobstructed for its entire length. In a few the ostia are recessed into the heart to form valvelike flaps which divide the heart into segmental chambers.

In addition to the heart, a varied assortment of structures exist to aid the blood flow through the appendages or its distribution in the body

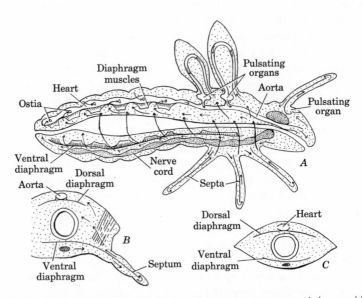

Fig. 118. Circulation accomplished by heart and accessory structures. *A*, insect with fully developed circulatory system, schematic; *B*, transverse section of thorax of the same; *C*, transverse section of abdomen. Arrows indicate course of circulation (based largely on Brocher). (Redrawn from Wigglesworth)

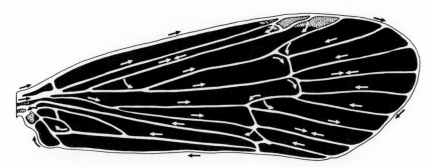

Fig. 119. Blood circulation in front wing of the caddisfly *Limnephilus rhombicus*. Direction of flow shown by arrows. (After Arnold.)

cavity. In rare cases the aorta may discharge into vessels which carry the blood in different directions. In many insects the antennae and legs are divided by longitudinal membranes or septa so that the blood enters on one side, flows the length of the appendage, and empties on the other side. Blood movements into the appendages are aided also by the respiratory movements, so that the "pulse" in the legs may synchronize with respiratory contractions and not with the heartbeats.

There are frequently supplemental blood pumps, or pulsatile organs, in the meso- and metathorax for sucking blood through the wings. In these instances the blood flows through certain veins of the wings, fig. 119, and is returned either directly to the aorta or to the body cavity. When well developed, the ventral diaphragm also assists blood flow; contractions of the diaphragm muscles drive the blood both laterally and backwards.

The diagrams in fig. 118 outline the direction of flow set up by these various methods.

The heart is so well supplied with nerves from both the visceral nervous system and the segmental ganglia that many investigators believe all its activities are controlled by nerve impulse. It is still a moot point, however, as to whether the automatic heartbeats are due to nerve stimulation or to muscle which possesses the ability to contract and relax periodically without nervous stimulation.

Metabolism

Metabolism is the total of all the chemical and physical processes which take place within the organism. This includes both the constructive phase (anabolism) and the destructive phase (catabolism). The metabolism of insects is affected markedly by both the activities of the insect itself and the external conditions which surround it, such as temperature, humidity, and atmosphere. These ecological influences on fundamental metabolism integrate with those on actions and habits, as discussed under Ecological Considerations, in Chapter 9.

Temperature Control. Insects are cold-blooded animals and their body temperature, in general, depends on that of the environment. Within certain limits, insects can change their body temperature. At high temperatures insects of sufficient size can reduce their temperature by evaporation of water from the surface. At low temperatures the chemical changes going on within the body may raise its temperature above that of the environment. Certain soil-inhabiting beetles, for example, have been observed to depress the body temperature 3.6°F.(= 2°C.) by tracheal evaporation. Evidence indicates that fairly high body temperatures are necessary for the extremely rapid muscular activity of flight. For instance, experiments with large hawk moths have demonstrated that they cannot fly at body temperatures below 86°F.(= 30°C.). Below this point the moth stands and vibrates its wings until this muscular activity has raised its temperature to 86°F., at which point it can fly. During flight its temperature may rise above 104°F.(= 40°C.), owing to the violent muscular action.

To what extent insects at rest increase their metabolism to maintain body temperature is not known.

Metabolic Rate. Within certain limits the metabolism of an insect increases with an increase in temperature. This variation in metabolic rate is correlated with the following physicochemical phenomena which automatically accompany a rise in temperature:

1. Chemical reactions increase in rate.
2. Solubility of solids in liquids increases.
3. Speed of diffusion of gases increases.
4. Solubility of gases in liquids decreases.

In insects an increase in temperature also induces an increase in activity, which in turn increases metabolism. Humidity also influences metabolism; it has been demonstrated in several insects that an increase in humidity decreases the rate of metabolism. These correlations obviously do not operate below or above temperatures detrimental to the insect's well-being or under inimical combinations of temperature and humidity.

Various attempts have been made to show definite mathematical relationships among temperature, humidity, basal metabolism, activity, and growth. So many variables have been encountered that few workers agree on the interpretation of results.

The resting metabolism of the bee is proportionate per unit of weight to that of man, consuming about 20 gram-calories of heat per kilogram of weight per minute. But in extremes of exercise the bee may increase this as much as 1300 times, whereas man in his greatest exertions is able to increase it only 10 or 12 times. Thus the bee in flight consumes as high as 26,000 gram-calories (= 26 standard or large calories) per kilogram of weight per minute.

In insects that are normally well fed, the respiratory quotient during activity is near unity, indicating that only carbohydrates are being oxidized. During starvation the quotient falls, a phenomenon associated with the oxidation of fats and proteins. This indicates that, under starvation conditions, insects follow the same procedure as other animals, burning their carbohydrates until these are exhausted, and only then using the fats and proteins for energy.

Oxygen Requirements. Insects are remarkably resistant to oxygen deficiency. The rate at which they are able to extract oxygen from the atmosphere remains the same down to a very low level of oxygen pressure. The exact level differs in various species. Below the critical oxygen pressure, the rate at which it is absorbed drops rapidly.

Certain insects possess a peculiar *anaerobic tolerance;* they are able to exist for long periods in the complete absence of oxygen. In these circumstances, the insect stops its activities and becomes quiescent. Only a minimum of metabolism goes on, and it probably consists chiefly of the involuntary muscular activities controlling circulation and digestion. The lactic acid, other unoxidized metabolites, and carbon dioxide from these processes accumulate in the body. When air is again available, the insect absorbs oxygen from it at an extremely high rate and oxidizes the waste products accumulated during the anaerobic period. This anaerobic tolerance has been demonstrated experimentally in many insects. The larva of the horse bot *Gasterophilus* (parasitic in the stomach of the horse) normally enjoys a cycle of aerobic and anaerobic conditions correlated with digestive activities of its host. Experimentally it has been found that the bot larva can survive as long as 17 days without oxygen. Diurnal aerobic and anaerobic cycles undoubtedly occur in certain lake-inhabiting insects that may hide below the lake's thermocline during the day and feed near the oxygen-carrying surface layer of water at night.

Temperature Resistance. In the course of their lives insects may be exposed to great extremes of temperature. Insects that have lost water by desiccation are more resistant than normal ones to inimical temperatures. The exact reason for this is not known. It has been shown that desiccated individuals of the beetle genus *Leptinotarsa* may withstand 1° to 8° of temperature (F.) above the point which is lethal to undesiccated individuals.   Similarly desiccated individuals are more resistant to cold.

There is as yet no complete explanation of these phenomena related to partial desiccation.   The cold resistance is thought to be due in part to a lowering of the freezing point of the cell contents by simple concentration of dissolved substances.   Another reason given is that more of the free body water goes into combination, or is "bound," with body colloids.   It is thought that such a depletion of free water by its conversion to bound water lowers the freezing point of the body contents and thus effects a greater degree of cold hardiness.   But the exact chemical and physical changes, and their significance, are not known.   The explanation of increased resistance to heat probably hinges on similar changes.

Color-Pigment Metabolism.   The color pigments of insects (see p. 111) show interesting reactions to both metabolism and external conditions.   Experiments on many insects, including the potato beetle *Leptinotarsa* and its predator, the stink bug *Perillus,* have shown that, as the metabolism increases with higher temperature or lower humidity, more of the pigments are oxidized, resulting in lighter-colored insects.   Both the black melanin and the orange carotinoids have been found to react in this fashion in some insects.   In these cases (for example, the Colorado potato beetle), individuals reared at a high temperature and low humidity may be almost colorless; others raised at medium temperatures and humidity may be orange; and still others raised at low temperatures and high humidity may be black.

The extent of these effects is not uniform.   In certain insects only some parts of the body show these reactions.   In others the color pattern may be rigidly fixed and apparently not influenced by effects of external conditions.

Physiology of Development

The growth from egg to the functional reproducing stage is considered development.   It represents the total of metabolic processes in the growing-up period.   The descriptive treatment of development is treated in the following chapter, but certain physiological features associated with development are of interest at this point.

Metamorphosis.   In insects with gradual metamorphosis the physical changes leading to adulthood are spread more or less evenly throughout the entire life.   In insects with complete metamorphosis the assumption of adult characters is accomplished suddenly, during the pupal stage.   There is at this period an apparent physical revolution from larval to adult characteristics. There is, however, no accompanying all-inclusive physiological revolution. The epidermis and tracheal system are reconstructed simply by the normal secretion of their matrix cells, but the secretions are cast in a different "mold." The nervous system enlarges rapidly by growth of the constituent parts, sometimes accompanied by the fusion of certain ganglia. The heart grows without marked change. The digestive tract is changed by the growth or reduction of some parts and the remodeling of others, fig. 81.

Not represented in larval structures are certain features of adults, such as the wings and reproductive system. Certain other adult features are usually radically different in size or organization from their larval counterparts, notably the legs and the musculature, especially muscles controlling flight and reproductive activities. These adult parts are built up from larval fat bodies, blood sugar, and muscles, in a series of conversion processes which are grouped into two phases, *histolysis* and *histogenesis*. Histolysis is a breaking-down process, essentially catabolism; leucocytes and enzymes convert the larval fat body, much of the muscle tissue, undoubtedly parts of other tissue also, and later the leucocytes themselves, into a nutritive matrix transportable by the blood to growing tissues. Histogenesis, representing anabolism, is the construction of adult tissues from the products of histolysis. Both phases go on simultaneously.

Before pupation, larvae enter a quiescent stage lasting one to several days. It is during this period that the conversion processes begin. They continue through the pupal stage until the adult structure is complete. During these processes, the stores of fat and glycogen accumulated by the larva during its feeding period are drawn upon for a food source.

Hormonal Regulation of Development. Experiments in ligaturing insects, fig. 120, and in exchanging body fluids and tissues of individuals

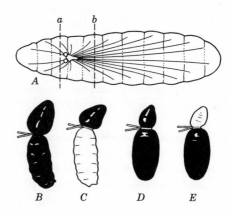

Fig. 120. Muscid larva (*A*) showing the ganglion and nerves and the positions *a* and *b* at which ligatures were applied. *B-E*, larvae ligatured at different levels and times and showing different combinations of pupating and nonpupating parts. (From Wigglesworth, after Fraenkel)

in different stages of development have shown that retention of immature or juvenile characteristics, development of adult structures, and molting are controlled by hormones secreted by the brain and several bodies associated with it. The principal hormones and their sources are:

ECDYSONE. A small gland (variously shaped in different insects) in the prothorax, the prothoracic gland, secretes this hormone which is necessary for molting and which promotes growth and development. The hormone is necessary for development of adult structures during pupation.

JUVENILE HORMONE. The corpora allata, fig. 89, secrete this hormone that retards development of adult structures and is responsible for the maintenance of nymphal or larval characteristics during pre-adult life. In many species the juvenile hormone is necessary for the development of ovaries.

BRAIN HORMONE. The brain secretes a hormone which activates the prothoracic gland to produce ecdysone.

Actually the brain controls the activation of both the prothoracic glands and the corpora allata, the former by hormone secretion, the latter by direct nerve connection.

It appears that, in response to a hereditary pattern of tissue and organ behavior, the brain causes first one of the growth hormones and then the other to be secreted into the individual's blood stream. These cause the rhythmic occurrence of molting and development of the proper immature or adult structures. Other hormones are involved in certain phases of development. There is evidence, for instance, that the developing ovaries themselves secrete a hormone affecting the brain or the corpora allata. The rabbit flea *Spilopsyllus cuniculi* requires a hormone-like substance from pregnant rabbits before its own eggs will mature. This requirement has the effect of synchronizing the reproductive cycles of both the flea and its rabbit host. Possibly other instances of such relationships occur, but there are few detailed studies in the area.

Of unusual interest is the discovery that the brain hormone and ecdysone (the prothoracic gland hormone) are steroids and hence may explain the insects dietary need for cholesterol (a possible steroid precursor). The juvenile hormone appears to be a complex alcohol or its derivative.

Suspended Activity: Diapause. In the life of many insects there are more or less prolonged periods of quiescence, during which visible activity and many physiological processes are suspended. These periods may occur in the egg, nymph, larva, pupa, or adult. They are characterized chiefly by a cessation of growth in immature stages and by cessation of sexual maturation in adults.

In some instances activity ceases only at the onset of some unfavorable condition, such as cold or drought, and resumes at the termination of the unfavorable condition. Thus chinch bugs breed continuously if kept at favorable temperatures, but the adults enter a quiescent condition if kept at a low temperature. In other instances the quiescence is a hereditary characteristic triggered by an internal "timing" mechanism which brings about a cessation of activity in advance of the unfavorable condition; some unfavorable condition, however, is usually a necessary stimulus to break this quiescence. Such a quiescent period is called diapause. An example is the cecropia moth (*Platysamia cecropia*). When full grown, usually toward the end of summer, its larva constructs a cocoon, and transforms into a pupa, which enters diapause. If chilled to $3°-5°C$. for about six weeks, then returned to room temperature, the pupa will resume development. If kept constantly at room temperature it will be six months to a year before diapause is broken.

Experiments on a large variety of insects have led to interesting general ideas about diapause. Initiation of diapause is linked with seasonal changes in day length, or photoperiodism. The exact relationship differs in various species. In many the critical period may be considerably prior to the onset of diapause. In the moth *Polychrosis* pupal diapause is predetermined by the photoperiodic conditions of embryonic development. In the beetle *Leptinotarsa*, on the other hand, a long photoperiod experienced by late larvae may cause them to enter diapause as soon as they dig into the ground. In the silkworm *Bombyx* a lengthening (vernal) photoperiod triggers diapause; in most diapausing insects it is a shortening (autumnal) photoperiod that does it.

The principal condition bringing about a termination of diapause seems to be exposure to temperatures below those favoring normal development. The actual mechanism involved, however, may be obscure. Certain unchilled grasshopper eggs and moth pupae may break diapause after being immersed in xylol or other organic fat solvents. In the case of the grasshopper eggs, the xylol dissolves a water-proofing wax cap covering the micropyle, water then enters the egg, and this water uptake breaks the diapausing condition. Other insects, however, may absorb a complete supply of water before entering diapause.

The length of diapause varies to a tremendous extent, ranging from little more than a week to ten or twelve years. In certain species having several generations a year, all members of each generation do not act alike with respect to diapause. A portion of the population from each of several generations may enter diapause while the remainder continues normal development, fig. 121. This differential behavior is linked with genetic factors, but their interplay with factors of the environment has proved to be complex.

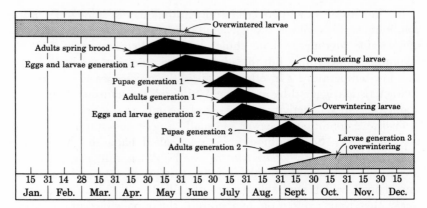

Fig. 121. Generations of the tobacco moth *Ephestia elutella,* showing portions of generations 1 and 2 entering diapause and overwintering. (After Tenhet and Bare)

The internal regulation of diapause release appears to be through the liberation of hormones by either the brain, the suboesophageal ganglion, or the prothoracic glands, depending on the species involved. Hormone secretion by the two latter structures may also be under control of the brain, either directly by innervation, or indirectly by hormone secretion. The details of this mechanism are not well understood, especially the exact response elicited in the brain by the external stimuli concerned with diapause, and the manner in which responses presumably made at one time can direct actions which do not occur until several instars later.

REPRODUCTION

In insects reproduction is the function of the sexual reproductive system. Normally insect reproduction is bisexual, in that the egg produced by the female will not develop unless fertilized by spermatozoa produced by the male. Except in a few species, only one sex is represented in any one individual. In most insect species, therefore, the physiology of reproduction deals with the development and maturation of spermatozoa in the male and of eggs or ova in the female and the manner in which they are brought together.

Development of Spermatozoa. Spermatozoa are produced in the follicles of the testis, fig. 122. The upper portion of the follicles contain primary germ cells called spermatogonia. These divide repeatedly to form cysts, which move to the base of the follicle due to the pressure of their own increase in size. At the base of the follicle each cell in a cyst undergoes repeated division and may increase in number 5 to 250 times.

In the next cell division following this multiplication stage there occurs the reduction division of the chromosomes. This is followed by a transformation period in which the round cells develop into slender flagellate

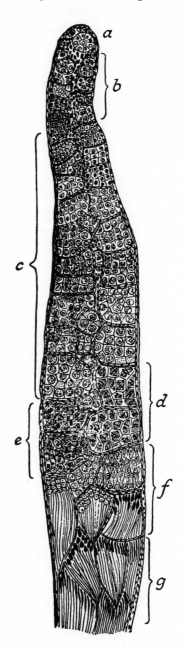

Fig. 122. Longitudinal section of testis follicle of a grasshopper, semischematic. *a*, apical cells surrounded by spermatogonia; *b*, zone of spermatogonia; *c*, zone of spermatocytes; *d*, cysts with mitoses of second maturation division; *e*, *f*, zone of spermatids; *g*, zone of spermatozoa. (From Webber, after Depdolla)

spermatozoa. These mature sperms escape from the duct of the follicle (*vas efferens*) into the genital ducts (*vas deferens*); they are stored in an enlarged or coiled portion of this duct, the seminal vesicle, until mating. At the time of mating the spermatozoa are transferred to the spermatheca of the female, where they are stored until needed for fertilization.

Development of Eggs. The eggs are developed in the ovarioles of the ovary. The tip of the ovariole, the *germarium*, contains primary germ cells which divide to produce developing eggs, or oocytes. These usually appear in successive stages of growth down the length of the ovariole. The oocytes derive nourishment for their growth from either the follicular epithelial cells forming the ovariole, fig. 123*A* or from special "nurse cells" present in the ovariole, figs. 123*B* and *C*. Below the oocyte at the end of the ovariole is a plug of epithelial cells which seals the duct leading from ovariole to oviduct. When the oocyte is fully developed, this plug breaks down, and the oocyte or egg is released into the oviduct. The portion of the ovariole which contained the released egg shrinks, and a new plug forms below the next oocyte. As this oocyte matures and its chamber enlarges, it assumes the same position as the previously discharged egg.

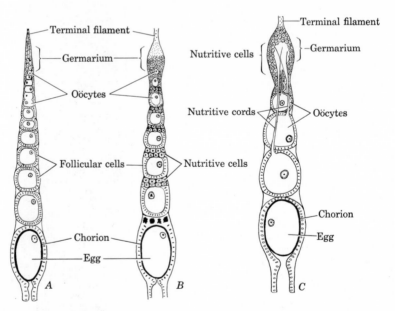

Fig. 123. Longitudinal section of ovarioles. *A*, simple or panoistic type having only oöcytes; *B*, polytrophic type having oöcytes and nurse or nutritive cells alternating; *C*, teleotrophic type having nurse cells connected to oöcytes by nutritive cords. (Redrawn from Weber)

The eggs at time of discharge into the oviduct are surrounded by the eggshell or chorion. The eggshell is perforated in one or more places by minute pores or micropyles. It is through these minute openings that the spermatozoa gain entrance to the interior of the egg.

Fertilization. As the eggs pass down the oviduct (by peristaltic action of the oviduct muscles) into the vagina, they come to lie at the opening of the spermathecal duct. From this duct spermatozoa emerge and enter the egg micropyle. After the spermatozoa enter the egg, the egg nucleus undergoes two divisions, one a reduction division, with the production of the female pronucleus and polar bodies. The spermatozoan loses its tail and changes to the male pronucleus. The male and female pronuclei unite to form the zygote.

From this usual sequence of events there are many deviations. The following are interesting examples. In the bedbug *Cimex* the spermatozoa migrate from the spermatheca, into the follicular structure of the ovarioles and from there into early-stage oocytes. Fertilization is thus accomplished before the eggshell is formed. In parthenogenetic species such as the European spruce sawfly *Diprion hercyniae,* fertilization does not occur, but the diploid chromosome count is restored by fusion of a polar body with the female pronucleus.

Mating. At each mating a large number of spermatozoa are transferred from the male to the female. The female stores and controls the spermatozoa so that only a small number are liberated at a time as successive eggs pass down the oviduct. In this way a separate mating is not necessary for the fertilization of each egg. As a consequence a large number of insects mate only once in their lifetimes, and most of the remainder mate only a few times.

Mating is induced by stimuli of many kinds; by peculiar movements, such as the dancing of swarms of male mayflies; by sound, as the chirping of crickets and grasshoppers; by color reactions, as in some butterflies; and chiefly by a wide variety of scents. The gonads seem to have no marked influence on mating behavior, since many species mate before the female ovaries are well developed, and castrated males will mate normally but without transfer of spermatozoa.

The mechanics of spermatozoa transfer may be divided into several distinctive types. In many forms, such as some of the true bugs, the penis is inserted into the female spermatheca and the spermatozoa are placed directly in this storage chamber. In many moths, grasshoppers, and beetles, the penis discharges the spermatozoa into the female bursa copulatrix; after mating, the spermatozoa become transferred from this structure to the spermatheca. The mechanism of this transfer is not known.

In many insects of this group having a bursa copulatrix, the spermatozoa are transferred in a membranous sac or *spermatophore,* formed by the secretion of the male accessory glands. This sac of spermatozoa is deposited in the bursa or vagina, and its contents are transferred to the spermatheca. After this transfer the empty spermatophore is ejected by the female.

Longevity of Spermatozoa. Apparently the secretions of the female spermatheca or its associated glands can keep spermatozoa viable for a considerable period. The honeybee can sustain its spermatozoan store for several years. In moths the spermatozoa remain alive in the spermatheca for several months. In females of a few insects, such as the bedbug *Cimex,* the spermatozoa not utilized in a few weeks are digested and absorbed by the body tissues, and mating occurs from time to time, replenishing the supply.

IRRITABILITY

A characteristic of living organisms is their ability to respond to stimuli, a property called irritability. In a general way irritability is the protective function by which the organism can move away from harmful environmental conditions or toward more favorable conditions. From the standpoint of mechanical performance, three definite functions are embodied in irritability. These are *sensitivity,* the ability to detect or perceive stimuli; *conductivity,* the transmission of stimuli from the point of reception to various parts of the body; and *contractility,* the power of contraction, on which depends the organism's ability to make a response to the original stimulus.

In primitive forms of unicellular life all three of these functions are performed by the same cell. Sensitivity lies in the cell membrane; conductivity and contractility are apparently properties of the general protoplasm. In highly organized animals each of these functions is performed by special structures or tissues. Insects have a well-developed system for accomplishing these various components of irritability. Sensitivity is seated in sense organs, some simple and some complex, distributed over various parts of the body. These receive stimuli. Conductivity is performed by the nervous system, which "telephones" notice of stimuli from the sense organs to reacting tissues. Contractility is accomplished by various cells or tissues especially modified for this purpose, notably muscle tissue and certain glands. When activated by "messages" coming over the nerve fibers, contractions of the muscles or secretion of hormones by the glands cause a reaction, or response, to the stimuli detected by the sense organs.

In observations on behavior, the terms *reception* and *response* are usually employed. These terms pertain to the two ends of the irritability chain, emphasizing the stimulus received and the response given by the organisms.

Sensitivity

Sensitivity of the organism as a unit is centered in specialized cells or groups of cells termed sense organs. These serve for the reception of many external types of stimuli, including tactile, auditory, gustatory (taste), olfactory (smell), thermal, and visual. In addition, insects respond definitely to temperature changes, hunger, and internal physiological conditions; no special receptors are known for these senses.

Nerve fibers have no selectivity. They transmit only an abstract impulse. The central nervous system can identify the nature of different stimuli because each sensory area is responsive to only one type of stimuli, and each one has its separate nerve endings. The nervous system is therefore able to identify types of stimuli by the sensory stations from which the impulses come.

The actual structures which serve as sense receptors vary from simple types to extremely complex organs such as the compound eyes.

Structure of Sense Receptors. Hairlike sense organs are the simplest type. They are typical setae to which have been added nerve cells and nerve endings, fig. 124. The nerve ending is set at the base of the seta in

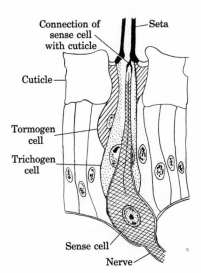

Fig. 124. A simple hairlike sense organ. (Redrawn from Snodgrass)

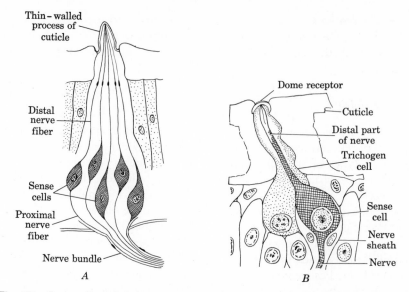

Fig. 125. Sense cells and their receptors. *A*, chemoreceptor having thin-walled peg and multiple sense cells. *B*, domelike sense receptor on cercus of a cockroach. (Redrawn from Snodgrass)

such a manner that the movements of the seta change the pressure on the tip of the nerve ending. Such changes of pressure cause a definite impulse to be transmitted along the nerve fiber. These hairlike sense organs usually serve for tactile stimuli. A great variety of organs for reception of stimuli relating to taste, smell, or humidity are similar in general structure to these hair organs. They differ in that the hair has been replaced by a thin-walled peg or plate with which the tip of the nerve ending is in contact, fig. 125*A*. Some of these sense organs may have a group of sense cells associated with the peg or plate, allowing the accommodation of several nerve endings in the same receptor, fig. 125*A*.

Stress and Sound Receptors. Domelike sense receptors, called the *campaniform sensilla,* occur on many areas of the integument, including that of the appendages. In these structures, fig. 125*B*, the distal process of the sense cell connects with the dome. Changes in the mechanical stresses of the integument surrounding such a sensillum cause the domelike plate to bow or flatten, and this action presumably lengthens or shortens the nerve apex, causing stimulation of the nerve. These organs respond to body movements and may be important in geotropic responses, coordination of leg movements, and other unknown functions.

In sense organs of a somewhat similar type, called *scolopophorous sensilla,* the axial fiber of the nerve cell is surrounded by an enveloping cell and surmounted by a cap cell. These scolopophorous organs occur in many parts of the body and are also thought to respond to changes in mechanical pressure. The exact function of many of these organs is not known, but certain of them are receptors of sound. Those that are sound receptors are called *chordotonal organs* and are associated with some sort of drumlike surface or tympanum.

Temperature Receptors. In the order Orthoptera a series of paired specialized areas on the head, thorax, and abdomen have been found which appear to be thermoreceptors. These areas differ in cuticular structure and texture from surrounding epidermis. Different groups of Orthoptera have different arrangements of thermoreceptor areas.

Eyes. Visual organs or eyes occur in most insects and consist of aggregations of photoreceptive cells. Photoreceptive sense cells are extremely varied in histological detail. They differ from other types of sense cells in two features: (1) The cuticle overlaying them is transparent, forming a *cornea,* and (2) the sense cells have no definite tip, but instead contain fine surface striations which are apparently the sensitive receptive elements of the cell, fig. 126*A.*

Insect eyes may be divided conveniently into two types: simple and compound. The simple types have a single lens for the entire eye, fig. 127.

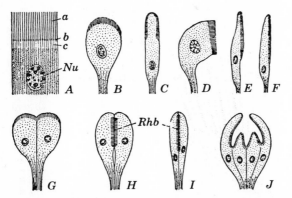

Fig. 126. Diagrams of light receptor cells and the development of a rhabdom. *A,* receptive pole of a sense cell; *a,* striated zone (rhabdomere) formed of ends of neurofibrillae; *b,* basal bodies; *c,* clear zone (from Hesse). *B–J,* different positions of the striated zones on ends of sense cells. *H, I,* union of the striated zones of adjacent cells to form a rhabdom (*Rhb*). (*B–J* from Weber) (After Snodgrass)

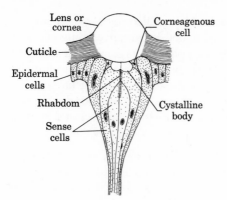

Fig. 127. The simple eye of a caterpillar. (Redrawn from Snodgrass)

The lens is specialized cuticle secreted by a layer of epithelial cells called
the corneagenous cells, which are themselves transparent. Nerve cells
form a retina beneath the corneagenous layer. In most eyes the striated
sensitive elements of the sense cells have migrated to form a line down one
side of the cells. The cells are frequently oriented so that these "lines" of
adjacent cells are together. The linear compound sensitive element so
formed is called a *rhabdom*, fig. 126*H, I.*

Compound eyes have the same basic parts as simple eyes, but the sense
cells are grouped into concentric units called *ommatidia*. Each ommatidium, fig. 128*B*, has its own lens (distinguished externally as a *facet*), some-

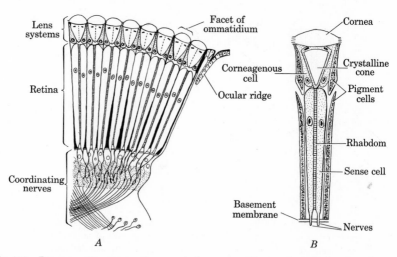

Fig. 128. Diagram of a compound eye, and of an ommatidium. *A*, vertical section of part of
eye. *B*, typical structure of an ommatidium. (Redrawn from Snodgrass)

times a lenslike cone, below this a rosette usually of eight sense cells with a central rhabdom, and pigment cells around both cone and rosette. The pigment cells contain colored granules which can move up and down in the cell. This movement is synchronized in all the pigment cells surrounding an ommatidium and controls the amount of light reaching the sensory portion of the ommatidium.

In adult insects the ocelli are simple eyes, and the large lateral faceted eyes are compound. Larvae have only simple eyes; sometimes several of these form a cluster. Both types of eye connect as a unit directly with the brain.

Conductivity

In insects, as in all the higher Metazoa, the basis of conductivity is the nerve cells. But these cells have become sufficiently developed into different types that they act also in the capacity of coordination or association.

Nerve Cells or Neurones. A neurone is simply an elongated cell capable of receiving and transmitting stimuli. Each neurone, fig. 129, consists of three principal parts: (1) a cell body or *neurocyte*, (2) one or more receptor fibrils, and (3) a transmitting fiber or axon which ends in a group of fibrils called arborizations because of the treelike pattern of branching.

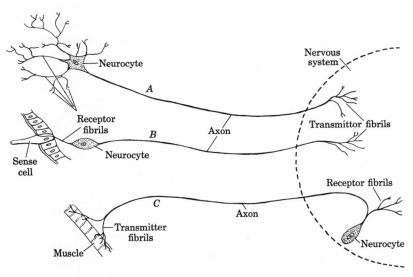

Fig. 129. Diagrams of neurones. *A, B,* sensory types; *C,* a motor type. (Adapted from Wigglesworth)

At least one end of each neurone is situated in the central nervous system or associated ganglia. There are two chief types of neurones. In one type, fig. 129*A*, *B*, one or more receptor fibers arise directly from the neurocyte. These are sensory neurones in which the receptor fibers are connected with sense cells and the axon runs to and terminates in the central nervous system. In the other type, fig. 129*C*, the receptor fibrils are situated on what appears as a branch of the axon called the *collateral branch.* This type includes motor neurones, in which the receptor fibrils and neurocyte are situated in the nervous system and the axon forms a nerve fiber running to muscle tissue; and association neurones, all parts of which are situated within the central nervous system.

When considering neurones one is inclined to think only of their highly specialized transmission function. The ultimate basis of this activity is a process of physico-chemical change within the cell requiring relatively large amounts of energy. It is therefore certain that the neurone is specialized in some fashion for a high rate of internal respiration in order to keep the nervous "motor" going.

In all cases the stimuli are received by the receptor fibrils and travel to the tips of the axon. This direction is not reversible. Impulses may be passed from one nerve cell to another through a *synapse,* an area of the central nervous system in which are intermingled the end fibrils of the axon of one neurone and of the collateral branch of another. In a generalized reaction, fig. 130, an external stimulus causes some change in a sense cell, this in turn stimulates the associated sensory neurone; the impulse from this neurone passes through a synapse to an association neurone and from this through another synapse to a motor neurone; and this motor

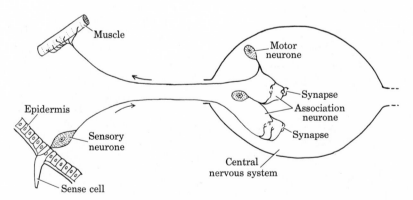

Fig. 130. Diagram of a simple reflex circuit of sensory, association, and motor neurones. Direction of impulse journey is shown by arrows. (Adapted from Wigglesworth)

neurone transmits an impulse through its axon to a muscle fiber, which contracts as a result of the stimulus received.

Coordination. The synapses of each body segment are grouped together to form the ganglia of the central nervous system. Thus the sensory nerves all "report" to these centers, and the "orders" go out from these to the reactive tissues. Association cells run from ganglion to ganglion, and into the brain. They also may link the same motor cell to several sensory cells or several motor cells to one sensory cell. This whole communications system coordinates responses in different parts of the body with stimuli received at only one station. Thus a touch on a cockroach's cercus (reporting to the terminal abdominal ganglion) will cause the animal's legs (motivated by the thoracic ganglia) to respond with running movements.

Contractility

The function of contractility in insects includes muscle reaction, resulting in movement; glandular reaction, resulting in secretion; and a combination of the two, resulting in the ejection of secretions. Muscle tissue is the more conspicuous type, and its contractions are responsible for movements of the insect as a whole and its internal and external parts individually.

1. Muscular Reaction. Insect muscles are similar to those of other animals in that they are elongate and have the power to contract when stimulated by an impulse from a nerve cell. The period of contraction is followed by a period of recovery in which the muscle cell returns to its original shape.

All insect muscles are made up of striated fibers, although in some visceral muscles the striations may be difficult to detect. Each fiber is composed of a number of parallel threadlike fibrillae laid down in a matrix of plasma (called sarcoplasm) containing nuclei. The arrangement of fibrillae and sarcoplasm is different in different insects, fig. 131, or in different muscles of the same insect.

Muscular contraction results in the movement of some part. For every movement there is a countermovement, in which the part regains its normal position. This countermovement is frequently brought about by the action of a second muscle, as in fig. 132C. Here the tibia articulates at each side with the femur at the point *l*; contraction of the elevator muscle *lv* raises the tibia for the initial movement; contraction of the depressor muscle *dpr* lowers the tibia for the countermovement. In other cases only a single muscle in involved. In these the countermovement is

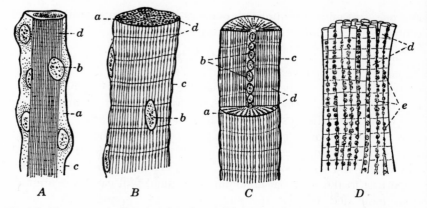

Fig. 131.   Insect muscle fibers.   A, from larva of honey bee; B, from leg muscles of a scarab beetle; C, from leg muscles of honey bee (tubular muscles); D, indirect flight muscles of honey bee (a group of sarcostyles from fibrillar muscle).   a, sarcoplasm; b, nuclei; c, sarcolemma; d, fibrils or sarcostyles; e, sarcosomes.   (From Wigglesworth after Snodgrass)

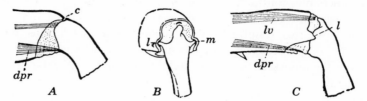

Fig. 132.   Movement of an adult insect leg.   A, monocondylic (single socket) joint (c). B, C, dicondylic (double socket) joint, (l and m), end view and side view with levator (lv) and depressor (dpr) muscles.   (After Snodgrass)

brought about either by the pressure of the blood or by tension of flexible membranes.   In countermovement by blood pressure, the hydraulic pressure of the blood simply forces back the part when muscle tension is released.

FLIGHT.   The muscular mechanism for insect flight is of unusual interest because nothing at all like it is found in any other animal group.   Flight is attained by the stroking of the wings, which are essentially outgrowths of the lateral margin of the meso- and metathoracic tergites.   In the dragonflies, damselflies, lacewing flies, and other primitive orders, the two pairs of wings move independently.   In the damselflies, for example, as one pair of wings goes up, the other pair goes down.   In the moths and butterflies, bees and wasps, and certain others, the two wings on a side are held together or coordinated by various types of hooks, bristles, or folds, so that both pairs work in unison. In most beetles and some of the true bugs, the front wings form hard armor plates, and only the hind wings function in flight. In the twisted-winged insects (Stylopoidea), the front pair is reduced, and in the true flies (Diptera), the hind pair is reduced, so that each of these two groups has only a single pair of flight wings.

Wing movements producing flight are controlled by about six to a dozen pairs of muscles. The exact muscle arrangement differs in various orders. These muscles act on the notum, basalar and subalar sclerites, and the axillary sclerites, but are not connected directly with the wing itself. In spite of the seeming complexity which would appear to attend insect flight, the wing movements are produced by a very simple seesaw or leverage arrangement.

The principal points and parts involved are these, fig. 133*A*:

1. The wing is attached intimately to the edge of the central portion of the notum. A hinge joint *j* is present along the line of attachment.

2. The center of the notum can be moved up and down by muscle action, carrying the base of the wing with it.

3. The wing, just beyond its base, passes over the pleural process *p*, which is the pivot or fulcrum of the wing seesaw. As far as a consideration of flight is concerned, this pivot point is stationary.

Thus, fig. 133*B*, when the set of powerful muscles connecting the

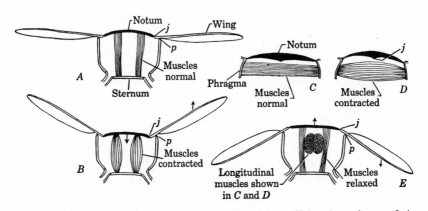

Fig. 133. Diagrams of muscles and movements producing insect flight. *j*, attachment of wing to notum; *p*, pleural process which acts as a pivot point for the "seesaw" of the wing. *A* and *B* are cross sections that show only the vertical muscles connecting notum and sternum; when these muscles contract, as in *B*, the notum is pulled down, depressing the edge of the wing, which pivots at *p*, forcing upwards the major portion of the wing. *C* and *D* are longitudinal sections of the mesonotum that show the longitudinal muscles; when these contract, as in *D*, they "bow" the notum, causing the center to raise. The effect of this is shown in cross section *E*; the raised edges of the notum at *j* carry up the ends of the wing, which pivots again at *p*, and this forces downward the projecting wing.

sternum and notum contract, the notum is pulled down and pulls the wing base with it. The pivot point does not move, so that the expanded part of the wing beyond the pivot is raised at an angle proportionate to the lowering of the wing base. This causes the upstroke. As the sternonotal muscles relax, another set of muscles contract. These latter muscles, fig. 133*C, D*, connect the front and hind margins of the notum, like a bow-string. When they contract, fig. 133*D, E*, the "bow" or notum is arched, causing the central portion to elevate, carrying the base of the wing upward with it. This causes the downstroke and produces the other half of the seesaw motion. Figures 133*A–E* are greatly exaggerated to illustrate the points made; in reality the distances involved are minute, and the accessory musculature is complex.

The deflection of the wing is caused by other muscles which, acting on the axillary sclerites, can pull down the front margin of the wing. In this movement the pleural process is used again as the fulcrum or pivot point.

FLIGHT SPEED AND DIRECTION.   In flight three types of wing movements are produced: (1) an upstroke and downstroke, (2) a deflecting or tilting, and (3) a forward and backward swing. All three types occur simultaneously but to different degrees, depending on speed and wing size.

The up-and-down stroke alone produces little more than elevation. Thus in butterflies which rely chiefly on this stroke, flight is mostly fluttering, with very little speed in a forward direction. Addition of the deflecting movements decreases the air pressure in front of the wing and increases the air pressure behind it, in the same manner as an airplane propeller. The pressure behind pushes the insect forward at the same time that the partial vacuum in front pulls it forward. Deflection is therefore the principal agent in forward movement. Insects with highly developed deflection, such as dragonflies, are able to maintain a steady forward flight.

Rate of stroke varies exceedingly with different species. The slowest group is exemplified by the butterflies; the cabbage butterfly makes about 10 strokes per second. Faster rates are made by the honeybee, with 190 strokes per second, the bumblebee with 240, and the housefly with 330.

Greatest speed is attained by a combination of great deflection, a long narrow wing, and usually at least a fairly high rate of stroke. Fastest recorded flight of insects is developed by the sphinx moths (Sphingidae), attaining a rate of over 33 miles per hour; the horsefly (*Tabanus*) with a rate of over 31 miles per hour; and the dragonfly, which cruises at about 10 miles per hour but can speed up to about 25.

In forward flight, the path of the wing makes a figure "8" which leans forward at the top of the wing stroke. Many insects such as the hover flies (Syrphidae), bees, and many flower-visiting moths, are able to fly backwards; this is done by reversing the path of the wings so that the figure "8" leans back at the top of the stroke.

2. Glandular Reaction. In addition to regulating the internal chemical machinery of the organism as explained in the preceding section on control of development and diapause, hormonal secretions also play a part in bringing about certain immediate responses of the organism to its environment. In butterfly pupae of the genus *Pieris,* for instance, it appears that color reception by the pupal eye controls a hormone which affects pigmentation of the epidermis and brings about a pupal color tone which matches that of its surroundings. Another example is provided by the stick insect (genus *Dixippus*), which is normally pale during the day, changing to dark at night. Experimental use of ligaturing and moist chambers indicate that the color changes are in response to high and low humidity. The brain responds to these changes by secreting a hormone which is the active agent causing the epidermal changes that result in differences in color.

In discussing diapause (p. 144) we mentioned that changes in temperature or other external factors may cause changes in the hormonal activity of the brain and so bring about changes in the state of diapause. Although there is a considerable time lag between stimulation and visible response on the part of the organism, these cases also exemplify a hormonal type of contractility.

3. Emission Mechanisms. Various insects have glandular organs that emit defensive secretions. The morphological variety of these organs is fantastic, but most consist basically of special secretory cells and a cuticular sac in which the secretion is stored. Usually the contractile function is double—the glandular cells produce the secretion when the sac is empty and the secretion is emitted from the sac by muscular contraction as a response to an external stimulus, such as an ant bite. The muscular contraction may involve muscles in the organ itself, as in the walkingstick *Anisomorpha,* or various body muscles contract producing an increase in internal body pressure that causes an emission of the secretion. The organs usually open directly to the exterior, but in a few insects the products are discharged through certain spiracles.

Many kinds of insects have remarkable abilities to aim these defensive emissions in the direction of the predator. This is often accomplished by either twisting the body or aiming a special ejection nozzle so that the secretory spray goes in the direction from which the discharge stimulus came. Over thirty defensive chemical compounds have been identified, most of which are organic acids, alcohols, aldehydes, or paraquinones.

PHEROMONES. Insects also secrete to the outside substances capable of eliciting behavioral or developmental responses in another of the same species. These include sex attractants, some determinors of caste differentiation in bees and termites, and alarm-producing and trail-marking chemicals in ants. Certain of these are produced in saclike glands opening on the mandibles. These compounds also represent the contractility function.

Behavior

Over the millions of years they have been in existence, insects have evolved ways of responding to their environment that tend to insure their successful propagation. This total pattern of responses is known as their behavior pattern. A great amount of work has been done on the study of insect behavior. Most of these studies probe the question of behavior from one of two approaches. One is the individualistic approach. Experimentation, especially training criteria, has established the reaction of many insects to various qualities and quantities of light, odors, tastes, sounds, touch, and other stimuli. The reactions and adjustments of an individual to abnormal situations have resulted in information on learning ability and the orbit in which instinct and intelligence overlap.

The second approach deals with the behavior characteristics of the species. The entire life cycle of an insect (or any other organism) is a succession of definite behavior patterns. Even in the same species they change constantly and regularly. Thus the larva of the June beetle shuns light, but the adult is attracted to it; certain cutworms move down into the soil during the day but climb plants at night for feeding. Cocoon formation and mating represent behavior patterns which appear and persist for only a very short period of the insect's life cycle. In the life history of any one species the total of this succession of behavior traits is so constant in both sequence and relative timing that the whole is spoken of as the "biological clock."

The basic units of inherited behavior are reflex actions, that is, automatic responses to the same stimulus. When the entire organism orients itself automatically in relation to a given stimulus, this is known as a tropism. For instance, at night certain species of moths will always fly towards a flame, exhibiting a positive tropic response to light. Tropisms that are inherited and therefore operate without benefit of experience are called instincts. Insect behavior in general is predominantly instinctive, but it is not a simple addition of reflexes, tropisms, and instincts. In the first place, these may be modified, inhibited, or coordinated as a result of experience or learning. In the second place, many investigators believe that the entire perceptual panorama of the organism forms a sort of pattern and that the organism responds to general changes in this pattern and not to the individual stimuli of which the pattern is composed.

REFERENCES

Arnold, J. W., 1964. Blood circulation in insect wings. *Mem. Ent. Soc. Canada,* **38**:1–60.

Fraenkel, G. S., and D. L. Gunn. 1962. *The orientation of animals.* 2nd ed. New York: Dover Publications. 376 pp.

Harvey, E. Newton, 1952. Luminescent organisms. *Am. Scientist,* **40**:468–481.

Khalifa, A., 1950. Spermatophore production in *Galleria mellonella. Proc. Roy. Entomol. Soc. London* (A), **25** (4–6):33–42.

Lees, A. D., 1955. *The physiology of diapause in arthropods.* New York: Cambridge University Press. 151 pp.

——— 1956. The physiology and biochemistry of diapause. *Ann. Rev. Entomol.,* **1**:1–16.

Lipke, H., and G. Fraenkel, 1956. Insect nutrition. *Ann. Rev. Entomol.,* **1**:17–44.

McFarlane, John E., 1952. Chordotonal organs of the lesser migratory grasshopper. *Can. Entomologist,* **85**:81–103.

Pepper, J. H., 1937. Breaking the dormancy in the sugar-beet webworm, *L. sticticalis* L., by means of chemicals. *J. Econ. Entomol.,* **30**:380–381.

Pierce, George W., 1949. *The songs of insects.* Cambridge, Harvard University Press. 329 pp.

Roeder, K. D., 1953. *Insect Physiology.* New York: John Wiley & Sons. 1100 pp.

Roth, L. M., and T. Eisner. 1962. Chemical defenses of arthropods. *Ann. Rev. Ent.,* **7**:107–136.

Roth, L. M., and E. R. Willis, 1951. Hygroreceptors in Coleoptera. *J. Exptl. Zool.,* **117**:451–487.

——— 1952. Cockroach behavior. *Am. Midland Naturalist,* **47**:66–129.

Slifer, Eleanor H., 1951. Some unusual structures in *Locusta migratoria* and their probable function as thermoreceptors. *Proc. Roy. Soc.,* B, **138**:414–437.

Snodgrass, R. E., 1935. *Principles of insect morphology.* New York: McGraw-Hill Book Co. 667 pp.

Watanabe, M. I., and C. M. Williams, 1953. Mitochondria in the flight muscles of insects. *J. Gen. Physiol.,* **37**:71–90.

Wigglesworth, V. B., 1939. *The principles of insect physiology.* London and New York: E. P. Dutton and Co. 434 pp.

Chapter Six

The Life Cycle

THE start of growth within an egg signals the beginning of a long series of changes leading ultimately to the attainment of adulthood and to production of eggs or young for another generation. This chain of events, from egg to completed adulthood, constitutes the life cycle of the individual. In the Insecta there are many types of life cycle, involving different methods of development, different relations of one generation to another, and a possible alternation of food or abode between different divisions of a single life cycle. In social insects, contemporaneous members of a population living in the same colony may have different features of the cycle.

DEVELOPMENT

The life cycle of the individual usually has two phases, development (from egg to adult), and maturity or adulthood. Development is a period of growth and change which is fundamentally gradual and continuous throughout its course. On the basis of external manifestations, however, it is broken into definite segments or eras. Because most insects start as eggs, the most universally important division point of insect development is the phenomenon of hatching from the egg. The development period within the egg is *embryonic development;* the period after hatching is *post-embryonic development.* Change of form during this latter period is termed *metamorphosis.*

Embryology

The Egg, or Ovum. Insect eggs are of many shapes, fig. 134; many of them are simple smooth ellipses; others may be ribbed or sculptured in various ways; others are provided with processes of different kinds, such

164

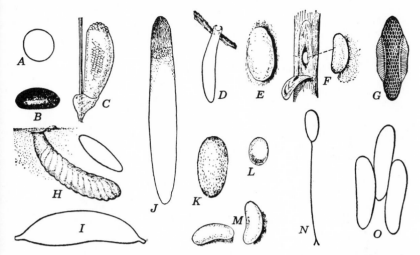

Fig. 134. Eggs of insects. *A*, a collembolan *Sminthurus viridis; B*, an aphid *Toxoptera graminum; C*, the sucking cattle louse *Solenopotes capillatus,* attached to a hair; *D*, apple mirid *Paracalocoris colon,* in plant tissues; *E*, the ladybird beetle *Hyperaspis binotata; F*, a weevil *Sphenophorus phoeniciensis; G*, a malarial mosquito *Anopheles maculipennis; H*, grasshopper egg pod in the soil and a single egg; *I*, an ichneumon fly *Diachasma tryoni; J*, a damselfly *Archilestes californica,* removed from water plant; *K*, webbing clothes moth *Tineola biselliella; L*, dog flea *Ctenocephalides canis; M*, pear thrips *Taeniothrips inconsequens,* removed from tissues of plant; *N*, a lacewing *Chrysopa oculata; O*, house fly *Musca domestica.* (After Essig from various authors)

as the lateral floats of *Anopheles* eggs, fig. 134*G*, which keep them afloat in water.

A typical egg, fig. 135*A*, is a cell encased in two coverings. The outer covering is a tough shell, the *chorion,* which has one to several minute pores, or *micropyles,* through which the spermatozoa enter the egg. Within this chorion is a delicate enveloping membrane, the *vitelline membrane,* surrounding the large nucleus and mass of cytoplasm. The cytoplasm consists of a large central area of yolk (essentially a food store), and a peripheral or cortical layer which is denser than the central part and relatively free from yolk.

Early Cleavage. In the order Collembola the entire egg divides during the early cleavages. As far as is known, this order is the only one in the Insecta which has holoblastic cleavage. In other insects only the nuclei divide during the early cleavages (meroblastic cleavage); because this is the common method of cleavage in insects, it is the one chosen to illustrate generalized insect embryology.

The nuclei produced by the early cleavages are at first scattered

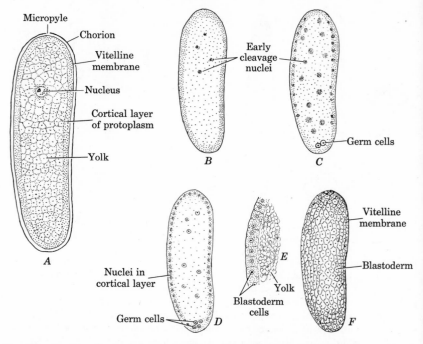

Fig. 135. Sections of a typical egg, showing its internal structure and early cleavage stages. *A*, fertilized egg; *B*, after a few divisions of the nucleus; *C*, after many cleavages (note migration of many nuclei to periphery); *D*, after cleavage nuclei form a definite layer at periphery; *E* and *F*, internal and surface views after cortical protoplasm of egg has condensed around the peripheral nuclei to form a surface layer of cells, the blastoderm. Note the early segregation of special germ cells in the posterior pole of the egg. (Redrawn from Snodgrass)

throughout the yolk, fig. 135*B* and *C*, and frequently several of them cluster together in nuclear aggregates. After many nuclei have been formed, most of them migrate from the yolk into the cortical layer, fig. 135*D* to *F*, where each nucleus becomes invested with cytoplasm and

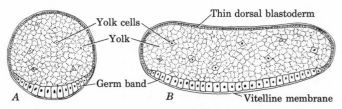

Fig. 136. One type of formation of the germ band on the ventral side of the blastoderm. *A*, cross section; *B*, longitudinal section. (Redrawn from Snodgrass)

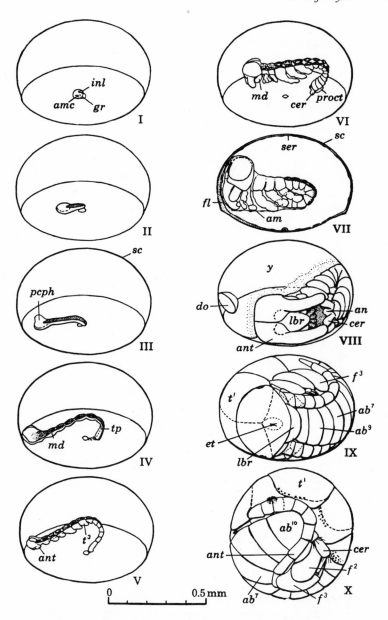

Fig. 137. Embryonic growth and segmentation in a stonefly. *ab*, abdominal segment; *am*, amnion; *amc*, amniotic cavity; *an*, anus, *ant*, antenna; *cer*, cercus; *do*, dorsal organ; *et*, egg tooth; *f*, femur; *gr*, grumulus; *inl*, inner layer; *lbr*, labrum; *md*, mandible or mandibular segment; *pcph*, protocephalon (= prostomium + first postoral segment); *proct*, proctodeum; *sc*, serosal cuticle; *ser*, serosa; *t*, thoracic segment; *tp*, tail piece; *y*, yolk. (After Miller)

a cell wall.   These cells make a wall around the egg, one cell thick.   In the ventral region, the cells are crowded together to make a thicker area, the ventral plate or *germ band*, fig. 136.   This is the first organized form of the embryo.   In some cases it is quite extensive, extending over a considerable area of the egg.   In others it forms only a small platelike area that may be termed a germ disc rather than a germ band, fig. 137, I.

Growth of the Embryo.   The germ band or disc grows by cell multiplication and differentiation.   At first growth is largely an increase in the size of the germ disc, but this is followed rapidly by a surface partitioning whereby the body segmentation and appendages are set out.   This is illustrated in fig. 137 which portrays graphically the gross changes leading to the formation of the completed embryo and the first-stage nymph, fig. 138.

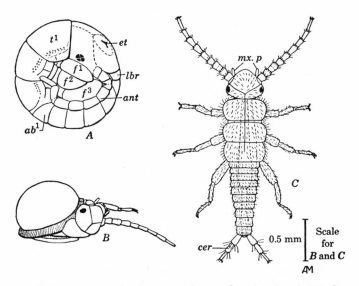

Fig. 138.   End of embryological development in the stonefly.   *A*, embryo in egg; *B*, completed embryo escaping from egg shell; *C*, newly hatched nymph.   Abbreviations as for fig. 137. (After Miller)

Segmentation and Appendages.   The development of these two features in the embryo parallels to a considerable extent the supposed evolutionary history of the insect group.

The body segments are first formed in the embryo by a series of transverse incisions, fig. 137, IV.   The segments which ultimately bear the mouthparts and fuse with the head structure appear originally as similar to the posterior segments. It is not until the appendages are well developed that the mandibular, maxillary, and labial segments become fused with the head structure, fig. 137, VII.

Appendages begin to develop soon after segmentation is evident, fig. 137, V. Typically each segment develops a pair of ventral appendages, but most of the abdominal appendages become only poorly defined. In many groups of insects all the abdominal appendages except the cerci are never more than small rudiments which are reabsorbed at an early stage. The anterior segments and appendages develop more rapidly than the posterior ones, so that as a rule the embryo at this stage appears as in fig. 139.

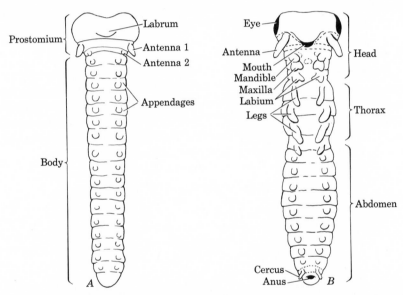

Fig. 139. Embryo at early stage of appendage development. *A*, early embryo having only small appendages on each but the last segment; *B*, later stage in which the head and thoracic areas and their appendages are better developed. (Redrawn from Snodgrass)

Body Shape. In the early stages only the appendages and the ventral portion of the body are formed. Thus, in fig. 137, V and VI, there are no sides or dorsum, the embryo representing the appendages and the sternal region. In essence, the body is open on top. During later growth the sides grow out and up, first over the anterior and posterior ends. Figure 137, VIII, shows a stage when the head is closed dorsally and also the posterior four of five segments of the abdomen; the embryo rests so that the open "top" of the intervening segments is pressed against the yolk, which fills the rest of the egg. From this stage the lateral margins of the open segments grow out and up along the sides of the egg until they meet dorsally, inclosing the yolk and completing the body closure.

Germ Layers; Tissue Determination. At first the germ band consists of only a single layer of cells, but early in embryonic life it forms a second layer. This is usually formed by gastrulation, that is, the infolding of a section of the germ band. In insects the common methods by which this is achieved are shown diagrammatically in fig. 140. Gastrulation usually begins as a longitudinal groove of the germ band, *A*; the outside edges of the groove grow toward each other, and the future second layer proliferates inward, *B*; finally the edges of the groove meet and fuse to form an outer layer or ectoderm. The inner layer or mesoderm spreads out above it, *C*. In fig. 137, I, to V, the mesoderm is shown as a darkened dorsal area. At each end of the mesoderm an invaginated cluster of cells forms, these clusters being the endoderm, or third germ layer.

The groups of cells which later develop into specific body segments, organs, or appendages are determined very early in embryonic life. Extensive experiments have shown that most of this determination is established by the time the germ layers are formed. A group of such apparently unspecialized cells, which are nevertheless destined to grow into a particular structure, is called an *anlage* (pl. *anlagen*).

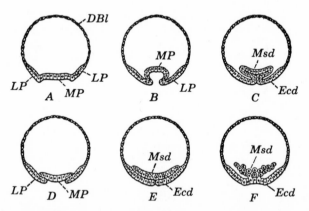

Fig. 140. Development of mesoderm by simple gastrulation methods. *A*, cross section of egg with germ band differentiated into lateral plates (*LP*) and middle plate (*MP*). *B*, later stage of same with middle plate curved in to form a tubular groove, edges of lateral plates coming together below it. *C*, still later stage, with edges of lateral plates united, forming the ectoderm (*Ect*), and middle plate spread out above the latter as internal layer of cells, the mesoderm (*Msd*). *D*, *E*, second method of mesoderm formation: middle plate, separated from edges of lateral plates, becomes mesoderm (*Msd*) when lateral plates unite beneath it. *F*, third method, in which mesoderm (*Msd*) is formed of cells given off from inner ends of middle plate cells. (After Snodgrass)

The ectoderm gives rise to the body wall, the stomodeum and proctodeum of the digestive tract, the nervous system, tracheal system, and many glands. The mesoderm gives rise to the muscular system, the gonads, the heart, and the fat body, and the endoderm gives rise to the mesenteron.

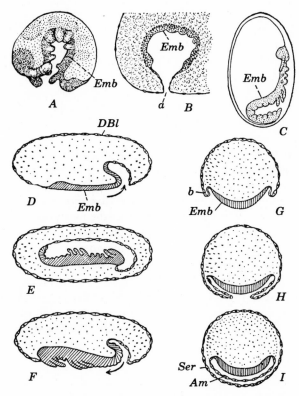

Fig. 141.   Diagrams of position and movement of embryo within the egg, illustrating three methods.   *A*, embryo (*Emb*) of a springtail *Isotoma cinerea*, curved into the yolk on under side of egg.

B, C, embryo of a silverfish *Lepisma:* first (*B*), at early stage when deeply sunken into yolk near posterior end of egg, the opening of the cavity closed to a small pore (*a*); and second (*C*), in later stage when partially revolved to outside of egg, in which position it completes its development.

D–F, lengthwise sections of an egg in which the embryo revolves rear end first into the yolk (*D*), becoming entirely shut in the latter (*E*) in reversed and inverted position, and then again revolves to surface (*F*) in original position before hatching.

G–I, cross sections of an egg in which embryo becomes covered by membranes originating in folds of the blastoderm around its edges (*G, b*), the folds extending beneath the embryo (*H*), and finally uniting to form two membranes (*I*), the outer serosa (*Ser*), the inner the amnion (*Am*).   (From Snodgrass, *A* after Philiptschenko, *B* after Heymons)

Embryonic Coverings. During much of its development, the embryo becomes partially or entirely immersed in the yolk, presumably for protection, and a pair of membranes form around it. The two principal methods followed are illustrated in fig. 141. In the first, the embryo slides tail first into the yolk, pulling the membrane with it, fig. 141*D*. When the embryo is completely immersed, fig. 141*E*, the membrane grows over the end of the entrance cavity to form two final membranes; the outer one is the *serosa,* and the inner one (which incloses a space around the ventral aspect of the embryo) is the *amnion.* The second method by which immersion occurs involves only a sinking of the embryo into the yolk, fig. 141*G* to *I*; the membrane grows over the ventral area, and the opposing membrane edges unite to form the amnion and serosa. At a later stage of development the embryo changes position again, breaking through the two membranes and assuming a position with its back to the yolk, as in fig. 141*F*.

Formation of Digestive System. A detailed discussion of the origin and formation of the various insect organs is beyond the scope of this book. The formation of the digestive tract, however, is of unusual interest, and a brief outline of its early growth is illustrative of the general fashion in which organs arise. The successive stages in the formation of the digestive tract are shown diagrammatically in fig. 142. In fig. 142*A* there are two masses of endoderm cells, the anterior mesenteron rudiment, and the posterior mesenteron rudiment, growing inward from each end of the embryo. In fig. 142*B* each of these rudiments has begun the formation of a sac, open toward the middle of the body, and beginning to inclose the central yolk mass; at the same time the ectoderm at each end has invaginated to form the beginnings of the anterior and posterior parts of the

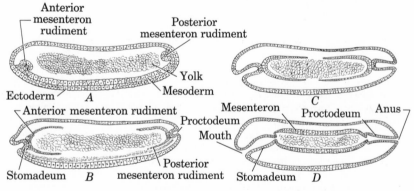

Fig. 142. Embryonic formation of the alimentary canal. (Redrawn from Snodgrass)

digestive tract. In fig. 142*C* these developments have continued to a further stage. The completed structure is shown in fig. 142*D*; the anterior and posterior sacs of the mesenteron have joined, completely inclosing the remains of the yolk; and openings have formed connecting the mesenteron with the anterior and posterior ectodermal invaginations. The digestive tract thus has three distinctive areas: (1) the anterior stomodeum, formed of ectoderm; (2) the central mesenteron, of endodermal origin; and (3) the posterior proctodeum, of ectodermal origin.

Hatching. When the embryo is full grown and ready to leave the egg, or hatch, it must force its way through the eggshell or chorion by its own efforts. Prior to hatching, the embryo may swallow air or the amniotic egg fluid to attain greater bulk or turgidity. In hatching, the embryo produces rhythmic muscular activity and presses against the shell or strikes it repeatedly with its head.

In some insects, such as grasshoppers, the embryo simply forces a rent in the anterior part of the eggshell. In others, such as many Hemiptera and certain stoneflies, fig. 138*B*, a portion of the egg forms an easily detached cap, which the embryo pushes open like a lid. In a third group, the anterior part of the embryo is armed with an egg burster, which may be a sclerotized saw, spine, or some blades which pierce the chorion to produce the initial tear.

Once the shell is broken, the embryo works its way out of the egg. In many cases the nymph is encased in an embryonic covering or pronymphal membrane which is molted when the nymph is partway out of the egg. The cast skin remains inside or protruding from the egg. The egg burster is a thickening of this pronymphal membrane. When free from the egg and its embryonic coverings, the embryo is considered as the first-stage nymph or larva of the postembryonic period, fig. 138*C*.

Polyembryony. The eggs of certain parasitic Hymenoptera frequently produce more than one embryo. In *Platygaster hiemalis,* a parasite of the hessian fly, each egg may develop two embryos; in *Macrocentrus gifuensis,* a member of the Braconidae, each egg may develop several embryos; and in other species each egg may produce one hundred to three thousand embryos. Each of these embryos develops into an active larva. The division into multiple embryos takes place before any other embryonic development occurs. The segmentation nucleus divides by mitosis into the required number of daughter nuclei, and each of these then develops into an embryo. The embryos may form either irregular groups or long chains. By means of polyembryony, a small parasite can insert a single egg into a large host and by that one egg produce enough offspring to take advantage of the great food possibilities of the host. An example of this

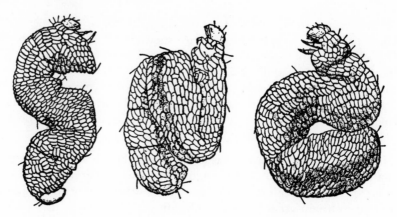

Fig. 143. Polyembryony in the chalcid *Litomastix;* the 2000 parasite larvae in each host caterpillar all developed from one egg. (From Clausen, after Sylvestri, *Entomophagous insects,* by
permission of McGraw-Hill Book Co.)

is the minute chalcid, *Litomastix truncatellus,* which parasitizes large Lepidoptera larvae, fig. 143; from only a small number of eggs over two thousand larvae usually develop in each caterpillar, and these allow no part
of the host to go to waste.

Postembryonic Development; Metamorphosis

From the time of hatching from the egg to adulthood the individual
passes through a period of growth and change. The insect integument has
no stretch, and so to accommodate increase in size the insect periodically
sheds its old skin and replaces it with a larger one. The mechanics and
physiology of molting are discussed on page 113. Most insects molt at
least three or four times, and in some cases thirty or more molts occur
during normal development. The average is five or six molts.

The process of molting is sometimes called *ecdysis.* The old skins cast
off by the insect are called *exuviae.*

Instar and Stadium. With few exceptions the molts for each species
follow a definite sequence as to number, duration of time between them,
and the increase in size accompanying them. The total period between
any two molts is called a *stadium.* The actual insect during a stadium is
termed an *instar.* Thus from time of hatching until the first molt is the
first stadium. Any individual which is in this period of development is
called the first instar. To put it another way, we might say of a species that
the first stadium is 5 days and that the first instar is slender and yellow.
In all but the primitive wingless orders, no further molts occur after the
functional adult stage is reached.

Adulthood. The *adult*, or *imago*, is the stage having fully developed and functional reproductive organs and associated mating or egg-laying structures. In winged species it is the stage bearing functional wings. The only known exception to this latter is the mayfly order, Ephemeroptera, in which the stage before the winged reproductive also has wings and uses them; this curious flying pre-adult instar is called a *subimago*.

Metamorphosis. In the primitive wingless orders, known as the Apterygota, the newly hatched young grow into adults with little change other than in body proportions and development of the reproductive organs. In the winged insects or Pterygota, the newly hatched young are still relatively simple creatures but the adults have wings and special body sclerites associated with them, so that the adults are quite different in appearance from early stages of the young. This condition of exhibiting different forms in different stages of the life history is known as metamorphosis.

GRADUAL METAMORPHOSIS. In the more primitive winged insects the wings first appear in about the third instar as slight backward outgrowths of the second and third thoracic nota. These wing pads increase in size with each subsequent molt, fig. 144, in the last immature instar forming

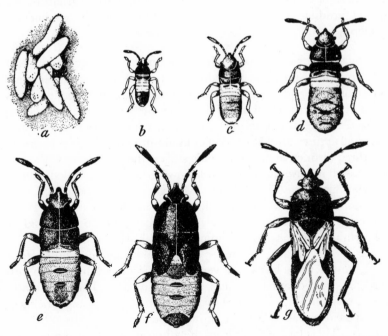

Fig. 144. Gradual metamorphosis; instars of a hemipteron, the chinch bug. (From U.S.D.A., E.R.B.)

pads which may extend over several segments of the abdomen. Up to this point no change occurs in the sclerites of the thorax. The complete adult structures develop inside the last immature instar, and the adult emerges fully formed. At emergence the adult wings are somewhat soft and pleated but these soon straighten and harden into their functional condition.

This type of development, in which the wings develop gradually as external pads is termed gradual metamorphosis. It is also called simple or partial metamorphosis, and paurometabolous development. The immature stages are called nymphs.

The cockroaches, grasshoppers, stoneflies, leafhoppers, and bugs are among the simplest examples of gradual metamorphosis. In these forms the nymphs resemble the adults in body form except for the wings, reproductive organs, and structures associated with them. In wingless species of these groups the nymphs can be distinguished from the adults chiefly by the incompletely formed genitalia. In the two orders Ephemeroptera (mayflies) and Odonata (dragonflies) the nymphs, which are aquatic, have evolved many specializations for life in the water, such as peculiar mouthparts, well-developed lateral or anal gills, or peculiar body shape. As a result the nymphs are considerably more different from the adults than those of cockroaches and stoneflies. This variant from the simpler type of gradual metamorphosis is sometimes termed hemimetabolous development.

COMPLETE METAMORPHOSIS. In many insects the wing rudiments develop internally until the pre-adult instar, in which stage the wings are everted as large pads. This is a nonfeeding stage and is chiefly a quiescent one in which structures of the adult are rebuilt from tissues of earlier stages. Externally it seems that in these insects the wings appear suddenly as well formed structures toward the end of immature growth. There are, therefore, three distinctive postembryonic stages: the early form without wingpads, called a *larva;* the quiescent form with wingpads, called a *pupa;* and the *adult,* fig. 145. This type of development is termed complete metamorphosis (sometimes termed holometabolous development). Examples of insects having this type include moths, bees, flies, beetles, and thrips, and certain members of the Hemiptera, notably the scale insects.

One of the simplest instances of complete metamorphosis occurs in thrips, in which the larvae are very similar to the adults in body proportions and shape of antennae, legs, and mouthparts. The thrips and scale insects, incidentally, each have two pupal stages, both stages being quiescent, typical pupae. Another simple type occurs in dobsonflies (Megaloptera). Their larvae differ from the adults in structure of antennae and

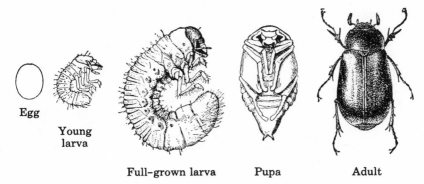

Fig. 145. Complete metamorphosis; life stages of a beetle, *Phyllophaga* sp. (From U.S.D.A., E.R.B.)

mouthparts more than do those of thrips, but the legs and many other structures are remarkably like those in the adults. In both dobsonflies and thrips, as is true in insects with gradual metamorphosis, the adult thoracic sclerites differ greatly from those of the immature stages.

In other orders with complete metamorphosis the larvae have more structures unlike those of the adults. In some groups, such as moths and sawflies, the larvae have pairs of lobate abdominal legs. In larvae of all these orders the antennae, eyes, and thoracic legs are simplified or completely undeveloped. These structures, like the wings, seem to appear suddenly in the pupal stage in roughly the outlines of the adult structures. Actually the adult antennae, legs, eyes, and other structures are not formed by remodeling the reduced larval structures but, like the wings, are built up from discrete areas of dormant embryonic tissues called *anlagen.* The anlagen develop in the larva to some extent as internal rudiments or *histoblasts.* Wing development is a good illustration of this process.

Histoblasts begin to develop early in larval life, sometimes even in the late embryo. Typical stages in the growth of a wing histoblast are illustrated in fig. 146*A–G*; these are diagrammatic. At first a histoblast is only an area of thickened epithelial cells, *A.* This area enlarges, *B,* and begins to pull away from the cuticle, until it forms an internal pocket, as in *C.* One portion of the pocket wall then begins to enlarge into the pocket, as in *D,* gradually infolding to form a double-walled sac, as in *E* and *F.* Just before pupation this sac, the rudimentary wing, is usually pushed out, and the cavity flattens out, so that the sac lies directly underneath the cuticle, as in *G.* When the cuticle is shed during the molt to the pupal instar, the wing is finally an exposed external structure. If rudi-

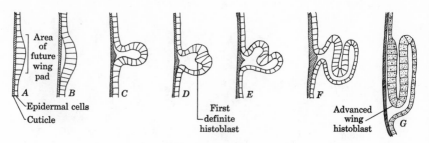

Fig. 146.   Stages in the growth of an internal wing pad or histoblast.

mentary wings are dissected out of the histoblasts, they look very much like wing pads of hemimetabolous nymphs.

The histoblastic development of so many adult structures allows the development of specialized larval structures which are not carried over into the adult, since they are cast off with the last larval skin or dissolved during the catabolism of early pupation.   This arrangement has provided a mechanism by which the larvae and adults of the same species can evolve in entirely different directions.   This they have done.   In general the larvae have specialized for better food gathering, the adults for better dispersal and reproduction.   The culmination of this trend is found in the muscoid flies (Diptera), whose larvae (maggots) are legless and appear to lack eyes, antennae, all conventional mouthparts, and head capsule, whereas the adults are typical, fast-flying insects.   Although fly larvae appear to be extremely simple creatures, their mouthparts, tracheal system, enzyme system, and musculature are specialized to a remarkable degree.

There are certain instances in which both larvae and adults have become specialized for food gathering, but in different ways.   Mosquito larvae, which are aquatic, have chewing mouthparts, and most of them feed on microorganisms; the adults have piercing-sucking mouthparts, the males feeding on nectar and most of the females on avian or mammalian blood.   The order Siphonaptera (fleas) offers another striking example.   Flea larvae feed as scavengers on inert organic material, and the adults suck blood.

HYPERMETAMORPHOSIS.   In the majority of cases all the larval instars of a species are similar in feeding habits and general appearance, differing chiefly in size.   Some groups, however, may have two or more quite distinct types of larvae in the life cycle.   When this sort of development occurs, it is called hypermetamorphosis.   Among the best examples are many of the parasitic Hymenoptera, fig. 147.   The first-instar larva is a

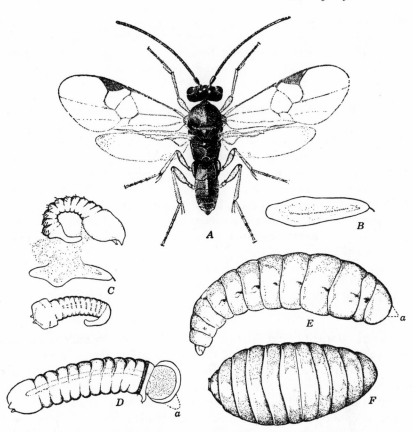

Fig. 147. Stages of the life cycle of a parasitic hymenopteron, *Apanteles melanoscelus*. *A*, adult; *B*, egg; *C*, first-instar larvae, the upper one with yolk still attached to it; *D*, second-instar larva; *E*, third-instar larva in feeding stage; *F*, same, but ready to spin cocoon (prepupal stage). *a*, anal segment. (From U.S.D.A., E.R.B.)

motile form bearing bristles, tails, or other processes; this instar either penetrates the host integument or migrates through host tissues. The later instars are sedentary and have none of the modifications of the first instar. Another striking example is the peculiar Stylopoidea (illustrated on p. 348). The young larvae are active legged forms furnished with bristles and tails; the succeeding instars are grubs. The same sort of development occurs in the blister beetles Meloidae, fig. 272, although in these the difference between larval instars is chiefly general shape accompanied by a few structural changes.

Transitional cases are frequent, in which various instars differ considerably in shape and habit but have few morphological differences. For instance, certain parasitic rove beetles have slender active first-instar larvae and grublike succeeding instars. A few genera of caddisflies (Trichoptera) have free-living slender first-instar larvae and casemaking stout-bodied later instars.

Paedogenesis. This is a precocious reproductive maturity which occurs in a few insects, resulting in the production of eggs or living young by larvae or pupae. In the life cycle of the rare beetle, *Micromalthus debilis,* some larvae lay eggs or produce young. Midge larvae of the genera *Miastor* and *Oligarces* produce young but no eggs. Pupae of the midge genus *Tanytarsus* may produce either eggs or young. Paedogenesis is a freak type of metamorphosis and growth involving a maturation of the reproductive organs without a similar maturation of other adult characteristics. It often is associated with unusual generation cycles, discussed on page 191.

MATURITY

Sexual Maturity and Mating. Adult insects are seldom sexually mature immediately on emerging from the preadult stage. In most cases the males require a few days to mature, and the females longer. Mating occurs in many forms before the females have mature eggs; the sperms are stored in the spermatheca of the female until the time for their use.

In the case of certain short-lived insects, such as many mayflies, both sexes are fully mature sexually within a matter of hours after completing the last molt. Mating takes place a day or so after emergence, and oviposition occurs shortly afterwards.

Parthenogenesis. This ability to reproduce without fertilization is possessed by certain insects. In some species parthenogenesis occurs only irregularly. Generally in the order Hymenoptera unmated females lay eggs which produce only males, while the eggs of inseminated females produce either males or females. Normally no males are produced at any time in some other parthenogenetic insects. The females lay unfertilized eggs, and these produce females. The pear sawfly *Caliroa cerasi* and the rose slug *Endelomyia aethiops* are examples of permanent parthenogenesis.

Oviposition. Most insects are oviparous; that is, they lay eggs; but the various kinds of insects differ tremendously in egg-laying habits. The Phasmidae (walkingstick insects) drop their eggs singly onto the ground; butterflies glue theirs to leaves; sawflies and some crickets, fig. 148, saw

Fig. 148.   Egg laying of the snowy tree cricket.   (From Metcalf and Flint, *Destructive and useful insects,* by permission of McGraw-Hill Book Co.)

out a cavity in leaf or stem, forming a recess for each egg.   Eggs may be laid separately, or they may be grouped together in large masses.   Extruded eggs of stoneflies and mayflies collect as a mass at the end of the body; this mass is deposited as a unit.   In the cockroaches this tendency is greatly developed.   The eggs are glued together as they emerge from the body and, cemented by glandular secretions, form a compact capsule or oötheca, which is deposited.   In forms such as the mayflies, oviposition is completed with the deposition of a single large mass of eggs.   Oviposition in bedbugs occurs at a slower rate but continues for a period of months.   These are only a few examples; additional material is incorporated in the synopsis of the orders in the following chapter.

Viviparity.   Not all insects lay eggs.   Species of various groups are viviparous; that is, they deposit living young instead of eggs.   In viviparous insects the eggs develop in the oviducts or vagina until at least the completion of embryonic growth.   This phenomenon occurs in many groups scattered throughout the insect orders.   There are several kinds of viviparity in adults, some cases being only slight modifications of the oviparous condition and others involving the development of special structures.

Precocious hatching of the embryo within the egg passage occurs in the flesh flies, Sarcophagidae. Eggs remain in the vagina until mature and hatch just as they are deposited. The young larvae pass through the ovipositor like eggs and at birth are in an early stage of larval development, corresponding to the point of hatching from the egg in oviparous species. In Stylopoidea (Coleoptera) the eggs hatch inside the body of the female and the minute larvae crawl out through the genital openings.

Uterine development of larvae occurs in the ked flies, Pupipara, and the African fly genus *Glossina* (both members of the Diptera). In these the uterus is a large chamber, fig. 149, provided with glands producing nourishment for the larva, which develops to maturity in the body of the female. The maggot pupates as soon as it is discharged from the vagina.

One of the more common examples of viviparity occurs in the parthenogenetic generations of aphids. The unfertilized eggs develop within the ovaries and are liberated as active nymphs equivalent to first-instar nymphs of oviparous generations.

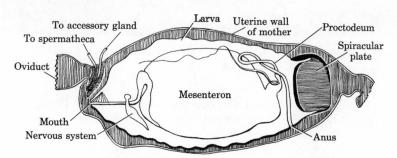

Fig. 149. Larva (unshaded except for its spiracular plate) of the tsetse fly *Glossina* in uterus of its parent. (Redrawn from Snodgrass)

Longevity of Adults. Adult insects have a normal life span ranging from a few days to several years, depending on the species. The length of life is correlated with fecundity, death usually occurring a short time after the completion of mating or oviposition activities. Thus the females of certain species of psocids (Psocoptera) have an adult life of about 20 days and die 5 or 6 days after final oviposition. Unmated females of these species rarely oviposit and live about 20 days longer than mated females that lay the normal number of eggs. The hibernating forms in many species are adults, and in these cases they may have a life span of nearly a year. Many adult leaf beetles, for instance, mature in July, feed for the rest of the summer, and then hibernate. They become active the follow-ing spring, mate, lay their eggs through May and June, and die soon after.

FOOD HABITS

Food is essential to the growth of any organism and therefore is an important consideration in the life cycle of an insect. A wide range of organic substances, living and dead, are used by insects as food. Accord-ing to the type of food utilized, insects may be grouped in the following manner with an example given for each category.

1. *Saprophagous*—feeding on dead organic matter.
   General scavengers—Dictyoptera (cockroaches).
   Humus feeders—Collembola (springtails).
   Dung feeders, coprophagous—some Scarabeidae (dung beetles).
   Restricted to dead plant tissue—Isoptera (termites).
   Restricted to dead animal tissue—Dermestidae (larder beetles).
   Carrion feeders—Calliphoridae (flesh flies).
2. *Phytophagous*—feeding on living plants.
   Leaf feeders—Saltatoria (grasshoppers).
   Leaf miners—Agromyzidae (flies).
   Stem and root borers—Cerambycidae (beetles, round-headed borers).
   Root feeders—some Scarabeidae (beetles, white grubs).
   Gall makers—Cynipidae (gall wasps).
   Juice suckers—Leafhoppers and aphids.
   Mycetophagous, fungus feeders—Mycetophagidae (fungus beetles).
3. *Zoophagous*—feeding on living animals.
   Parasites (living on another animal).
       Living on warm-blooded vertebrates—Anoplura (sucking lice).
       Living on other insects—Ichneumonidae (ichneumon flies).
   Predators (seeking out and killing prey)—Reduviidae (assassin bugs).
   Blood feeders—Culicidae (mosquitoes).
   Entomophagous—either parasites or predators on other insects.

Of these food categories, two involve relations between insect and host which are quite unusual and merit further discussion. These are the gall makers and the parasites.

Gall Makers. Many insects cause plants to develop abnormal out-growths or disfigurements called galls, fig. 150, and live within the shelter of these structures. Insect galls are formed by abnormal growth of par-ticular tissues of a plant and may occur on leaves, buds, stems, or roots. Each insect produces a particular type of gall and always on the same region of the plant. The sawfly *Euura salicisnodus* always makes a gall on

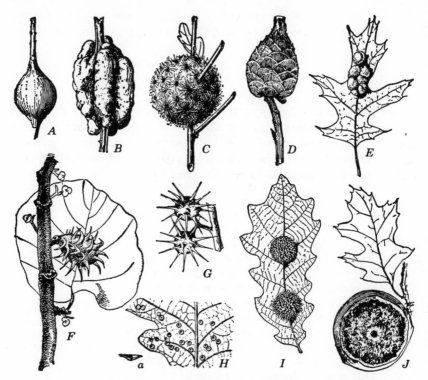

Fig. 150.   Examples of insect galls.   *A*, goldenrod ball gall, caused by a fly *Euaresta solidaginis;* *B*, blackberry knot gall, caused by a gall wasp *Diastrophus nebulosus; C*, wool sower gall on oak twig, caused by a gall wasp *Andricus seminator; D*, pine-cone gall, common on willow, caused by a gall fly *Rhabdophaga strobiloides; E*, oak leaf galls, caused by a gall wasp *Dryophanta lanata;* *F*, spiny witch hazel gall, caused by an aphid *Hamamelistes spinosus; G*, spiny rose gall, caused by a gall wasp *Rhodites bicolor; H*, oak spangler, caused by a gall fly *Cecidomyia poculum*, one gall shown in section at *a; I*, spiny oak gall, caused by a gall wasp *Philonix prinoides; J*, large oak apple, caused by a gall wasp *Amphibolips confluens*.   (From Metcalf and Flint, rearranged from Felt)

the stems of willow, and another sawfly *Euura hoppingi* always makes a gall on the leaves of willow.   In certain insects that have alternation of generations, each generation may cause the formation of a differently shaped gall.

The cause of gall formation is not known exactly, but a gall is presumably excessive growth of the plant tissue brought about by either irritation caused by the presence of the insect, or hormones or growth-stimulating substances secreted by the insect.   When sucking insects cause a gall, such secretions may be injected with saliva when the insect feeds.   When chewing insects are involved, the secretions may be freed on the lacerated tissue of the plant and absorbed by it.

Fig. 151.   Diagram of open and closed galls.   *A* and *B*, open galls; *C*, closed gall.   (*A* and *B* redrawn from Wellhouse)

Galls are of two types, open and closed, fig. 151.   Open galls are essentially pouchlike and have an opening to the outside.   Aphids cause galls of this type.   A gall begins to form around the feeding station of a single aphid.   The plant tissue around this area enlarges and gradually becomes twisted or bowed to form a purselike cavity with the aphid inside.   The edges of these galls are usually tightly appressed until a generation of aphids within have completed their growth; at this point the edges separate and allow the migrating aphids to escape from the gall.   The galls of *Phylloxera* and mites are also of this type.

Closed galls are formed by several groups of Hymenoptera, including a few genera of sawflies, certain groups of chalcid flies, and the family Cynipidae.   In these groups the females insert each egg beneath the surface into the plant tissue.   The larva never leaves this haven, feeding on the inner tissues of the gall which forms around it.   The insect must eat its way out.   In some sawflies the larva leaves the gall and pupates elsewhere; in others the larva pupates within the gall, and the adult eats its way out.

Parasites.   In the zoological sense, a parasite is an animal that lives in or on another animal, known as its *host,* from which the parasite derives its food for at least some stage in its life history.   Several insect groups are true parasites.   On the basis of habits and hosts, insect parasites fall into two categories—parasites of warm-blooded vertebrates, and parasites of insects or other small invertebrates such as spiders and worms.

The insects parasitizing warm-blooded vertebrates do not kill their host, so that many individuals or many generations of the parasite may live on the same host animal.   Sucking lice (Anoplura) and chewing lice (Mallophaga) are examples of external parasites.   They spend all their lives on a bird or mammal host, frequently occurring in large numbers on the same animal, and having continuous generations during the life span of the host.   Seldom does the host die from these attacks, although its general health and resistance to other maladies may be greatly impaired.   Bot flies and warble flies furnish excellent examples of internal parasites.   The larvae of these flies live and mature in the nasal passages, stomach, or back of the host, and when full grown leave the host and pupate in the soil. Many individuals of the parasites may infest one host individual. Again the host is not killed, although harmed, and is attacked by successive generations of the parasites.

Insects which parasitize other insects differ from those parasitizing vertebrates in two particulars: (1) Usually only a single parasite attacks a host individual, and (2) the parasite usually kills the host. As with the internal parasites of vertebrates, only the larvae actually live entirely on the host. This type of parasitism occurs in many families of Hymenoptera, several families of Diptera, and a few genera of Coleoptera.

Parasites of insects usually infest a host in one of three ways. The most common manner is for the female to lay an egg on the host or to insert her ovipositor through the host integument and lay an egg in the host tissue. Most of the parasitic Hymenoptera and Diptera use this method. The second manner is an indirect approach. The parasitic female deposits her eggs on leaves of the plant on which the insect host feeds. If the host eats some of the egg-infested leaves, the parasite eggs are unharmed and hatch within the digestive tract, the larvae making their way into the host tissue. The hymenopterous family Trigonalidae and many species of the Tachinidae, or tachina flies, use this method. In the third manner the eggs are laid more or less in random places, and the first larval instar finds its way to a host, as in the Meloidae (p. 338) and Stylopidae (p. 347).

Most of these parasites are internal, but in a number of parasitic Hymenoptera the larvae feed externally on the body of the host. In these instances the circumstances approach closely the condition of predatism rather than parasitism.

The habit of parasitism is an extremely specialized one. Thousands of species of Hymenoptera and Diptera are parasitic; yet each species usually parasitizes only a single host species or a group of closely related species. The parasites, called primary parasites, are often the host to other parasites, called secondary parasites, and these may be the host to tertiary parasites. Usually secondary and tertiary parasites are much less specific in host selection than are primary parasites.

Feeding and Life-Cycle Stages. In general, the life of the adult insect is concerned primarily with reproduction. The adult feeds to maintain its metabolic losses due to activity and life processes, or to furnish nutriment to the eggs or sperms developing in its body. In such groups as Orthoptera, Hemiptera, Siphonaptera, and blood-sucking members of the Diptera, the adult requires a large amount of food to supplement the reservoir from nymphal or larval stage.

In many groups sufficient stores of fats or other nutrients are carried over from the immature stages so that the adults need to do little or no feeding. This is true of caddisflies (Trichoptera), many Hymenoptera, Lepidoptera, and Diptera. In extreme cases such as mayflies (Ephemeroptera), the eggs are practically ready to deposit by the time the adult emerges, and no feeding is done in this stage. This last example is an extreme in the direction of non-feeding on the part of adults, and there are also extremes in the opposite direction. The most outstanding instance in our fauna is the sheeptick group, Pupipara (Diptera). In these the larvae develop to maturity in the body of the female, and she does all the active feeding in the life cycle. A somewhat parallel situation is found in the bees and many wasps, which gather food for the larvae.

SEASONAL CYCLES

Whereas the *life cycle* is the development of the individual from egg to egg, the *seasonal cycle* is the total successive life cycles or generations normally occurring in any one species throughout the year, from winter to winter.

The life cycle of many species consists of a single generation each year. In this case the life cycle and seasonal cycle are the same.

In cases such as the housefly there are continuous generations produced throughout the warmer months, followed by a hibernating period or diapause (see p. 144). Thus the seasonal cycle consists of several life cycles.

There are some species in which the life cycle is longer than a year in duration, as, for instance, many June beetles, whose larvae require 2 or 3 years to mature, and the 17-year locust, a cicada having a 17-year developmental period. In these insects the seasonal cycle includes only a portion of the life cycle. In most cases, however, the generations overlap so that adults of each species are produced every year; the seasonal cycle is used here to include the activities of the combined generations of the species for the year.

Seasonal cycles made up of more than one life cycle are of two types, those having repetitious life cycles, and those having alternation of generations.

Repetitious Generations

In this category successive life cycles are fundamentally the same. One generation of houseflies, for instance, lays eggs which develop into another generation just like the first, having the same morphological characteristics, food habits, and reproductive habits.

Interruptions in the development of certain generations, due to estivation, hibernation, or other types of diapause, are not considered as altering fundamentally the general pattern of the life cycle. To cite the housefly again, it has successive generations during the summer, the life cycle of each varying from 4 to 5 weeks, depending on weather conditions; adults produced in autumn, however, hibernate during the winter and in spring resume normal activities. A life cycle interrupted by onset of winter differs from that of summer in no feature other than the time element.

Alternation of Generations

There are several groups of insects in which succeeding generations are quite different in method of reproduction and sometimes in habits.

Forms with Reproduction Only by Adults. Two well-known groups belong in this category, the aphids and the gall wasps. These forms are all plant feeders.

The aphids (Aphididae, Hemiptera) have varied and complicated seasonal cycles involving sexual oviparous and parthenogenetic viviparous generations, winged and wingless generations, and frequently migrations between definite and different summer and winter host plants. A fairly simple seasonal cycle is exemplified by the cabbage aphid, *Brevicoryne brassicae*. Members of this species overwinter as eggs, laid in autumn on the stems of cruciferous plants. These hatch the following spring and develop into the wingless parthenogenetic viviparous form, the stem mothers. It is interesting to note that all the eggs develop into these stem mothers. These produce parthenogenetic viviparous generations which may be winged or wingless. Similar viviparous generations are continuous throughout the summer. As a rule parthenogenetic individuals live about a month and produce 50 to 100 young. When the days become shorter in autumn, the viviparous forms produce the sexual generation, wingless females and winged males. After mating, each female lays one to several eggs, which pass through the winter.

More complicated seasonal cycles have several additional specializations. Many species migrate to summer hosts. In these the winter or primary host is usually a tree or shrub. The eggs are laid on such hosts in autumn, hatch in spring, and develop into wingless stem mothers. These normally produce winged viviparous females that fly to summer or secondary hosts. Continuous viviparous generations, either winged or wingless, are produced on these hosts until autumn. At that time migratory forms are produced which fly to the winter host and there produce young of the sexual generation. In certain species the winged males are produced on the summer host, migrate to the winter host, and mate with the wingless oviparous females.   In other species both males and females are wingless, in which case they are both produced by winged viviparous females which have previously migrated to the winter host.

There are radical differences in habits among various generations of some aphids.   In the genus *Pemphigus,* for example, the stem mother makes a gall on the leaf or petiole of poplar; her progeny migrate to the roots of Compositae and other plants and initiate a series of root-inhabiting generations.

Table 4.   Chart of alternation of generations in aphids, illustrating a form with no host alternation, such as the cabbage aphid; and one having an alternation of hosts, such as the plum aphid.

	Non-Migratory Species	Typical Migratory Species	
Season	All forms on one host	Forms on primary host	Forms on secondary host
Winter	Eggs	Eggs	
Early spring	Stem mothers Apterous viviparous females	Stem mothers Apterous vivipa- rous females	
Late spring	Alate viviparous females (spring migrants)	Alate viviparous females (spring migrants)	Spring migrants from primary host
Summer	Alate and apterous viviparous females (these migrate from plant to plant of the same host species, or compatible related species)	A few strays	Alate and apter- ous viviparous females
Early fall		Fall migrants from secondary hosts	Alate viviparous females, some- times alate males (fall migrants)
Late fall	Sexual forms: males and oviparous females	Sexual forms: males and oviparous fe- males	
Winter	Eggs	Eggs	

The phylloxerans (Phylloxeridae, Hemiptera), closely related to the aphids, also have an alternation of parthenogenetic and sexual generations (some winged and others wingless) and complicated migration habits. They differ from aphids, however, in that all forms are oviparous. A common example is the grape phylloxera, fig. 152. A series of parthenogenetic generations form leaf galls, migrating in autumn to the grape roots, where the following spring another series of parthenogenetic

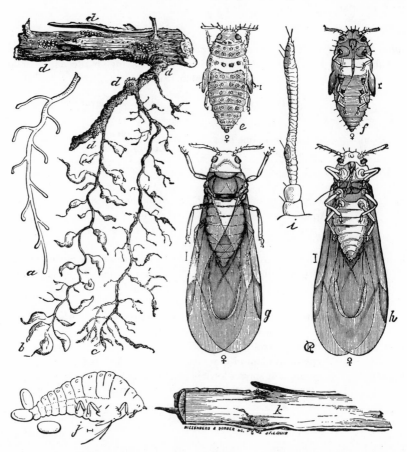

Fig. 152. Root-inhabiting form of the grape phylloxera. *a*, shows a healthy root; *b*, one on which the insects are working, representing the knots and swellings caused by their punctures; *c*, a root that has been deserted by them, and where the rootlets have commenced to decay; *d, d, d*, show how the insects are found on the larger roots; *e*, agamic female nymph, dorsal view; *f*, the same, ventral view; *g*, winged agamic female, dorsal view; *h*, same, ventral view; *i*, magnified antenna of winged insect; *j*, side view of the wingless agamic female, laying eggs on roots; *k*, shows how the punctures of the insects cause the large roots to rot. (From Riley)

generations form enlargements and galls on the small grape roots. Some of the root forms give rise to parthenogenetic winged migrants in the autumn; these crawl out of the ground and fly to the grapevines, there producing wingless males and females. After mating, the females each lay one egg in a crevice of the bark. The eggs overwinter and hatch the next spring, initiating a series of leaf-infesting generations.

The gall wasps (Cynipidae, Hymenoptera) contain many species having alternation of sexual and parthenogenetic generations. An example is an oak-feeding species *Andricus erinacei* which overwinters as eggs laid in leaf or flower buds. These hatch in spring, and each larva becomes surrounded by a soft bud gall produced by the plant. The inner layer of the galls provides food for the larvae. In early summer, when those larvae mature, winged males and females emerge. The females lay their eggs in the veins of oak leaves. Larvae from these eggs cause pincushion-like growths on the leaf veins, called hedgehog galls. Larvae in the hedgehog galls mature in autumn, and all emerge as short-winged females, which reproduce parthenogenetically, laying the overwintering eggs in the oak buds.

The benefits of alternation of generations are undoubtedly concerned with food supply. The habit allows the species to feed on more species of hosts, or more individuals, or on different parts of the same host. This in turn permits larger populations of the insect species to be produced without danger of the food supplies being curtailed.

Forms with Paedogenesis. Paedogenesis involving reproduction by larvae is always combined with a complex and irregular cycle of generations. The cases studied indicate that there may be successive generations of paedogenetic larvae, with irregular production of larvae which mature normally and pupate. Adults from these pupae mate normally and produce fertilized eggs. The beetle *Micromalthus debilis* has a complex cycle of paedogenetic and normal generations. The most completely studied examples are in the midge family Cecidomyiidae, especially the European *Oligarces paradoxus*. For this species a typical cycle of generations, shown diagrammatically in fig. 153, is as follows: A paedogenetic larva *a* produces larvae *b* which may grow up to be one of four types: (1) another paedogenetic larva like *a*, or (2) a female-producing larva *c*, or (3) a paedogenetic larva *d* which gives birth only to male-producing larvae *e*, or (4) a paedogenetic larva *f* which gives birth to both male-producing larvae *e* and other paedogenetic larvae like *a*. The male- and female-producing larvae pupate, and normal adults emerge. The females lay fertilized eggs which develop into the paedogenetic larvae *a*.

In *Oligarces paradoxus* adults are produced more frequently as the colony

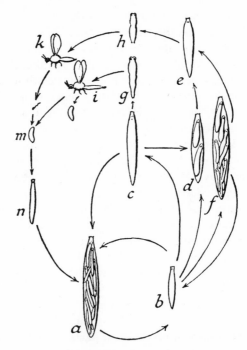

Fig. 153. Succession of generations in the midge *Oligarces paradoxus*. *a*, paedogenetic larva giving rise to *b*, undetermined daughter larva which may develop into *a* again or into *c*, female-imago-producing larva, or into *d*, which gives rise to male-imago-producing larvae (*e*) or into *f*, which gives rise both to undetermined daughter larvae and to male-imago-producing larvae; *g*, female pupa; *h*, male pupa; *i*, female imago; *k*, male imago; *l*, sperm; *m*, egg; *n*, young larva from egg. (From Wigglesworth, after Ulrich)

becomes overcrowded or the food supply less ample. In the paedogenetic species of the midge genus *Miastor*, temperature changes have a determining effect on the type of generation produced. Thus in these midges the cycles of paedogenetic generations are correlated with day-to-day conditions as well as a more inclusive seasonal cycle.

SOCIAL INSECTS

In mode of life the great majority of insects are solitary. Each individual lives to itself, and members of a species have no marked attraction for each other except at times of mating. Beyond placing the eggs or young on or near their food (by instinct rather than by design) the parents usually take no interest in their offspring. The parents usually die before

their progeny mature, and consequently there is no opportunity for prolonged parent-offspring relationship. There are certain groups of insects,
however, whose species have a social mode of life. In the termites, ants,
social wasps, and social bees social life is well developed and complex,
embracing almost all phases of the individual's activities. Other insects
show interesting tendencies toward the beginnings of social life, through
such phenomena as maternal care, social larvae, and community
development.

Maternal Care. The females of certain earwig species deposit their
masses of eggs in a sheltered chamber and guard them, driving away
predators. After the eggs hatch, this watch is continued for a short time
until the young nymphs are active enough to leave the brood chamber.
At this point maternal care is discontinued, and each nymph goes its own
way. Similar observations have been recorded for a few other insects,
among them the mole crickets *Gryllotalpa*.

Social Larvae. Among the moths (Lepidoptera) are species in which
the larvae, hatching from the same egg mass together, construct a silken
weblike nest, and all use it as a common abode. The nest is built around

Fig. 154. Maternal care exhibited by female earwig. (After Fulton)

Fig. 155.   Communal home or tent of tent caterpillar.   (After U.S.D.A., B.E.P.Q.)

a fork or branch of a tree, all the larvae contributing to its construction. The larvae leave the nest during the day to feed on the foliage of the tree, and all return to it for resting.   Larvae of the tent caterpillars (*Malacosoma*) live in the nest for their entire larval period, leaving it finally to pupate.   Larvae of fall webworms (*Hyphantria*) spin a similar nest but leave it and follow solitary existences for the last larval instar.

Community Development.   In the order Embioptera or web spinners, some species are gregarious.   They live in colonies that consist of interlocking silken tunnels in soil, surface cover, or the base of plants.   The females exercise maternal care to a high degree, watching over the eggs and young nymphs.   Some of these colonies are described as forming a solid silken mat over many square yards of ground and must contain hundreds and possibly thousands of individuals.   To date, however, no intimate relationship has been observed among individuals of a colony, and therefore the gregarious nature of the forms may mean little more in a social sense than the clustering of aphids or scale insects on or about their parents and grandparents.

From the standpoint of development toward social life, a more significant type of colony occurs in the cockroaches. Wood cockroaches of the genus *Cryptocercus* live together as family colonies in rotten logs. The cockroaches eat wood, which is digested by a specialized symbiotic protozoan fauna in their digestive tracts. When young cockroaches molt, they completely empty the digestive tract, so that after molting they have no symbiotic fauna, and the nymphs would soon starve if this situation were not remedied. The newly molted nymphs replenish their supply of Protozoa by eating some of the fresh excrement of another member of the colony. This necessary interchange of Protozoa requires that groups of individuals live together in a colony.

Social Life. The ants, termites, certain wasps, and a few kinds of bees have developed social life to a high level. They live in family colonies that have a division of labor among individuals and an interchange or sharing of food and other things among members of the colony. In each of the four groups where it occurs, social life arose independently. This is one of the most striking known examples of parallel evolution in animal habits. The actual details of metamorphosis, feeding habits, and colony formation are different in the four social groups, often radically so; yet the final organization attained in each is remarkably similar. Because of the differences among them, it is illuminating to present a brief sketch of the social features of the four groups.

TERMITES (the order Isoptera) form a definite colony, either hollowed out of wood or built from masticated products, and populated by several different forms. At certain times of the year swarms of winged sexual forms issue from the old colonies and disperse. After their flight, these forms alight, and the wings fall off. Males and females pair off and together begin a small excavation for a new nest. At this stage mating occurs, and later the female deposits and watches over her first brood of eggs. She feeds the first young with saliva and other secretions. Thus a new colony is founded. Soon after hatching, the nymphs are self-reliant and feed themselves and their parents also. From this time on the original male and female, called the royal pair, perform only the function of reproduction. In the early stages of the colony, the nymphs develop into three castes, fig. 156, all of them wingless: (1) a worker caste, which is simple in structure, feeds on wood or fungus products, and by regurgitation feeds the young and other castes; (2) a soldier caste that is large-headed and has

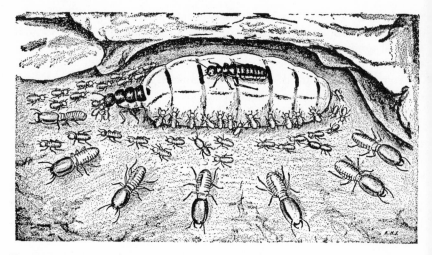

Fig. 156. A nest of termites, diagrammatic. The greatly enlarged queen rests in her royal cell with the king on her abdomen. Surrounding the queen are numerous workers grooming and feeding her. Guarding the workers are the soldiers with their greatly enlarged heads, while, among the workers, smaller soldiers (traffic police) regulate the movement of the workers. (From Matheson, *Entomology for introductory courses,* by permission of Comstock Publishing Co.)

a protective function in the colony, guarding the nest entrances and the royal pair; and (3) a substitute reproductive caste that may become fertile and replace the royal pair if the latter die. There are usually two kinds of substitute reproductives, one with well-developed wing pads (but never wings), called second-form queens, and one without wing pads and very similar to the worker caste, called third-form queens. The nonreproductive castes contain both males and females, but their reproductive organs are vestigial. In some species the soldiers may be replaced by a long-headed form having a long snout. This caste, called nasutes, fig. 180, emits a disagreeable odor that is apparently designed to keep away enemies. After the colony is flourishing, periodic broods of winged reproductives are produced which disperse to found new colonies.

Nearctic termite species nest in cavities hollowed out in the ground or in wood. Some neotropical species build elaborate nests in trees, and certain species in Africa and Australia build mounds on top of the ground. The mounds of a particular species have a distinctive shape and size and range in height from a few inches to 20 feet, fig. 157. These nests or houses are built by the workers, using a "plaster" of saliva and earth or wood masticated together. It is remarkable that these structures are so uniform in shape and size, made as they are by thousands of workers who never see the nest from the outside. The instinctive behavior pattern responsible for this and other activities is one of the most amazing phenomena displayed by animals.

Among the various members of the colony there is a constant exchange of materials. Workers give food to soldiers and reproductives and in return obtain from them secretions from mouth or anus. The queen is thought to secrete desirable substances at many points on the body, because other castes lick her body as well as obtaining oral or anal secretions from her. This exchange of substances is called *trophallaxis*.

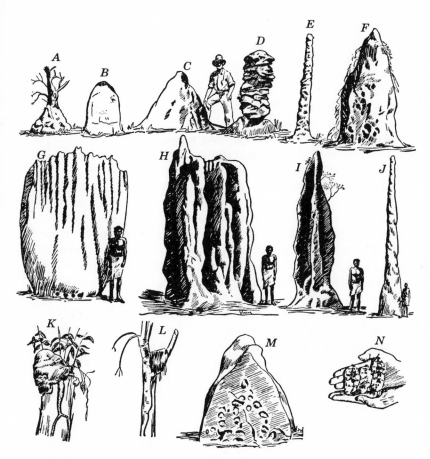

Fig. 157. Termite mounds or termitaria. *A, B, C, D, J*, from South Africa; *D, Nasutitermes lamanianus*, Belgian Congo; *E*, vent of subterranean nest, South Africa; *F* and *M*, exterior and cross section of mounds of *Termes redemanni*, Ceylon; *G, Amitermes meridionalis*, Australia; *H, Nasutitermes triodiae*, Australia; *I, Nasutitermes pyriformis*, Australia; *K* and *L*, arboreal nests of *Nasutitermes corniger*, Panama; *N*, queens of *Macrotermes bellicosus*, Nigeria. (From Essig, *College entomology*, by permission of The Macmillan Co.)

Termites, like the wood cockroaches, have in their intestines a symbiotic protozoan fauna which predigests cellulose eaten by the termites. Without these symbionts the termites would be unable to subsist on their wood or mycelium diet. The Protozoa are passed from termite to termite by way of the secretions of which the termites are very fond. It is probable that social life in this group originated in family colonies centered around the dissemination of symbionts, as we find them today in the wood cockroaches.

The actual mechanics by which the different castes are produced has been a subject of research and speculation by many investigators. Except for the primary sexual caste, all forms in the colony are individuals that, even when mature, fail to develop complete adult characteristics. More specifically, the mature second-form reproductives develop functional reproductive organs, but their wings never grow beyond the wing-pad stage, in this latter respect resembling the last nymphal stage of the perfect form; the mature third-form reproductives have the functional reproductive organs but lack any trace of wings, in this last character being similar to an early nymphal stage; and the workers and soldiers fail to develop either functional reproductive organs or any trace of wings.

This situation implies that a control of growth is responsible for the differences among the castes. The control is differential or qualitative and probably consists of a set of complex hormones, each of which may affect one part of the insect and yet not interfere with others. Evidence based on study of embryos indicates that certain phases of growth regulation are imposed on the individual before it hatches from the egg. But some control is exercised during the entire life of the individual. For example, the substitute reproductives do not become functional until the queen dies or is removed; then several individuals of the substitute caste begin to lay eggs. It has been suggested that in this and similar cases growth hormones are exchanged from caste to caste by trophallaxis. Thus the functional queens may secrete a hormone that inhibits the growth of certain adult characteristics and prevents maturation of the substitute reproductives. When the queen dies, the inhibition would be removed and allow the maturation of the reproductive system of the substitute reproductives. There is a further possibility that substances eaten by the queen, offered by various castes, may contain hormones influencing the quality of eggs and young and aid in keeping fairly constant the numerical ratio of the different castes in the nest.

These speculative hypotheses are based on little in the way of experimental proof but have considerable justification in colony observations. For there is no doubt that some pliable mechanism exists that allows each colony to adjust itself to misfortunes and depredations, an adjustment which could not be made with a behavior pattern controlled completely by inflexible blind instincts.

In contrasting social life in the termites with the habits of solitary insects, it is apparent that several features of termite habits are of special significance in making possible their type of social life. These features are:

1. Care of eggs and young by parents during the founding of a colony.

2. Greatly extended life span of the sexually mature adults, over a period of several years, during which many generations of progeny are matured.

3. The feeding of parents and young by offspring of the reproductives which began the colony.

4. Control of individual growth leading to a development of different castes, correlated with a division of labor within the colony.

ANTS (Formicidae, Hymenoptera) exhibit a type of social life closely paralleling that of the termites in organization. A typical ant nest, or colony, is usually in a log or cavity, or in the ground, and often has a mound of earth above it. Colonies vary in number of individuals, from a few dozen to many thousands. Each colony is founded by a migrant winged queen. After the mating flight, the male dies, and the female loses her wings. When well developed, the colony contains the original queen, a large number of wingless sterile workers, frequently wingless sterile large-headed soldiers, and the brood or young (see p. 310). The wingless sterile workers are not true neuters but are modified females. In respect to metamorphosis ants are quite different from termites. The latter are hemimetabolous, and the nymphs are active and able to feed themselves soon after hatching. But ants are holometabolous. Their immature stages are legless, helpless grubs, incapable of moving about, and they must be fed for their entire period of development. After this the larvae pupate, in some groups spinning a cocoon and in others lying naked in the nest. In the early stages of the colony, only sterile workers are produced, these having no wings and only vestigial reproductive organs. When the colony is well established, periodic broods of winged males and females are produced, and these disperse.

The four habit features listed for termites apply also to the ants, although with modifications regarding the manner in which each is executed:

1. The life span of only the female queen is prolonged, since one mating suffices for her life of several years, and the males die after the nuptial flight.

2. The maternal care of the first eggs and young by the queen when founding a new colony covers a more protracted period because the larvae

must be nourished until they are mature. During this period the queen's wing muscles undergo histolysis to provide the source of both the queen's own nourishment and the oral secretions she feeds her first brood of larvae.

3. The workers take over feeding of the queen and brood after the nest is established.

4. There is the same control of development, with production of different castes and a division of labor.

Ants have an exchange of foods between individuals as have termites, the workers feeding queens, soldiers, and larvae, and obtaining exudates or anal secretions from each. Ants are omnivorous and apparently have no peculiar intestinal fauna as have the termites. Trophallaxis in the ants, therefore, seems to be a simple "reward" system by which the receiver pays the giver. The role played by the possible exchange of growth hormones in ants is unknown.

Whether an individual develops into a sexual form or a neuter is thought to be determined in the egg. Among the neuters, however, are many cases of polymorphism due to heterogony. In heterogony different parts of the body grow at different rates, so that in small individuals all parts may appear well proportioned, but in larger individuals some part or parts may appear unusually large. In the case of ants, the head is usually the region of disproportionate growth. Thus in *Pheidole*, fig. 158, small neuters have heads and abdomens about the same size; in larger individuals the head becomes disproportionately larger, and in the largest individuals the head is enormous. The larger-headed forms perform special services such as cracking seeds or acting in a soldier capacity and are usually considered as an additional caste. The amount of food given to the larva determines the size of the adult developing from it; hence, the workers are able to control the production of certain castes by feeding.

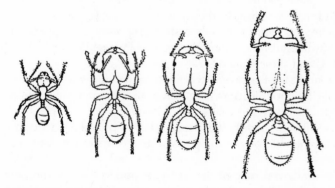

Fig. 158. Heterogony in the ant *Pheidole*. Workers of *Pheidole instabilis* showing increase in the relative size of the head with absolute size of the body. (From Wigglesworth after Wheeler)

SOCIAL WASPS, comprising *Vespa, Polistes,* and certain related genera of the Vespidae, and *social bees* in several genera of the family Apidae, are social in habits, living in colonies and having a caste of sterile workers. All these differ from both termites and ants in certain respects. The workers are winged and differ in appearance from the queens chiefly in their smaller size. Both wasps and bees are holometabolous insects and have legless grubs which must be fed for their entire growth period, as is the case with ants. The bees and wasps, however, construct an individual cell for each larva, and the cell is sealed during pupation. Wasps feed on insects; bees on nectar and pollen. Except for honeybees, the nearctic wasps and bees make only annual colonies. These are started each year by a fertilized female. She begins the nest, lays eggs, and forages food for the first brood of young. These mature into workers, who take over the duties of nest making and foraging for the colony. In fall only males and females are produced. These disperse, and mate, the males dying soon after mating, but the fertilized females hibernate for the winter and emerge the following spring to start new colonies. At the approach of winter old queens and workers die.

HONEY BEES are much more specialized socially than the other social bees. Their colonies are perennial, made possible by storing during the summer enough food to keep the colony alive through the colder months when there are no flowers to provide essential nectar and pollen. Beginnings of this storage habit are found in more specialized species of bumblebees (*Bombus*), close relatives of the honey bees. In nests of these bumblebees a few cells are constructed for storage of food, which is used later in the fall when flowers are less abundant. In a honey bee nest (normally built in a hollow tree) there are vertical rows of wax cells, or combs. A large number of the combs are used for raising the brood, but many are used for storage of food, in the form of honey. During the cold months, individuals of the colony are active, though sluggish, within the nest and by oxidation of foodstuffs in their bodies keep the nest temperature well above freezing. Unfertilized honey bee eggs produce males or drones that do no work. They stay around the nest for mating purposes but after a few weeks are driven out by the workers. Fertilized eggs develop into either queens or workers, depending on the diet given the larvae. Those fed "royal jelly" (a milky product of glands in the heads of workers) and afterwards pollen and nectar develop into workers; those fed royal jelly for their entire larval period develop into queens. Since the workers do all the feeding, they determine the production of queens.

The queens have lost the power of founding new colonies alone. This is accomplished by swarming, in which a queen and part of a worker colony leave the nest together, settle in a new site, and start a new colony.

In this feature the honey bee differs from all other social insects except some ants.

Social Life Cycles. Social insects differ from others in that the *colony* and not the *individual* is the reproductive unit. The productivity of the reproductive caste is made possible by the home building, foraging, and protection performed by sterile workers and soldiers. The instincts and behavior pattern which will result in new generations of these sterile castes in a new nest must be represented in the genes carried out of the nest by migrating reproductives.

We have seen how certain insect species have a cycle of dissimilar generations, each generation specialized to aid in some way the welfare of the species. In social insects we have the same principle applied, except that the various generations (reproductives, workers, and soldiers) all occur contemporaneously and work side by side in a common abode. In all four groups in which this social phenomenon occurs there is a division of labor or biological functions: One caste performs the reproductive function, a second caste the feeding function, and sometimes a third the function of protection. This is carrying to individuals the same kind of specialization that occurs in the cells of the metazoan body.

REFERENCES

Balduf, W. V., 1935. *The bionomics of entomophagous Coleoptera.* St. Louis, by the author. 220 pp.
1939. *The bionomics of entomophagous insects,* Pt. 2. St. Louis, by the author. 384 pp., illus.
Clausen, Curtis P., 1940. *Entomophagous insects.* New York: McGraw-Hill Book Co. 688 pp., illus.
Frost, S. W., 1942. *General entomology.* New York: McGraw-Hill Book Co. 524 pp., illus.
Johannsen, O. A., and F. H. Butt, 1941. *Embryology of insects and myriapods.* New York: McGraw-Hill Book Co. 462 pp., illus.
Wheeler, W. M., 1928. *The social insects, their origin and evolution.* New York: Harcourt, Brace & Co. 378 pp.

Chapter Seven

The Orders
of Insects

ONE of the greatest marvels of the living world is that insects, starting from a single simple form, evolved into the multitude and diversity of species found in the world today. The story of this evolution is the drama of the fortuitous development of structures which endowed their possessors with added advantages in the struggle for existence. A sufficient number of early, primitive insect types have persisted to the present time, and enough fossils are known to allow us to reconstruct the main trends in the evolution of these fascinating creatures, fig. 159.

The earliest insects differed from their centipede-like ancestors chiefly in having a 3-segmented thorax with its three pairs of legs, and the abdominal legs greatly reduced or absent. In these early forms the young differed little in appearance from the adults and wings had not yet evolved. Five orders of these early wingless types are known, collectively termed the Apterygota. The most primitive of these appears to be the order Diplura, fig. 162. Considerable evidence suggests that two other orders, the Collembola and Protura, arose from a Diplura-like ancestor in which the tibia and tarsus fused and the abdominal spiracles atrophied. In spite of their many features in common, the Collembola and Protura have evolved into remarkably dissimilar animals. In the Protura the antennae are nearly atrophied and the front legs have become somewhat antenna-like; in the Collembola the abdominal segments are reduced in number and the vestigal legs of the fourth abdominal segment have fused and developed into a forked leaping organ or spring. In all three of the orders Diplura, Protura, and Collembola, the sides of the mouth cavity have fused with the sides of the labium and grown out to form a cavity surrounding the functional mouthparts, the mandibles and maxillae.

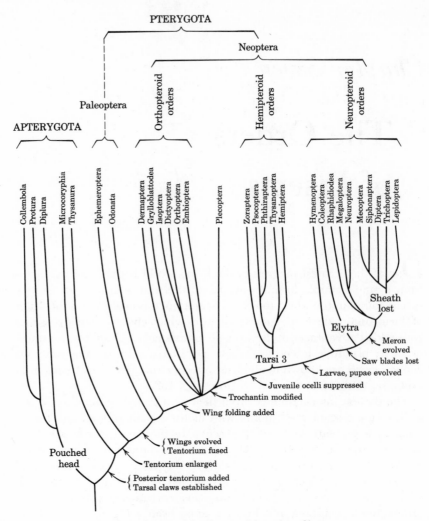

Fig. 159. Suggested family tree of the orders of insects.

In mouthparts and certain body characters, the other two apterygote orders are more primitive than any of the three mentioned above but they have developed structures which later evolved into distinctive features of winged insects. Most important of these structures are the longer and stronger thoracic legs and the development of the dorsal and posterior arms of the tentorium. The Microcoryphia (machilids) is the more primitive of these two orders, preserving styli on all the abdominal segments. In the Thysanura (lepismatids), the styli are lost on the first six abdominal segments, the parts of the tentorium are much more developed, and the body is wider and flatter.

There is little doubt that some offshoot ancestral to the Thysanura developed planing habits and structures and eventually wings and flight. The first successful wings were sharply and deeply pleated like a fan, but could not be folded and laid over the back in repose. Insects with wings of this type are called the Paleoptera ("ancient wings"). The first insect fliers had primitive mandibles with only a single socket. The order Ephemeroptera (mayflies) contains the only living survivors of this type. The next advance seems to have been the development of the kind of two-socketed mandibles found in most winged insects. The order Odonata (dragonflies) are the only modern representatives of this stage in insect evolution.

In the more rapidly changing insect line a set of mechanisms developed by which the wings could be folded and laid over the back in repose. Concurrently much of the wing fluting was reduced, resulting in larger flat areas on parts of the wing. Insects with wings of this type are called the Neoptera ("modern wings").

The first abundant members of the Neoptera may have been the extinct order Protorthoptera, a diverse group of fairly large, running insects having completely veined wings and long cerci. From the Protorthoptera or similar early forms arose probably five and possibly more lineages represented by living species. In one line, from which arose the Grylloblattodea (grylloblattids), the ovipositor remained the same primitive type as that of the Thysanura, but our only persisting members are wingless. In a second line the ovipositor became highly modified, evolving into the grasshopper type; from this line evolved the Dictyoptera-Isoptera (cockroach-termite) line and the Orthoptera (grasshopper) line. Difficult to place but probably representing another separate lineage is the order Dermaptera (earwigs), in which the wings became reduced to short, square wing covers. The peculiar order Embioptera (web-spinners) probably also arose near this point in the family tree.

In these five orders just mentioned (and presumably in the Protorthoptera also) the wings developed as external wing pads and, except for the lack of wings and difference in size, the nymphs look much like the adults and usually live and feed the same way. These five living orders are collectively termed the orthopteroid orders.

In the other neopteran line, one of the mesopleural sclerites, the trochantin, became reduced to a thin strap forming a movable leg articulation, and certain adult characters became suppressed in the immature stages. It is possible that the order Plecoptera (stoneflies) is an early offshoot of this line for it has the narrow trochantin but no appreciable suppression of adult characters in the nymph. Beyond the possible point of origin of the stonefly line, however, a form arose in which the ocelli became suppressed in the nymphs. This ancestor gave rise to two vigorous groups, the hemipteroid orders and the neuropteroid orders.

The first members of the hemipteroid line were probably general feeders much like the order Psocoptera (psocids). The most primitive branch seems to be represented by the Zoraptera, a group of colonial insects resembling psocids but preserving distinct cerci, fig. 195. In the Psocoptera the lacinia of the maxilla evolved into a slender, chisel-like piece which, in two other orders, forms the basis of sucking-type mouthparts. The first step in this development is exemplified by the order Thysanoptera (thrips) which have evolved a rasping-sucking set of mouthparts. The end stage is seen in the Hemiptera (bugs) in which the mouthparts form a slender piercing-sucking organ. The other branch of the hemipteroid line evidently arose directly from a Psocoptera-like ancestor, that became established as skin scavengers of animals and evolved into the ectoparasitic order Phthiraptera (chewing and sucking lice).

In the neuropteroid line the juvenile suppression of adult characters continued, the immature stages or larvae having greatly reduced eyes, antennae, and body sclerotization. The last immature instar developed into a transformation stage or pupa in which the various parts are re-formed into those of the adult.

The primeval neuropteroid insect can be reconstructed with considerable reliability by adding together primitive characters still persisting in living orders. It was probably of medium size for an insect. In the adults, the two pairs of wings were about the same size and each had a fairly large number of veins and crossveins, the coxae were of a simple type, the abdomen was well sclerotized, and it possessed a sawlike ovipositor much like that of a sawfly. The saw sheath possessed a stylus at the apex (see fig. 70). The larva was elongate, predaceous, fast-running, and resembled present-day snakefly and some beetle larvae. This primeval neuropteroid apparently gave rise to three lines that are represented by present-day major insect groups. One line evolved into the sawflies or Hymenoptera, preserving the functional sawlike ovipositor but undergoing reduction in wing venation and evolving plant-feeding larvae. Another line evolved into the beetles or Coleoptera; in the adult of these the front wings became hard wing covers and the ovipositor atrophied, but the larvae changed little.

In adults of the third line, the meron of the coxae became enlarged but the bladelike saw atrophied, only the sheath remaining as an ovipositor (valvifer 2 plus valvulus 3, fig. 70.) This third line in turn gave rise to two main branches: (1) the Megaloptera branch in which the sheath persisted as a functional ovipositor and (2) the Mecoptera branch in which this structure atrophied completely. From the Megaloptera branch arose three orders: The Raphidiodea (snakeflies) that changed little, the Megaloptera in which the abdomen lost much of its sclerotization and the larvae became aquatic, and the Neuroptera in which the stylus of the ovipositor atrophied.

The Mecoptera branch appears to have divided early into the Trichoptera-Lepidoptera stem in which the longitudinal veins and especially the crossveins became reduced in number, and the Mecoptera-Diptera stem in which the lower edge of the pronotum fused with the anterior edge of the mesopleuron below the spiracle. In the Trichoptera (caddisfly) line arising from the Trichoptera-Lepidoptera stem, the larvae became aquatic and their spiracles atrophied; in the Lepidoptera (moth and butterfly) line, the larvae remained terrestrial and spiraculate. In the Mecoptera (scorpionfly) line arising from the Mecoptera-Diptera stem, little change seems to have occurred; in the Diptera (true fly) line, the venation became greatly reduced, the second pair of wings evolved into small, knoblike balancing organs called halteres, and the pronotum fused very solidly with the adjacent mesothoracic parts.

The order Siphonaptera (fleas) has no winged species and has become so modified for existence as an ectoparasite that only a few clues to its ancestry remain. Fleas have a large posterior plate on the coxa that is probably a large meron, suggestive of the Mecoptera branch; so also is the complete lack of ovipositor plates. Certain internal structures of the proventriculus are remarkably like those of the Mecoptera. Unlike the Mecoptera, however, the flea pronotum is not fused with the mesopleuron, but this is a connection that could readily have been lost in the evolution of the wingless condition. Because of the uniqueness of the proventricular structure, it seems probable that the fleas are a highly-modified offshoot of primeval Mecoptera.

During their evolutionary history insects have multiplied in number of species as has no other group of organisms. The twenty-eight living orders of insects comprise nearly a thousand families and many thousand genera, and up to the present about nine hundred thousand described species. This does not include a large number of fossil orders, genera, and species. Known from North America are all twenty-eight orders, over five hundred families, several thousand genera, and over one hundred thousand species. It is certain that many more species than this occur in the world. It has been estimated that nearly twenty-five thousand species remain to be discovered from North America and at least a million more from the entire world.

The suggested evolutionary paths outlined in fig. 159, it must be

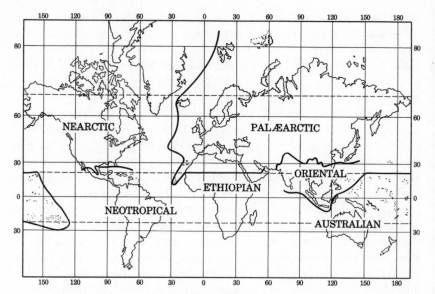

Fig. 160. Faunal realms or zoogeographic regions of the world. (After Sclater and Wallace)

remembered, are based on the limited amount of comparative data, both morphological and physiological, known at the present time, and on only the species, both fossil and recent, known to date. Further studies will undoubtedly shed much needed light on these problems.

Various large regions of the world have many distinctive taxonomic elements in their biotas. On the basis of relative similarities and differences between their animal components six major realms or zoogeographic regions of the world are recognized by zoologists, as shown in fig. 160. These regions are mentioned frequently in taxonomic literature.

NAMES OF ORDERS AND FAMILIES

In many cases more than one name is applied to the same order or family. Certain of these different usages are based on contentions regarding either the descriptive propriety or the date priority of the names in question. In this book the attempt is made in all such cases to employ the name in most common use. Notations are given concerning synonymous names which the beginning student is likely to encounter when using reference material.

General References on the Orders of Insects

After each order are listed references which will be helpful in obtaining a more thorough knowledge of the respective order. The following list will be found helpful in learning more concerning the rarer or exotic families of insects, or in obtaining information on collecting or rearing insects.

Borror, D. J., and D. M. DeLong, 1964. *An introduction to the study of insects,* 2nd. ed. New York: Holt, Rinehart & Winston.

Brues, C. T., A. L. Melander, and F. M. Carpenter, 1954. Classification of insects, *Museum Comp. Zool. Harvard Coll. Bull.,* **108**:1–917.

Comstock, J. H., 1936. *An introduction to entomology,* 8th ed. Ithaca, N. Y.: Comstock Publishing Co. 1044 pp., illus.

Essig, E. O., 1942. *College entomology.* New York: The Macmillan Co. 900 pp., illus.

Felt, E. P., 1940. *Plant galls and gall makers.* Ithaca, N. Y.: Comstock Publishing Co., Inc. 364 pp., illus.

Imms, A. D., 1951. *Insect natural history.* Philadelphia, Pa.: The Blakiston Co. 317 pp.

Jaques, H. E., 1941. *How to know insects.* Dubuque, Ia.: Wm. Brown & Co. 140 pp., illus.

Oman, P. W., 1948. Collection and preservation of insects. *USDA Misc. Publ.* **601**:1–42.

Peterson, A., 1948, 1951. *Larvae of insects.* Ann Arbor, J. W. Edwards. Pt. 1, 315 pp.; Pt. 2, 416 pp.

1953. *A manual of entomological techniques,* 7th ed. Ann Arbor, J. W. Edwards. 367 pp.

Ross, H. H., 1953. How to collect and preserve insects, 3rd. ed. *Illinois Nat. Hist. Survey Circ.* **39**:1–59.

Smith, Roger C., and others, 1944. *Insects in Kansas.* Topeka, Kan., Rep. Kansas State Board Agr. 440 pp., illus.

KEY TO THE ORDERS OF COMMON INSECTS

1. Wings well developed, fig. 167, the front pair sometimes forming short, hard wing covers, fig. 190 2
   Wings reduced to small pads, or no wings developed 32
2. Front wings hard, horny, opaque, and veinless, in repose lying over the body and forming wing covers, figs. 190, 253 3
   Front wings either transparent, membranous, or with definite veins 6
3. Front and hind wings both long, narrow, and the same shape; ventral region of head capsule prolonged into a beaklike structure at the end of which are situated typical chewing mouthparts, much as in fig. 287 (male *Boreus*)
   Mecoptera, p. 356
   Front and hind wings either broad or quite dissimilar in shape; head capsule not so produced 4
4. Mouthparts forming a needle-like piercing beak, fig. 207; venation usually indicated but indistinct . . . . **Hemiptera,** p. 264
   Mouthparts with mandibles, fitted for chewing, fig. 36 5
5. Abdomen terminating in a pair of external forcepslike appendages, fig. 190
   Dermaptera, p. 241
   Abdomen either without terminal appendages, or these pointed and stylelike, fig. 261 . . . . **Coleoptera,** p. 316
6. Having only one pair of wings, the hind pair at most forming small, clubbed, balancing organs, or halteres, fig. 292 7
   Having two pairs of wings, although the hind pair may be small 12
7. Wings leathery or parchment-like 8
   Wings membranous, sometimes dark in color 10

8. Mouthparts forming a piercing-sucking beak, fig. 207 . . . . . . **Hemiptera,** p. 264
   Mouthparts fitted for chewing, with generalized parts, as in fig. 36 9
9. Hind legs greatly enlarged for leaping, figs. 181–189 . . . . . . **Orthoptera,** p. 236
   Hind legs not greatly enlarged, fitted for running, figs. 174, 176
   Dictyoptera, p. 227
10. Abdomen without terminal filaments; hind wings represented by halteres
    Diptera, p. 361
    Abdomen with one, two, or three terminal filaments, as in figs. 167, 218 11
11. Halteres present; terminal appendages short, antennae long, fig. 218 (males of
    Coccidae) . . . . **Hemiptera,** p. 264
    Halteres not developed; terminal appendages very long, antennae short, as in
    fig. 167 . . . . **Ephemeroptera,** p. 219
12. Front wings, and usually also hind wings, clothed with overlapping scales,
    fig. 323, except for window-like areas, fig. 333 . . . . **Lepidoptera,** p. 395
    Wings bearing only scattered scales, having chiefly fine hair, bristles, or no
    vestiture 13
13. Tarsus ending in a round bladderlike structure, without evident claws, fig. 205;
    wings long and narrow, at most with two longitudinal veins
    Thysanoptera, p. 260
    Tarsus without a terminal bladder, sometimes having large pads and/or distinct
    claws, fig. 209C–E; wings not as above 14
14. Mouthparts forming slender stylets fitted together for piercing and sucking, and
    housed in a beak that is either triangular or rodlike, figs. 207A, 209G
    Hemiptera, p. 264
    Mouthparts not forming a beak, either vestigial or of the chewing or lapping
    type; mandibles not forming slender stylets 15
15. Abdomen having two or three terminal filaments as long as the body, and head
    having short hairlike antennae, fig. 167 . . . . **Ephemeroptera,** p. 219
    Either abdomen having short terminal filaments or none, or antennae long and
    slender, figs. 176, 179 16
16. Front wings leathery, hind wings membranous, the front ones when folded form-
    ing a protective covering over the hind pair, figs. 174, 181 17
    Both pairs of wings of about the same texture 18
17. Hind legs greatly enlarged for leaping, fig. 181 . . . . **Orthoptera,** p. 236
    Hind legs not greatly enlarged, fitted for running, figs. 174, 176
    Dictyoptera, p. 227
18. Both front and hind wings having many veins and many crossveins, forming a
    close network over part of the wing surface, figs. 179, 194, 284A 19
    Wings either with crossveins few in number, fig. 64, or longitudinal veins
    reduced, fig. 66, or both 28
19. Front and hind wing each with radius and its branches sclerotized and forming a
    heavy anterior band, remainder of venation semimembranous or subatro-
    phied, fig. 179 . . . . **Isoptera,** p. 233
    Venation sclerotized throughout 20
20. Antennae short and setalike, 5- to 8-segmented; wings long and exceedingly
    reticulate, figs. 170, 172 . . . . **Odonata,** p. 222

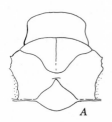

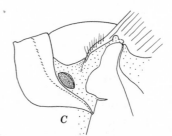

Fig. 161. Thoracic structures of insects. *A*, pro- and mesonotum of *Hemerobius* (Hemerobiidae); *B*, same of *Brachypanorpa* (Panorpodidae); *C*, anterior pleural region of *Brachypanorpa* (Panorpodidae).

30. Size less than 3 mm.; front wings with only three principal veins, fig. 195

    Zoraptera, p. 249

    Size 5 mm. or more; front wings with a much more extensive venation, fig. 194

    Plecoptera, p. 247

31. Mandibles sclerotized and large; wings clothed with minute setae, the venation either greatly reduced or forming a series of irregular cells, fig. 232

    Hymenoptera, p. 291

    Mandibles difficult to detect, subatrophied; either small species (less than 6 mm.) with extremely hairy wings, or venation composed of regularly branching veins, fig. 317 . . . . **Trichoptera, p. 391**

32. Abdomen ending in two or three long "tails," figs. 162A, 166 33

    Abdomen without long terminal tails 35

33. Having two terminal tails, fig. 162A . . . . **Diplura, p. 213**

    Having three terminal tails, fig. 166 34

34. Abdominal segments 2–9 having styliform appendages, fig. 165

    Microcoryphia, p. 218

    Abdominal segments 1–6 without styliform appendages, fig. 166

    Thysanura, p. 218

35. Abdomen ending in a pair of strong sclerotized pincerlike jaws or forceps, fig. 162B 36

    Abdomen without stout terminal forceps 37

36. Tarsus 1-segmented; head without eyes; fig. 162B . . . . **Diplura, p. 213**

    Tarsus 3-segmented; head having conspicuous eyes, as in fig. 190

    Dermaptera, p. 241

37. Tarsus 4- or 5-segmented 38

    Tarsus 1- to 3-segmented 48

38. Base of abdomen constricted to a narrow joint hinged to forward part of body, figs. 234D, 246, 249 . . . . **Hymenoptera, p. 291**

    Abdomen not constricted and hinged at its base 39

39. Antenna minute, flattened, and indistinctly segmented, fig. 289, or round, sometimes with a terminal hair, much as in fig. 297J 40

    Antenna slender and long, many-segmented, figs. 185, 338B 41

40. . Antenna with many indistinct segments; body bilaterally compressed, with distinct segmentation; head and pronotum often with ctenidia, or rows of stout spines, fig. 289 . . . . **Siphonaptera, p. 358**

    Antenna globular, appearing as one segment, sometimes with a terminal hair; body not greatly compressed from side to side, and often without distinct segmentation on abdomen; never having ctenidia on head or pronotum, fig. 316 . . . . **Diptera, p. 361**

41. Head prolonged into a beaklike projection at the end of which are located a set of chewing mouthparts, as in fig. 287 . . . . **Mecoptera, p. 356**

    Head not prolonged into a beak 42

42. Mouthparts vestigial or composed chiefly of a short coiled tube, mandibles indistinct; body densely hairy or scaly, fig. 338B . . . . **Lepidoptera, p. 395**

    Mouthparts having sclerotized and massive mandibles, mouthparts of simple chewing type, fig. 36; body never hairy 43

SUBCLASS APTERYGOTA

Order DIPLURA: Campodeids and Japygids*

Wingless, blind, slender insects of small size, with long, many-segmented antennae, well-developed legs, and a pair of conspicuous cerci

* Some authors use the name Entotrophi for this order.

which are either segmented or forcepslike. Mouthparts of chewing type hidden within the ventral pouch of the head. Metamorphosis not marked.

The young and adults differ chiefly in size and sexual maturity. The genae of the head and the labium form a ventral pouch in which the other mouthparts are situated. The legs are not so well developed as in the Thysanura; the abdomen has vestigial paired appendages but no caudal filament. In the Campodeidae, fig. 162*A*, the abdomen has a pair of many-segmented cerci; in the Japygidae, fig. 162*B*, the cerci are forcepslike.

The species of the order occur under leaves, stones, logs, or debris, or in

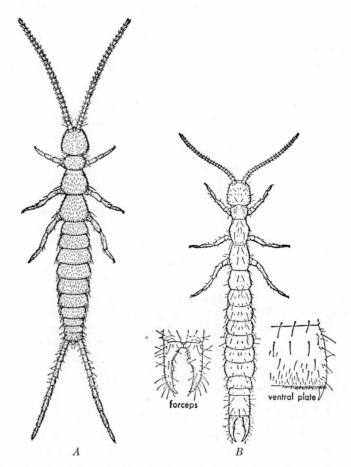

forceps

ventral plate

A   *B*

Fig. 162.   Diplura.   *A, Campodea folsomi; B, Japyx diversiunguis.*   (From Essig, *College entomology,* by permission of The Macmillan Co.)

the soil. Their movements are at most moderately rapid, and they seldom if ever come out into the light. Practically nothing is known about details of their life history; none of the species is of economic importance.

References

Smith, L. M., 1960. The family Projapygidae and Anajapygidae (Diplura) in North America. *Ann. Ent. Soc. Amer.*, **53**:575–583.
See also under Thysanura.

Order PROTURA: Proturans

The adults are small and slender, fig. 163, ranging from 0.5 to 2 mm. in length. The head is cone-shaped; it has no eyes; the antennae are reduced to minute, ocellus-like structures; and the mouthparts consist of stylet-like mandibles, small and generalized maxillae, and a poorly developed membranous labium that is fused at the sides with the cheeks. The three pairs of thoracic legs are similar in general appearance; the first pair serve as tactile organs.

The nymphs are similar to the adults in general appearance. In development they exhibit *anamorphosis*, that is, adding segments to the body at each molt. The abdomen of the first-stage nymph, the protonymph, has nine segments; the abdomen of the deutonymph has ten segments; that of the tritonymph has eleven segments; and, finally, the abdomen of the adult has twelve segments. The head and thorax are not affected in this manner.

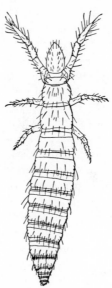

Fig. 163. A proturan, *Acerentulus barberi.*
(Redrawn from Ewing)

Proturans are moderately rare. They live in humus and soil, preferring damp situations. An ideal habitat for many species is old leaf mold along the edge of woods. Both adults and nymphs feed on decayed organic matter, and both may be found together during most of the year. Specimens may be collected either by examining leaf mold or by drying it in a Berlese funnel. Study specimens should be preserved in 70 per cent ethyl alcohol.

The order Protura has at times been considered as constituting a separate class, the Myrientomata.

Reference

Ewing, H. E., 1940. The Protura of North America. *Ann. Entomol. Soc. Amer.,* **33**:495–551, illus.

Order COLLEMBOLA: Springtails

Minute to medium-small wingless insects; with antennae and legs well developed; mouthparts of chewing type, but in some forms having the maxillae and mandibles long, sharp, and stylet-like; in the entire order the genae or cheeks have grown down and fused with the sides of the labium, forming a hollow cone into which the other mouthparts appear to be retracted. Abdomen frequently with a ventral jumping organ or furcula and a button-like structure, the tenaculum. Metamorphosis absent, both sexes usually similar.

The adults range in length from ⅕ mm. in the minute genus *Megalothorax* to over 10 mm. in the larger species of the Entomobryidae and Poduridae. In the suborder Arthropleona the body is long and cylindrical; in the suborder Symphypleona the abdomen is round and more or less globular, fig. 164. The antennae are four- to six-segmented, the last segment sometimes with many fine annulations. The eyes are either lacking or represented by a series of isolated ommatidia. Most members of the Collembola have a ventral springing organ or *furcula;* this is coupled with a ventral button or *tenaculum* when not in use. By means of the furcula these little animals can execute a leap of some distance, which has earned them their name of springtails. The young are similar to the adults in both appearance and habits, differing chiefly in size and sexual maturity. Many of the species are white or straw-colored, others are blue, gray, yellow, mottled, or marked with distinctive patterns.

Springtails are found abundantly in many types of moist situations, including deep leaf mold, damp soil, rotten wood, the edges of ponds or streams, and fleshy fungi. A few species attack plants, especially members of the family Sminthuridae, and may be of local economic impor-

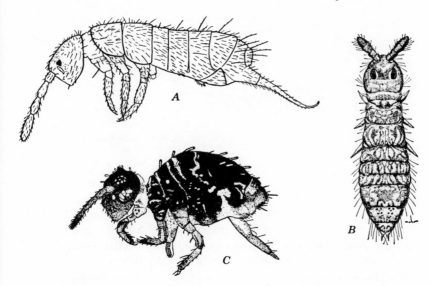

Fig. 164. Collembola. *A, Isotoma andrei, B, Achorutes armatus,* suborder Arthropleona; *C, Neosminthurus clavatus,* suborder *Symphypleona.* (*A, C,* after Mills, *B* from Illinois Nat. Hist. Survey)

tance. A small gray species, *Achorutes armatus,* fig. 164*B,* is sometimes destructive to mushrooms in commercial production. Egg-laying habits are known for only a few species, which lay their eggs singly or in clusters in humus or soil.

One of the most interesting features of the Collembola is their wide distribution. In the monograph by H. B. Mills, 132 species are listed from Iowa. Of these, 59 species, or 45 per cent, are holarctic or cosmopolitan.

KEY TO COMMON FAMILIES

1. Thorax and abdomen together comprising an almost globular or ovoid mass; all but anal segments of abdomen coalesced so that little external sign of segmentation remains, fig. 164*C* (suborder *Symphypleona*). . . . . **Sminthuridae**
   Thorax and abdomen tubular and more elongate, the segments of the abdomen indicated by external sutures, fig. 164*A, B* (suborder *Arthropleona*). 2
2. Dorsum of prothorax forming at least a semisclerotized plate similar in texture to that of mesothorax, fig. 164*B*. . . . . **Poduridae**
   Dorsum of prothorax completely membranous, in contrast to sclerotized or semisclerotized dorsal plate of mesothorax, fig. 164*A*. . . . . **Entomobryidae**

References

Maynard, E. A., 1951. *The Collembola of New York.* Ithaca, N. Y.: Comstock Publ. Co. 339 pp.
Mills, H. B., 1934. *A monograph of the Collembola of Iowa.* Ames, Iowa: Collegiate Press, Inc. 143 pp.

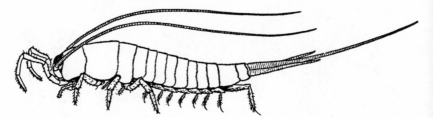

Fig. 165. Microcoryphia. *Machilis* sp. (Adapted from Snodgrass)

Order MICROCORYPHIA: Bristletails

Wingless, soft-bodied insects of small to medium size, with large eyes, long, multisegmented antennae, long cerci, and a long caudal filament, fig. 165. Mouthparts of chewing type. Abdomen with paired stylets on each segment, fig. 69. Comprises the family Machilidae.

The young and adults are extremely similar in shape and habits, differing chiefly in size and sexual maturity. The body is deep and tapers posteriorly. The abdomen has ten complete segments; the eleventh segment forms the caudal filament. The legs are moderately stout. All species are very swift runners and agile dodgers. A common American genus is *Mesomachilis,* which feeds on humus and is found among leaves and around stones. Little is known regarding their life history. It is thought that they lay eggs singly in cracks and crevices. The insects molt continuously throughout their life.

References

See under Thysanura.

Order THYSANURA: Silverfish, Firebrats

Similar in general to members of the previous order, differing in having a flatter, wider body, styliform appendages on only abdominal segments 7–9, and small or no eyes, fig. 166.

Most of the American species are domestic and belong to the family Lepismatidae. Like the Microcoryphia, they are extremely fast runners. They lay eggs singly in cracks, crevices, and secluded places. The young grow slowly, maturing in 3 to 24 months, and have a large and indefinite number of molts. Molting continues after adulthood is reached. The rare family Nicoletiidae contains a few small, ovate forms that live in ant nests and a few elongate subterranean species.

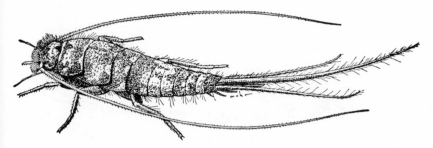

Fig. 166. Thysanura. *Thermobia domestica,* the firebrat. (From Illinois Nat. Hist. Survey)

Economic Status. Silverfish, *Lepisma saccharina,* firebrats, *Thermobia domestica,* and other domestic species feed commonly on starch. They cause considerable damage to books and clothing by chewing off the starch sizing; other articles containing glue or sizing are attacked.

References

Adams, J. A., 1933. Biological notes upon the firebrat, *Thermobia domestica* Packard. *J. N. Y. Entomol. Soc.,* **41:**557–562.
Escherich, K., 1904. Das System der Lepismatiden. *Dem Gesamtgebiete der Zoologie,* **43:**1–164.
Lubbock, J., 1873. *Monograph of the Collembola and Thysanura.* London: Royal Society. 276 pp.
Remington, C. L., 1954. The suprageneric classification of the order Thysanura. *Ann. Entomol. Soc. Amer.,* **47:**277–286.
Slabaugh, R. E., 1940. A new thysanuran, and a key to the domestic species of Lepismatidae . . . in the United States. *Entomol. News,* **51:**95–98.

SUBCLASS PTERYGOTA

Order EPHEMEROPTERA: Mayflies*

Small to large, soft-bodied, slender insects with gradual metamorphosis. Adults having two pairs of net-veined wings, the metathoracic pair small, completely atrophied in a few genera; legs usually well developed; antennae inconspicuous and hairlike; mouthparts vestigial; eyes large; and abdomen with a pair of cerci and in many species with a median terminal filament, all very long and tail-like, fig. 167. Nymphs aquatic, varied in shape, and similar in general structures to adults, but with well-developed chewing mouthparts; and usually with series of tracheal gills on the abdomen, fig. 168.

Metamorphosis in the mayflies is characterized by a feature unique

* The names Ephemerida and Plectoptera are sometimes used for this order.

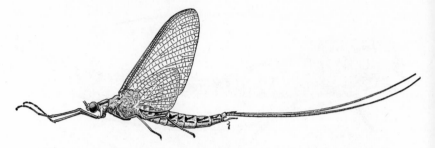

Fig. 167.   Ephemeroptera.   *Hexagenia limbata.*   (From Illinois Nat. Hist. Survey)

among insects.   The nymphs follow the usual type of gradual development with wings developing in external pads.   When full grown, they
swim to the surface of the water or crawl up on some support, and the
winged form escapes from the nymphal skin.   This winged form is capable
of flight and looks like an adult but in most species is not yet sexually

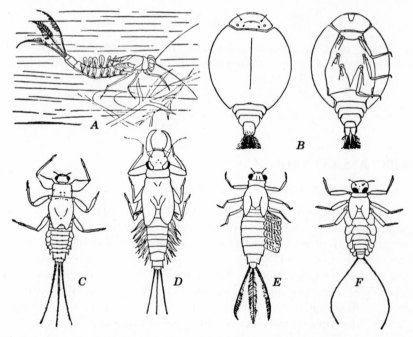

Fig. 168.   Nymphs of mayflies.   *A, Callibaetis fluctuans; B, Prosopistoma foliaceum,* dorsal and
ventral aspects; *C, Ephemerella grandis; D, Paraleptophlebia packii; E, Siphlonurus occidentalis; F, Iron
longimanus.*   (From Essig, after various sources, *College entomology,* by permission of The Macmillan Co.)

mature. The term subimago is applied to this stage. In a few genera, the subimago is the final form, the sexes mating and laying eggs in this stage. In most mayflies, however, the subimago molts again and produces the mature adult. The adults apparently take no solid food, probably imbibing only water during their short life. In certain genera, especially *Hexagenia* and *Ephemera,* mass emergence of adults may take place, resulting in the appearance of clouds of these insects over lakes and along streams. The adults mate in dancing swarms. In most genera, the female extrudes masses of eggs from the abdomen, swoops down to the water and releases the eggs into it. In the genus *Baetis,* the female crawls into the water and lays her eggs under stones. Each female may lay several hundred to several thousand eggs.

The complete winged life of many species of mayflies is extremely short, ranging from an hour and a half to a few days; mating normally occurs the same day adulthood is achieved, and the eggs are laid immediately. These eggs hatch in a few weeks or a month. In certain genera, such as *Callibaetis* and *Cloeon,* the adult females live much longer, from 2 to 3 weeks. In these longer-lived forms the eggs are fertilized and held in the body until the embryos are mature. When laid, the eggs hatch almost immediately on touching the water.

Nymphs, fig. 168, live in a great variety of lake, pond, and stream situations. The nymphs of some species mature in 6 weeks; others may require 1, 2, or 3 years to attain their full growth. Their food consists of microorganisms and fragments of plant tissue.

Nymphs of the Ephemeridae and its allies live in mud, burrowing through it by means of their large shovel-like front legs. Many nymphs of other families occur under stones and logs. Those which live in rapid mountain streams may have the entire venter of the body developed into a disclike suction cup which enables them to attach firmly to smooth surfaces. Nymphs of a few genera live in small pools or ponds and are free swimmers along the bottom or in the shallows.

Mayflies play an extremely important role in the fish-food economy of most North American waters. They are the most abundant insect group in many types of fishing waters. Studies of fish-stomach contents indicate that, by and large, mayflies and chironomids (midges) are undoubtedly the two most important insect groups from the standpoint of fish food.

References

Berner, L., 1950. The mayflies of Florida. *Univ. Florida Publ. Biol. Sci. Ser.,* 4(4):1–267.

Burks, B. D., 1953. The mayflies or Ephemeroptera of Illinois. *Illinois Nat. Hist. Survey Bull.* 26:1–216.

Day, W. C., 1956. Ephemeroptera. *Aquatic Insects of California,* 3:79–105. Berkeley, Univ. Calif. Press.

Edmunds, G. F., Jr., R. K. Allen, and W. L. Peters. An annotated key to the nymphs of the families and subfamilies of mayflies (Ephemeroptera). *Univ. Utah Biol. Ser.* 13(1):1–49.

Needham, J. G., J. R. Traver, and Yin-chi Hsu, 1935. *The biology of mayflies.* Ithaca, N. Y.: Comstock Publishing Co. 759 pp., 169 figs.

Order ODONATA: Dragonflies and Damselflies

Medium-sized to large predaceous insects with gradual metamorphosis. Adults slender or stout-bodied, with two pairs of nearly similar net-veined wings; legs well developed; antennae hairlike; mouthparts mandibulate, of the chewing type; eyes large; abdomen without long "tails." Nymphs aquatic; mouthparts of chewing type, with labium elongate and hinged to form a stout grasping organ for seizing prey; legs stout; three leaflike terminal gills present in the suborder Zygoptera.

All the adult Odonata feed on insect prey captured on the wing. They devour mosquitoes, midges, horseflies—in fact almost any insect that the odonate can tackle and catch successfully. The nymphs are aquatic, living chiefly in ponds, lakes, and backwaters of streams. They do not swim, but instead walk along the bottom or among debris or vegetation. The nymphs like the adults are predaceous, catching aquatic insects, crustaceans, and the like, trapping them with the extensile spined labium.

The eggs are laid in or near the water in a variety of ways. Some are thrust into aquatic vegetation or rotten wood; others may be deposited in masses on some object just beneath the water surface, or laid in ribbons or rings in the water, or thrust into wet mud near the water's edge. Females of many species dip down to the surface and wash the eggs off the end of the abdomen. Others crawl beneath the water to deposit eggs.

Nymphs of the smaller species mature in a year. In the case of larger species, development may take 2 to 4 years. Hibernation is passed in the nymphal stage. When full grown, the nymphs crawl out of the water and attach to a stick, stem, or other object for the last molt. The newly emerged adults harden and color relatively slowly, many of them requiring 1 or 2 days for the process.

A peculiar characteristic of the order is the method of mating, fig. 169. Before mating, the male bends the tip of the abdomen forward and transfers the spermatozoa to a bladder-like receptacle situated in the second abdominal sternite. In mating, the male, using its terminal claspers, grasps the female around the neck; the female then bends her abdomen forward to the second sternite of the male, at which place the actual transfer of spermatozoa is effected. This unusual procedure is known in no other order of insects.

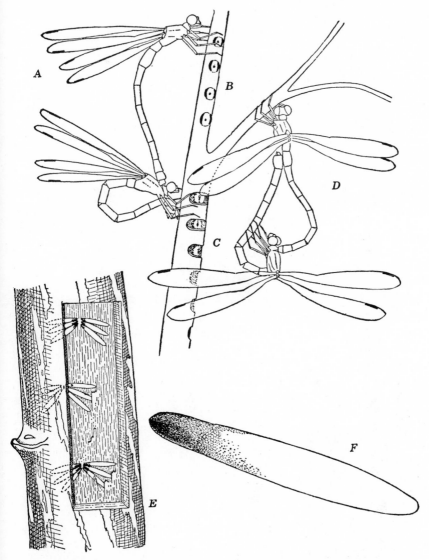

Fig. 169. Habits of the damselfly *Archilestes californica.* *A*, ovipositing; *B*, scars from oviposition, one year old; *C*, scars two years old; *D*, in copulation; *E*, bark cut away showing eggs in cambium; *F*, egg. (After Kennedy)

The Odonata includes three different types of insects which look and act strikingly different but which are separated by only a limited number of diagnostic characters. Present-day forms of one suborder, Anisozy-

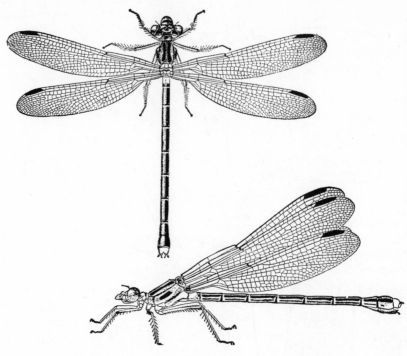

Fig. 170.   A damselfly *Archilestes californica*.   (After Kennedy)

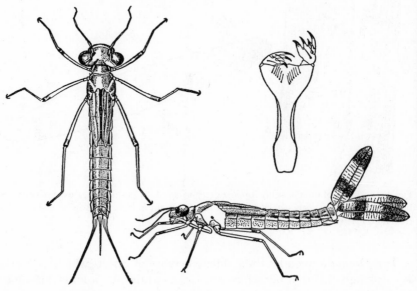

Fig. 171.   A damselfly nymph *Archilestes californica*.   Insert is labium showing grasping teeth. (After Kennedy)

goptera, are known to occur only as rarities in the oriental region. The two suborders which occur in North America may be separated by the following key.

DIAGNOSIS OF SUBORDERS

Nymphs provided with terminal leaflike tracheal gills, fig. 171. Adults having fore and hind wings of similar shape and venation, when at rest held together and extending parallel to the abdomen, fig. 170 . . . . .Zygoptera, damselflies

Nymphs without external gills, fig. 173. Adults with hind wings much wider than fore wings, especially at the base, extended outwards when at rest, fig. 172

Anisoptera, dragonflies

SUBORDER ZYGOPTERA

DAMSELFLIES. Damselflies are always slender and delicate, with a fluttering flight quite in contrast to the rapid and positive movements of the dragonflies. The damselfly adults have a thorax of very peculiar shape (fig. 170); the meso- and metathorax together are somewhat rectangular and tilted backward 70 to 80 degrees in relation to the linear axis of the entire body. The wings at rest are held together above the back at right angles to the upper margin of the meso- and metathorax. Because they are tilted to such a degree, the folded wings are nearly parallel to and held just above the abdomen.

Most of the adults are somber-hued, but a few have red or black banding on the wings or metallic green or bronze body and wings.

The nymphs (fig. 171) also are slender and possess three large caudal tracheal gills. They frequent the stems of aquatic vegetation more than the actual bottoms of ponds or streams.

SUBORDER ANISOPTERA

DRAGONFLIES. The adults of this suborder, fig. 172, are stout bodied, with strong, graceful, and superbly controlled flight. The thorax is not tilted as in the damselflies, and the wings at rest are extended to the side. Many species are gaudily colored and have conspicuous mottling or spotting on the wings. Older specimens frequently develop a pale-blue waxy bloom over the body and wings which may obscure the original colors and markings.

The adults, especially the larger ones, are great favorites with the out-of-doors enthusiast. Even if the dragonflies are not gaudily colored, their flight is of such speed and poise as to entrance the spectator. Each dragonfly has a regular beat. Up and down this it flies, patrolling the beat at regular intervals, and looking for flying insects as prey. When one of these is sighted, the dragonfly wheels from its course in pursuit of the prey; when the prey is captured, the dragonfly wheels back to its regular beat. Some-

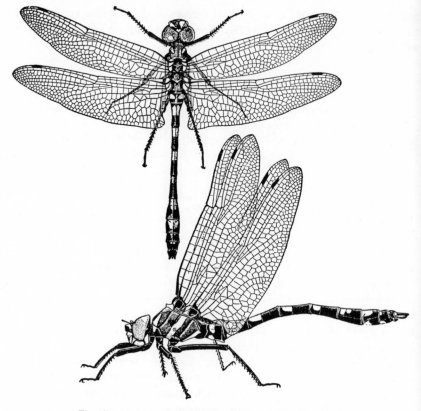

Fig. 172.   A dragonfly *Macromia magnifica*.   (After Kennedy)

times rivals clash, and there is displayed a real show of aerial acrobatics to the accompaniment of clicking of mandibles and rustle of wings.

The nymphs, fig. 173, are also stout, many of them frequenting the ooze or mud in the bottoms of ponds and lakes.   They have no external gills but have a rectal respiratory chamber (see p. 134) in which the gaseous exchange takes place.   Such a respiratory chamber is found in no other group of insects.

References

Borror, D. J., 1945.   A key to the New World genera of Libellulidae.   *Ann. Entomol. Soc. Amer.*, **38**:168–194.

Garman, Philip, 1927.   The Odonata or dragonflies of Connecticut.   Connecticut State *Geol. Nat. Hist. Survey, Bull.*, **39**:1–331.

Needham, J. G., and M. J. Westfall, Jr., 1955.   *A manual of the dragonflies of North America (Anisoptera)*.   Berkeley, University of California Press.   615 pp.

Smith, R. F., and A. E. Pritchard, 1956.   Odonata.   *Aquatic insects of California:* 106–153.   Berkeley, Univ. Calif. Press.

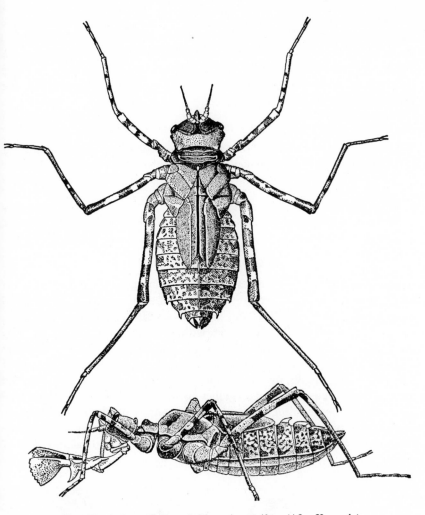

Fig. 173. A dragonfly nymph *Macromia magnifica*. (After Kennedy)

Walker, E. M., 1953–55. *The Odonata of Canada and Alaska.* Vol. 1 (Zygoptera), 2 (Anisoptera). Toronto, University of Toronto Press. 292 and 318 pp.

Order DICTYOPTERA:
Cockroaches, Mantids, and Walkingsticks*

In this order belongs a varied assemblage of insects, including wingless, short-winged, and long-winged forms. The winged forms have two pairs of net-veined wings: the fore wings, called *tegmina*, are leathery or parch-

* The name Cursoria is sometimes used for this order.

ment-like; the hind pair are membranous, larger, and folded beneath the tegmina in repose. The head bears antennae, eyes (rarely vestigial), and mouthparts of a generalized chewing type. Metamorphosis is gradual. Adults and nymphs are terrestrial. The hind legs are in about the same proportions as the middle legs, fitted for running. Classified in the Orthoptera by some authors.

KEY TO SUBORDERS AND COMMON FAMILIES

1. Front legs large, with series of strong teeth on opposing tibia and femur, fitted for grasping prey, fig. 176; pronotum elongate (Suborder *Mantodea*). . . .**Mantidae**
   Front legs similar to middle legs in general shape; pronotum wide, fig. 174, or body sticklike, fig. 178 2
2. Broad flat insects, fig. 174 (Suborder *Blattaria*). . . . .**Blattidae**
   Elongate, usually apterous insects having long slender legs, mimicking sticks, fig. 178 (Suborder *Phasmida*) . . . .**Phasmidae**

SUBORDER BLATTARIA

COCKROACHES. Cockroaches are rapid-running flattened insects, with long slender antennae, well-developed eyes, and chewing mouthparts having mandibles, maxillae, and labium very similar in type to figs. 44, 45, and 46, respectively. In species with well-developed wings, fig. 174, both pairs have many veins and a very large number of crossveins; the fore pair are narrower, thickened, and leathery or parchment-like, called *tegmina*, serving chiefly as a cover for the hind pair when not flying; the hind pair are thin, much larger, chief agent in flight, and are folded fan-like beneath the tegmina when not in use. Many species have only pad-like wings or no wings at all. The prothorax is large and conceals much of the head. The abdomen is large, many segmented, and bears a pair of apical cerci.

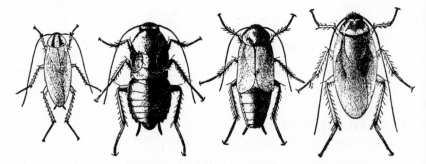

Fig. 174. Common cosmopolitan cockroaches. The German cockroach *Blattella germanica;* the Oriental cockroach *Blatta orientalis,* female and male; and the American cockroach *Periplaneta americana.* (From Connecticut Agricultural Experiment Station)

This suborder contains only one family, the Blattidae. In North America it is represented by about seventy species. More than two thousand species are known for the world.

Cockroaches frequent dark humid situations. Typically they belong to the tropics, where occurs a great variety of species, large and small. An extensive native fauna occurs in the southern portion of the United States, especially in humid regions. A limited number of outdoor species are found in the areas to the north, where they occur chiefly under bark of dead trees or fallen logs. The more conspicuous elements of the cockroach fauna of the northern states are not native to this continent but are a group of cosmopolitan species that are household pests. They find in human dwellings and heated buildings the semitropical conditions which enable them to thrive and multiply throughout the entire year. They are almost omnivorous in habit, eating a wide variety of animal and vegetable foods. The nymphs are similar to the adults in general structure and usually occur and feed along with the adults. In certain species in which wings are never developed, it is sometimes necessary to examine the genitalia to differentiate adults and nymphs.

The egg-laying habits of cockroaches are unusual. As the successive individual eggs are extruded from the oviduct they are grouped in an egg chamber and "glued" together by a secretion into a capsule or *ootheca*. These are definite in shape and sculpture for the species. The eggs in each usually number 15 to 40, arranged in symmetrical double rows, fig. 175. The ootheca is formed over a period of several days. In different species the hatching process is quite different. In some species the ootheca is deposited from a few to several days after it is formed, but long before hatching (e.g., *Periplaneta, Blatta*); in others, it is extruded but carried with its base firmly embedded in the brood chamber of the female until about hatching time (e.g., *Blattella*); and in others, the ootheca is carried within the brood chamber and the eggs may hatch there (e.g., *Pycnoscelus, Diploptera*). In the latter two categories the eggs obtain needed

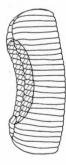

Fig. 175. Ootheca of *Blattella germanica*, with lateral portion cut away to show arrangement of individual eggs.

moisture from the mother's body. In at least the greenhouse roach *Pycnoscelus surinamensis* the species is parthenogenetic.

The nymphs are extremely active but grow relatively slowly. The smaller species may attain maturity in a few months, but the larger species may require a year or more. Many species are gregarious in habit, the adults and nymphs running together.

Of unusual interest is the wood roach *Cryptocercus punctulatus* found in the southeastern states. The species lives in colonies in decaying logs and has developed a close approach to true social life. The family unit forms the colony, several generations living together. The species feed on rotten wood. The first steps in the digestion of this cellulose material are accomplished by certain Protozoa, which are always abundant in the intestinal fauna of *Cryptocercus*.

Economic Importance. Cockroaches are one of the most disagreeable pests of human habitations. They get into many kinds of food, eat part of it, discolor and spot it with fecal material, and leave behind a disagreeable odor. In addition to the actual spoilage they cause, these scurrying insects are regarded as a general nuisance and a sign of unclean conditions. As a consequence the nation foots a large bill for the control of these insects in warehouses, eating places, and homes.

North of the frost line three cosmopolitan domestic species are most abundant, the small German cockroach *Blattella germanica*, the larger Oriental cockroach or "water bug" *Blatta orientalis*, and the American roach *Periplaneta americana*, the largest of the three and sometimes nearly as big as a small bat. In local areas the Australian cockroach *Periplaneta australasiae* is abundant; this is another cosmopolitan species as large as the American roach. To the south of the frost line other more tropical species invade buildings, some species attaining the size of a mouse.

SUBORDER MANTODEA

PRAYING MANTIDS. Predaceous insects of medium to large size, having an elongate prothorax and large spined grasping front legs, fig. 176. The middle and hind legs are usually slender. Otherwise the mantids are similar in general features to the cockroaches. Indeed, the structure of their mouthparts, internal organs, and genitalia indicates that the mantids are closely related to the cockroaches, in spite of the striking differences between the two in general appearance. The mantids comprise a single family, the Mantidae, represented in North America by only a few dozen species. As with the cockroaches the tropics support a larger fauna. Two species in the North American mantid fauna were introduced, *Mantis religiosa* from Europe and *Paratenodera sinensis* from the Orient. Both

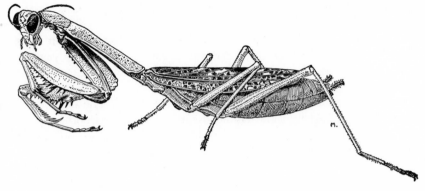

Fig. 176. A praying mantis *Stagmomantis carolina*. (From Illinois Nat. Hist. Survey)

species probably came into the United States as oothecae on nursery stock or packing.

The mantids may be long-winged, short-winged, or completely wingless. Many are green, brown, or mottled; a few species have brighter colors, and some have definite patterns.

All the species are predaceous in habit, feeding on other insects which they capture by means of the prehensile front legs. Cannibalism is not unusual, in fact, in certain species it is customary for the female to seize and devour the male after mating is completed.

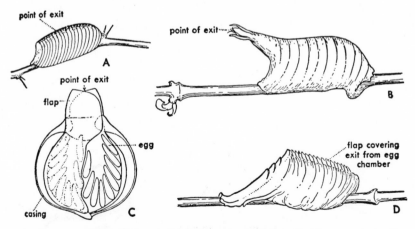

Fig. 177. Mantid oothecae or egg capsules. *A*, generalized type; *B*, *Oligonyx mexicanus;* *C*, sectional and *D*, exterior aspects of *Paratenodera sinensis*. (From Essig, *College entomology,* by permission of The Macmillan Co.)

Mantid eggs are deposited in large masses of definite pattern.   In these masses or ootheca the eggs are arranged in a series of rows, glued together with secretion, and the whole mass glued to a branch or other object, fig. 177.   In the northern areas there is single generation per year, and the winter is passed in the egg stage.   It is interesting to gather these ootheca in late winter or early spring and bring them into the laboratory, and see the young mantids emerge sometime later.   The eggs are frequently parasitized by some Hymenoptera which are as odd looking as the young mantids; the parasites normally emerge from the ootheca some time after the hatching date of the mantids.

SUBORDER PHASMIDA

WALKINGSTICKS.   Large sluggish insects which are either leaf mimics or stick mimics, fig. 178.   The North American species are wingless, except for a single Florida species, *Aplopus mayeri,* and belong to the family Phasmidae.   The resemblance of so many species to sticks has given the suborder the name "walkingsticks."   The head is round and has long slender antennae, small eyes, and simple chewing mouthparts.   The body and legs are long, sometimes thorny or extremely slender.   The smaller species may be only ½ inch long (over 12 mm.); the largest, the southeastern *Megaphasma dentricus,* attains a length of 6 inches (125 to 150 mm.). Some of the tropical species are broad and leaflike, with leaflike expansions on the leg segments.

All members of the phasmids are leaf feeders, most of them frequenting trees.   They sometimes are sufficiently abundant to defoliate large areas of woodland.   The insects themselves are never conspicuous.   Their sticklike appearance and green or brown coloring gives them almost perfect protection from observation without close scrutiny.   They move very slowly and feign death if disturbed.

Fig. 178.   A walkingstick, *Diapheromera femorata.*   (Original by C. O. Mohr)

The eggs are laid singly and simply dropped, falling to the ground. The winter is passed in this stage, the adults dying with the advent of cold weather. There is only one generation a year.

References

See under Orthoptera.

Order ISOPTERA: Termites, White Ants.

Termites are medium-sized insects having gradual metamorphosis, living in large colonies much like those of ants, and having several different social castes. A typical colony, for example, has three castes: sterile workers, sterile soldiers, and sexual forms (reproductives), fig. 179. The workers are white and sometimes appear to be translucent. They are wingless, and have round heads, long antennae, chewing mouthparts, and small eyes or none at all. The legs are well developed and all about equal in size. The soldiers have bodies similar to those of the workers, but their heads are enlarged and have massive mandibles. The reproductives are of two types: One type is white, wingless or with only short wing pads; the other type includes fully formed sclerotized winged males and females. These have round heads, long antennae, chewing mouthparts, well-developed eyes, and two pairs of similar transparent wings. After dispersal flights and mating, the wings fall off, each one leaving only a short stub or scale which persists for the life of the individual.

The termites in North America feed on cellulose, in almost all cases obtained from dead wood. The colonies, which may number several thousand individuals, are located in dead trees or logs or in the ground with covered runways connecting the nest to a log or stump which provides a food supply. The workers do all the foraging for the colony, feeding both the soldiers and reproductives. The soldiers afford protection from enemies from the outside, taking up strategic stations near the exits of the colony. The reproductives are the only fertile members of the colony and produce eggs almost continuously. The workers take care of the eggs until they hatch.

During most of the year only workers and soldiers are produced, but once a year, in spring or fall, a brood of winged males and females is produced by the more northern species. These are fully formed reproductive individuals called the *first reproductive caste*. They leave the nest in swarms, disperse, mate, and form new colonies.

A new colony is established by a single pair of winged individuals. The male and female lose their wings after the dispersal flight and, in our species, together eat out a small nest in a dead stump or log. They feed

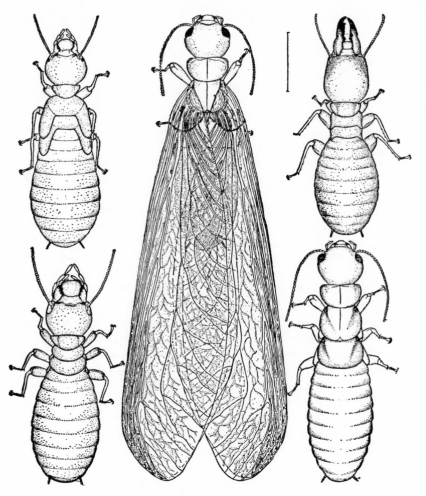

Fig. 179.   Castes in a colony of termites, order Isoptera.   Center, first form of winged reproductive; upper left, second form reproductive; upper right, soldier; lower left, worker; lower
right, first stage reproductive after the wings have broken off.   (From Duncan and Pickwell,
A world of insects, by permission of McGraw-Hill Book Co.)

as normal individuals, and the female produces eggs which develop into
workers and soldiers.   When a sufficient number of these have matured,
these neuter castes take over the activities of nest expansion and the feeding of both the female and male, called the *queen* and *king* of the colony.
If either of these die, their place is taken by the worker-like fertile forms
known as the *second reproductive caste*.   These are produced in small num-

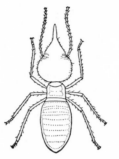

Fig. 180. Nasute of a termite.
(Adapted from Banks and Snyder)

bers in most colonies and appear to be held in reserve for substitution purposes.

In certain genera of North American termites there are no soldiers, but instead a caste called *nasutes,* fig. 180. These have a curious snout-like head; they produce a droplet of liquid with high deterrent quality and use it to repel enemies of the colony.

Economic Status. Every year termites cause a large loss to buildings and libraries. In their search for cellulose, several kinds of termites invade foundation woodwork of buildings and may spread from that point through the woodwork into upper parts of buildings. They may cross masonry or metal in their progress. Over these nonwood areas they build covered runways out of excrement, soil, and chewed wood and by this means always keep a contact with the soil from which they derive needed moisture. Books or wooden furniture may be attacked if these are stationary for long periods and in contact with wood. Freak cases of termite attack like the following are not uncommon. In an Urbana, Illinois, high school of concrete construction, an instructor was leaning against a corner of the desk. Suddenly there was a grinding of wood and the desk literally fell apart and crumbled to pieces. Out of the wreckage cascaded thousands of termites, which had eaten away the inner portions of the wood, never breaking through the wood surface but excavating the softer inner-ring fiber of every board. How the termites reached the desk appeared a mystery. Investigation finally showed that they had entered a piece of wood bracing embedded in the concrete foundation. This led from the ground to the next floor, which was of oak. The desk rested squarely on the oak floor. Along this unseen but direct route the termites had eaten their way, undetected until the desk was so thoroughly eaten away that a chance pressure caused it to break.

References

Banks, Nathan, and T. E. Snyder, 1920. A revision of the Nearctic termites. *U. S. Natl. Museum Bull.* **108**:1–226. 70 figs., 35 pls.

Emerson, A. E., 1933. A revision of the genera of fossil and recent Termopsinae (Isoptera). *Calif. Publ. Entomol.*, **6**:165–196, illus.

Snyder, T. E., 1935. *Our enemy, the termite.* Ithaca, N. Y.: Comstock Publishing Co. xiii + 196 pp., illus.

——— 1954. *Order Isoptera of the United States and Canada.* New York: Natl. Pest Control Assoc. 64 pp.

Order ORTHOPTERA: Grasshoppers and Their Allies

Medium-sized to large insects usually having the hind legs elongate, their femora enlarged for leaping, fig. 181. In almost all forms the pronotum is large and produced downward at the sides to form a large collar back of the head. The head is large, with long antennae, well-developed eyes, and chewing mouthparts of a simple type. In many species the wings are large and functional; the tegmina are invariably leathery, and the hind wings membranous, pleated fanwise in repose. Other species may be short-winged or completely wingless. To this order belong the grasshoppers, crickets, katydids, mole crickets, and pygmy locusts, altogether making up an array of forms varied in size, shape, color, and habits. Seven or eight families usually are recognized in the North American fauna, including several hundred species.

KEY TO COMMON FAMILIES

1. Front tibiae and tarsi enlarged for digging, the former having a group of large heavy sharp processes, the latter forming two or more heavy flanged knifelike processes, fig. 188 . . . . **Gryllotalpidae, p. 239**
   Front tibiae and tarsi lacking heavy black processes. 2

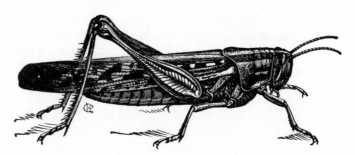

Fig. 181. The American grasshopper *Schistocerca americana americana*. (From Illinois Nat. Hist. Survey)

2. Hind tarsi minute or absent, but the tibial spurs forming large flat structures (used for jumping on mud), fig. 189 . . . . **Tridactylidae,** p. 239
   Hind tarsi well developed, projecting beyond tibial spurs, figs. 181, 186 3
3. Antennae much shorter than body and relatively heavy, fig. 181 4
   Antennae much longer than body, slender, fig. 184 5
4. Pronotum extending backward into a long shield covering all or nearly all of abdomen; tegmina short and ovate, fig. 182 . . . . **Tetrigidae,** p. 239
   Pronotum extending only over thorax, tegmina various but often extending beyond apex of abdomen, fig. 181 . . . . **Locustidae,** p. 237
5. Tarsi 4-segmented . . . . **Tettigoniidae,** p. 239
   Tarsi 3-segmented . . . . **Gryllidae,** p. 239

Locustidae. This family contains the grasshoppers and migratory locusts, fig. 181. The antennae are short, seldom half the length of the body, and, because of this characteristic, the family is often called the short-horned grasshoppers. Most of the group are grass or herb feeders, but a few feed on the foliage of trees. The eggs are deposited in masses in the soil. The female works the end of the abdomen down into the soil to form a chamber; into the end of this chamber she starts depositing eggs, and, as she gradually withdraws the abdomen from the chamber, more eggs are laid. When the chamber is filled with eggs, she secretes a weatherproof cap covering the opening and protecting the eggs from enemies and the elements.

Many members of the subfamily Oedipodinae have brightly banded hind wings of blue, red, pink, and black. In the field, males of many of these species attract attention by the crackling noise they make in flight.

To the Locustidae belongs the interesting and important group of grasshoppers known as migratory locusts. These are species which periodically develop populations of a size which staggers the imagination. Under these conditions the locusts soon completely denude the area in which they develop and after maturity migrate in huge swarms to other areas. These swarms may travel many hundred miles, eating all the foliage and visiting complete destruction on farm crops in their path. Every continent has its particular migratory species. In North America the most important are species of the *Melanoplus mexicanus* complex. In 1873 one of these, *Melanoplus spretus,* the Rocky Mountain grasshopper, swarmed from the Rocky Mountains eastward to about the Mississippi River.

Several other species of grasshoppers cause serious but less spectacular damage year after year. The most persistent are other species of the genus *Melanoplus,* including *femur-rubrum, bivittatus,* and *differentialis* and *Camnula pellucida.* Also local nonmigratory populations of *mexicanus* in the eastern part of its range cause some damage. All these species eat a

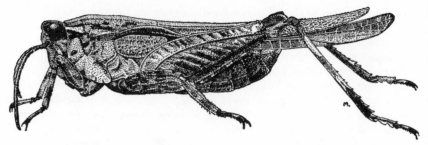

Fig. 182.   A pygmy grasshopper *Acridium ornatum.*   (From Illinois Nat. Hist. Survey)

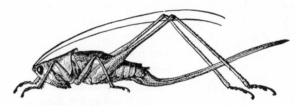

Fig. 183.   A female meadow grasshopper *Conocephalus strictus.*   (From Illinois Nat. Hist. Survey)

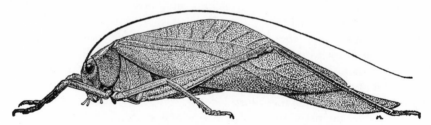

Fig. 184.   The bush katydid *Microcentrum rhombifolium.*   (From Illinois Nat. Hist. Survey)

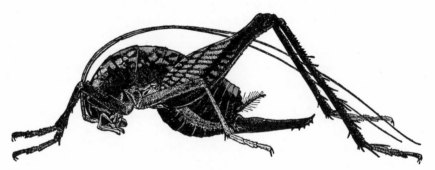

Fig. 185.   A camel or cave cricket *Ceuthophilus maculatus.*   (From Illinois Nat. Hist. Survey)

wide variety of crops, and the species occurring in the western states are extremely destructive to range land following overgrazing.

Tetrigidae. This contains the grouse locusts or pygmy grasshoppers. At first glance these appear similar to short-horned grasshoppers, but they differ from them in having the pronotum produced posteriorly into a long narrow shield extending over the entire length of the body, fig. 182. The North American species are few in number and all small, seldom more than 15 mm. long. They occur in a variety of situations, especially moist places near water. Certain species of the family display extraordinary variations in color pattern and have been employed in genetics research.

Tettigoniidae. Here belong the long-horned grasshoppers, those Saltatoria in which the antennae are long and slender, as long as or longer than the body, and the tarsi are four-segmented. The family is a large one, embracing the meadow grasshoppers, fig. 183; the cone-headed grasshoppers; the various types of katydids, fig. 184; and the cave or camel crickets, fig. 185. Best known are the katydids, which are large, usually green or pinkish insects with wide wings. The katydids produce a musical series of chirps and as insect musicians are as renowned as the crickets.

The most destructive member of the Tettigoniidae is *Anabrus simplex,* the Mormon cricket. This is a large wingless western species which often occurs in outbreak numbers in the Great Basin region of the Rocky Mountains and inflicts great damage on natural range and cultivated grain and grass crops.

Gryllidae. A varied assemblage of crickets comprise this family; these have long antennae, as have the Tettigoniidae, but the tarsi are three-segmented. A number of genera such as *Nemobius,* fig. 186, live in open fields or in woodland grasses. Other genera frequent shrubs or trees. One of these, *Oecanthus,* containing the tree crickets, has an awl-shaped ovipositor with which it drills holes into pithy stems and deposits its eggs in these holes, fig. 187. In local areas raspberry canes may be injured seriously in this manner by *Oecanthus* females.

The mole crickets represent two other families, the Gryllotalpidae and the Tridactylidae. The Gryllotalpidae, fig. 188, are about an inch (25 mm.) long and have large scooplike front legs used in digging. The species make burrows in fairly light soil and feed on small roots and insects which they encounter underground. The adults rarely emerge from their burrows and are seen only occasionally. The Tridactylidae, or pygmy mole crickets, are much smaller, at the most 5 mm. long, fig. 189. They occur at the edge of lakes and streams, where they may be found either burrowing in the sand or leaping about near the shore line.

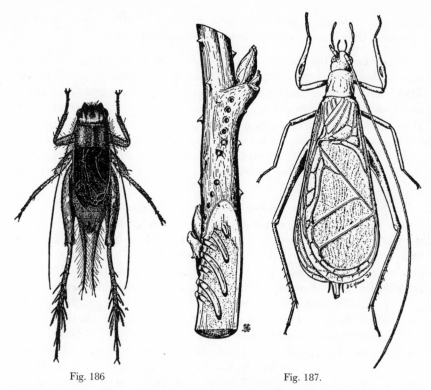

Fig. 186   Fig. 187.

Fig. 186.   A field cricket *Nemobius fasciatus*.   (From Illinois Nat. Hist. Survey)

Fig. 187.   The snowy tree cricket *Oecanthus niveus*.   Egg punctures and eggs exposed to view in a raspberry cane, and adult male.   The males are among the most fascinating insect musicians.   (From Essig, after Smith, *College entomology*, by permission of The Macmillan Co.)

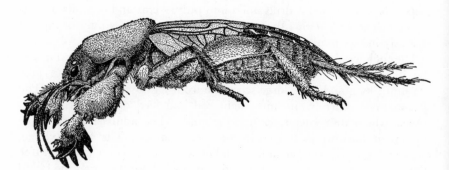

Fig. 188.   Mole cricket *Gryllotalpa hexadactyla*.   (From Illinois Nat. Hist. Survey)

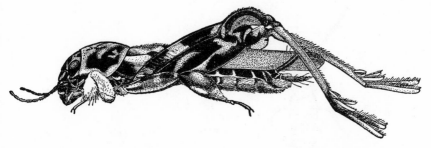

Fig. 189. Pygmy mole cricket *Tridactylus minutus.* (From Illinois Nat. Hist. Survey)

References

Ball, E. D., E. R. Tinkham, Robert Flock, and C. T. Vorhies, 1942. The grasshoppers and other Orthoptera of Arizona. *Univ. Ariz. Tech. Bull.* **93**:257–373.

Blatchley, W. S., 1920. *Orthoptera of northeastern America.* Indianapolis, Ind.: Nature Publishing Co. 784 pp. illus.

Gurney, A. B., 1951. Praying mantids of the United States. *Smithsonian Inst. Rept. for* 1950: 339–362.

Hebard, M., 1917. Notes on earwigs of North America, north of the Mexican boundary. *Entomol. News,* **28**:311–323, 5 figs.

——— 1917. The Blattidae of North America north of the Mexican border. *Mem. Am. Entomol. Soc.,* **2**:284 pp., 10 pls.

——— 1934. The Dermaptera and Orthoptera of Illinois. *Illinois Nat. Hist. Survey Bull.* **20**(3):125–279, 167 figs.

Morse, A. P., 1920. Manual of the Orthoptera of New England. *Boston Soc. Nat. Hist. Proc.,* **35**:197–556.

Order DERMAPTERA: Earwigs*

Medium-sized elongate heavily sclerotized insects in which the abdomen of American species has a pair of stout forceps, the modified cerci, fig. 190. The mouthparts are a simple chewing type; the compound eyes are large, the ocelli usually indistinct or lacking; the antennae are long, multi-segmented, and slender. Wings are sometimes lacking; if present, the first pair forms short usually truncate veinless hard wing covers, and the second pair is fan-shaped, with a peculiar radial venation, fig. 191. When not in flight, the second pair folds into a complicated compact mass almost entirely covered by the wing covers or elytra. Metamorphosis is gradual.

The earwigs in North America vary from about 5 to 15 mm. in length but are otherwise relatively uniform in shape and habits. They are nocturnal, roaming actively at night, and are omnivorous in food habits.

* Sometimes called the Euplexoptera.

Fig. 190.   An earwig *Labia minor*.
(From Illinois Nat. Hist. Survey)

Some species are apparently predaceous; others feed chiefly on decayed vegetation, or occasionally on living plant tissue.   During the day they hide in a wide variety of tight places—under bark and boards, in the soil, and in cracks and crevices of every sort.

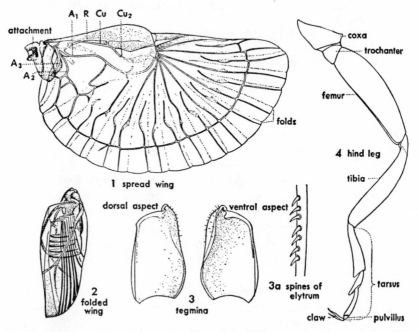

Fig. 191.   Structures of an adult earwig.   (From Essig, *College entomology*, by permission of The Macmillan Co.)

In temperate regions there is only one generation a year. The female lays a large cluster of white ovate eggs in a chamber in the ground in some protected spot. She watches over these for the few days required for hatching, and then extends her maternal care over the young for at least a short period, fig. 154. The young pass through four to six molts, maturing fairly rapidly.

The group is chiefly tropical, with a few representatives extending north into temperate areas. Less than twenty species occur in America north of Mexico, representing two families and several genera. The most widely distributed is the small *Labia minor* which is an introduced species, as are most of the nearctic earwigs.

Two curious families usually placed in this order occur only in the Old World—the viviparous Arixeniidae from southeastern Asia, associated with bats, and the Hemimeridae from Africa, ectoparasites on banana rats. This last family has 1-segmented cerci that are not forceps-like; it has often been considered a separate order, the Diploglossata.

Economic Status. In certain areas of North America the cosmopolitan European earwig *Forficula auricularia* has become a pest of great importance. It is especially abundant on the West Coast, where it is destructive to roses, dahlias, and other flowers, eating off the petals at the base and causing them to drop. Aside from this habit, it is chiefly a general feeder around the garden and home. Community poison-bait campaigns are often carried out in efforts to reduce its numbers.

References

See under Orthoptera.

Order GRYLLOBLATTODEA: Grylloblattids

This order is composed of one family, the Grylloblattidae, which contains only the single genus *Grylloblatta*. These are small wingless elongate insects, fig. 192, the head bearing long antennae, small eyes, and chewing mouthparts of generalized shape. The legs are slender, but well developed, and have five-segmented tarsi. The abdomen of the female bears at its apex a stout but primitive type of ovipositor and in both sexes a pair of eight- or nine-segmented cerci.

These are among the most interesting of all insects. In North America they have been found near the snow line of a few mountains in western Canada, California, Montana, and Washington. They live in soil or rotten wood, or under logs or stones, always in places which are covered with snow for much of the year. They feed on vegetation or dead organic matter. The females deposit black eggs in moss or soil.

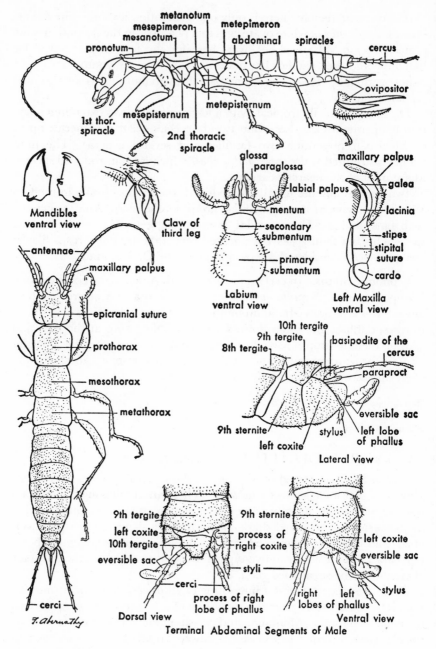

Fig. 192. *Grylloblatta campodeiformis*. (From Essig, *College entomology*, by permission of The Macmillan Co.)

Only a very few species of grylloblattids have been described. In addition to North America they have been found in Japan.

Reference

Gurney, A. B., 1937. Taxonomy and distribution of the Grylloblattidae. *Entomol. Soc. Wash. Proc.,* **50:**86–102.

Order EMBIOPTERA: Embiids, Web Spinners

Elongate flattened insects, fig. 193, with curious enlarged front tarsi, used for spinning silken webs in which the insects live. Mouthparts are the chewing type, primitive in structure; eyes well developed, ocelli absent; antennae many-segmented and elongate; legs short but stout, the tarsi three-segmented; cerci one- or two-segmented. Females always wingless; males usually with two pairs of long membranous similar wings with reduced venation. Metamorphosis gradual.

To this order belong a small number of peculiar tropical and semitropical insects, living in silken tunnels spun on their food supply. They feed on a wide variety of plant materials, especially dried grass leaves. Their tunnels may be found under loose bark, among lichens, or on the ground. The ground nets are often among matted leaves, or under dry cattle droppings or stones. Sometimes these nets are found around the bases of plants. In arid regions the insects may be active at the ground surface during the wet seasons and retire into the soil during the dry season. The embiids themselves are active and rapid in movement. The winged males fly readily and are frequently attracted to lights.

The web spinners live in large colonies, with numerous interlocking tunnels, and are gregarious. Most species have both males and females, but a few are parthenogenetic, and of these only females are known. The eggs are elongate and relatively large. They are laid in clusters attached to the walls of the tunnels. The female exhibits considerable maternal interest in both eggs and newly hatched nymphs, remaining near them and attempting to drive away enemies.

The nymphs are remarkably similar to the adults that are wingless. In those species with winged males, there is a noteworthy phenomenon. In the male nymphs the wing pads develop internally as imaginal buds until the penultimate molt and appear as typical wing pads only in the last nymphal instar. This is what hapens in holometabolous insects, so that this last embiid nymphal stage might well be called a pupa.

About seventy species of the order have been found in the Americas, representing seventeen genera and six families. Most of these occur in the tropical areas, but five species extend north into the southern portion

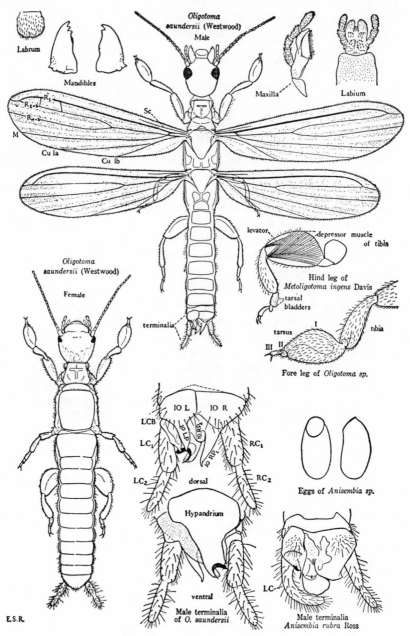

Fig. 193.   Structures and forms of Embioptera.   Letters on male terminalia refer to special parts used in embiid taxonomy.   (From Essig, *College entomology*, by permission of The Macmillan Co.)

of California, Arizona, Texas, and Florida. Two of these, *Oligotoma saundersii* and *nigra*, are tropicopolitan, and have been transported by commerce to most of the equatorial world. A few additional species are occasionally found by quarantine inspectors in shipments of material to the United States from other countries.

All species of the order are remarkably uniform in general appearance. In fact, to date few characters have been discovered to use for the identification of the females, and almost the entire classification of families, genera, and species is based on males.

References

Ross, E. S., 1940. A revision of the Embioptera of North America. *Ann. Entomol. Soc. Amer.,* **33**(4):629–676, 50 figs., bibl.
    1944. A revision of the Embioptera, or web-spinners of the New World. *USNM Proc.,* **94**(3175):401–504, 156 figs., 2 pls., bibl.

Order PLECOPTERA: Stoneflies

Moderate-sized to large insects with aquatic nymphs and gradual metamorphosis. The adults, fig. 194, have chewing mouthparts, frequently reduced in size and sclerotization; long many-segmented antennae; distinct eyes and ocelli; cerci ranging from short and one-segmented in some families to long and multisegmented in others. Two pairs of well-developed wings are almost always present. These are of similar texture and have only a moderate number of veins but frequently a large number of crossveins; the front pair is usually narrower than the hind pair. Several species have short wings, and in *Allocapnia vivipara* the males have no wings. The nymphs, fig. 194, of all stoneflies are aquatic. They have long antennae and a pair of long multisegmented cerci, chewing-type mouthparts, well-developed eyes and ocelli, and body proportions as in adults.

The nearctic stoneflies include about two hundred and fifty species, comprising about ten families and thirty-five genera. Their generalized mouthparts, antennae, and wings, together with their simple type of metamorphosis, indicate that the order is a primitive one allied to the orthopteroid orders.

The nymphs of this order are one of the abundant and interesting components of stream life. They range in body length from about 5 to over 20 mm. and present a varied appearance, including drab plain forms, spotted patterns, and forms striped with yellow, brown, or black. Many of them breathe by means of external finger-like gills. Sometimes the gills are filamentous. The gills are single in some and arranged in tufts

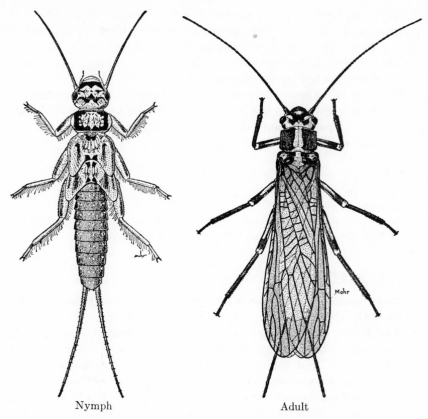

Nymph   Adult

Fig. 194. A stonefly *Isoperla confusa,* nymph and adult. (From Illinois Nat. Hist. Survey)

in others. Some nymphs have no external gills and simply use the cuticle for respiration. As a rule, the nymphs are found in cool unpolluted streams; a few species occur also along the wave-washed shore area of some of the colder lakes. The nymphs live in a variety of situations, frequently specific for the species. They are found under stones, in cracks of submerged logs, in masses of leaves that accumulate against stones or around branches trailing in the water, and in mats of debris. The majority of the nymphs are vegetarian, feeding on dead organic matter presumably incrusted with algae and diatoms. A number of species are predaceous, feeding on small insects and other aquatic invertebrates.

The females lay several hundred to several thousand eggs, discharging them in masses into the water. The eggs soon hatch. The smaller species and some large ones mature in 1 year, but other large species require 2 years to complete their development. When full grown, the nymphs crawl out of the water and take a firm hold on a stone, stick, tree trunk, or other object preparatory to the final molt. At molting a dorsal split occurs in the nymphal skin; then the adult emerges in about a minute or less. After another few minutes the wings have expanded and hardened enough for flight. The adults live for several weeks.

There is a peculiarity about certain groups of stoneflies which is only rarely encountered among insects. Winter signals the end of the active season and the beginning of the quiescent period for most insects. With many of the stoneflies the opposite is the case. Apparently the first-instar nymphs do not develop further during the warmer months of the year. With the approach of winter, nymphal development becomes accelerated, and the adults emerge during the coldest months of the year, beginning in late November or early December, and continuing through March. The adults are active on the warmer winter days and may be found crawling over stones and tree trunks, mating, and feeding on green algae. They show a decided preference for concrete bridges and may be collected in great numbers there. This group is called the fall and winter stoneflies and includes roughly the families Capniidae, Leuctridae, Nemouridae, and Taeniopterygidae. The latter three have members which appear later in the year, and their emergence overlaps that of the spring and summer species.

This peculiar growth behavior of the fall and winter stoneflies indicates a physiological adjustment to the warm and cold seasons quite different from that in most insects. When discovered, the controls and mechanisms for this adjustment will make an interesting story.

References

Claassen, P. W., 1931. Plecoptera nymphs of America north of Mexico. *Thomas Say Foundation, Entomol. Soc. Amer.,* 3:199 pp., 35 pls.

Frison, T. H., 1935. The stoneflies, or Plecoptera, of Illinois. *Illinois Nat. Hist. Survey Bull.* 20(4):281–471.

Frison, T. H., 1942. Studies of North American Plecoptera. *Illinois Nat. Hist. Survey Bull.* 22(2):235–355, 126 figs.

Jewett, S. G., Jr., 1956. *Plecoptera, aquatic insects of California.* Berkeley: Univ. Calif. Press. pp. 155–181.

Needham, J. H., and P. W. Claassen, 1925. A monograph of the Plecoptera or stoneflies of America north of Mexico. *Thomas Say Foundation, Entomol. Soc. Amer.,* 2:397 pp., 29 figs., 50 pls.

Ricker, W. E., 1943. Stoneflies of southwestern British Columbia. *Indiana Univ. Studies, Sci. Ser.,* 12:1–145.

Ricker, W. E., 1952. Systematic studies in Plecoptera. *Indiana Univ. Studies, Sci. Ser.,* 18:1–200.

Order ZORAPTERA: Zorapterans

Minute insects, 1.5 to 2.5 mm. long, with both winged and wingless adult forms, fig. 195. In both, the head is distinct and oval and has chewing mouthparts, long nine-segmented antennae, and one-segmented

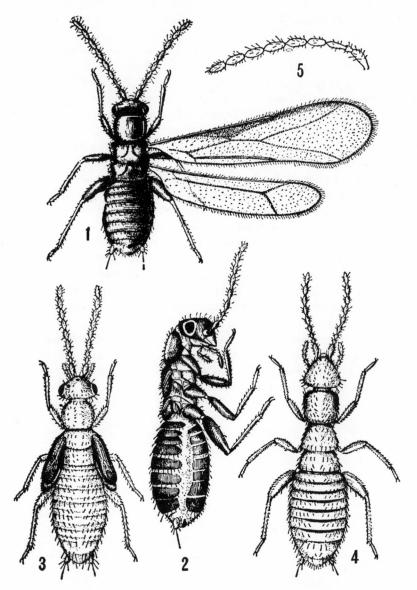

Fig. 195. Forms of *Zorotypus hubbardi*. 1, winged adult female; 2, adult female that has shed her wings; 3, nymph of winged form; 4, wingless adult female; 5, antenna of adult wingless *Zorotypus snyderi*. (After Caudell)

cerci. The wingless forms are blind, with only occasional vestiges of eyes or ocelli; the winged forms, or alates, have compound eyes and distinct ocelli. The alates have two pairs of delicate membranous wings, each with only one or two veins which may be branched. These wings are shed by the adults much as in the termites, leaving only small stubs attached to the body. Metamorphosis is gradual.

The order is one of the rarest among the insects. It contains one family, the Zorotypidae, in which there is only a single genus, *Zorotypus*. From the entire world less than twenty species are known, most of them found in the tropics. Two occur in North America, *Z. snyderi* described from Jamaica and Florida, and *Z. hubbardi* which has been collected in many localities in the southern states and as far north as Washington, D. C.

Zorapterans live in rotten wood or under dead bark and are usually found in colonies of a few to a hundred individuals. Their food, as far as is known, consists mainly of small arthropods, especially mites and small insects. Whether they are scavengers or predators has not been established, but observations on culture specimens indicate the former.

The wingless and winged adults have similar genitalia and reproductive habits. Eggs of only *Z. hubbardi* have been observed, laid without definite anchor lines or matrix in the runways of the colony. The creamy-colored oval eggs hatched in about 3 weeks. Collection observations over several years suggest that nymphs require several months to become adults.

Although the development of winged and wingless forms might indicate a forerunner of a caste system, no evidence of social life has been observed in the Zoraptera. There is apparently no division of labor, care of young, or social interrelationship between individuals. The gregarious nature of the colonies is very similar to the conditions found in many species of Psocoptera.

Reference

Gurney, A. B., 1940. A synopsis of the order Zoraptera, with notes on the biology of *Zorotypus hubbardi* Caudell. *Proc. Entomol. Soc. Wash.*, **40**(3):57–87, 56 figs., extensive bibl.

Order PSOCOPTERA: Psocids, Booklice*

Small insects, ranging in length from 1.5 to about 5 mm., with chewing mouthparts, long 13- to 50-segmented antennae, small prothorax, and no projecting cerci, fig. 196. Two pairs of wings are well developed in some forms, the front pair much larger than the hind pair, both of similar texture and with a reduced and simple venation. In other forms the wings may be small and scalelike or absent. Metamorphosis is gradual.

Most of the members of this order are either inconspicuously colored or exhibit marked protective coloration. For this reason they are seldom

* The names Corrodentia and Copeognatha are sometimes used for this order.

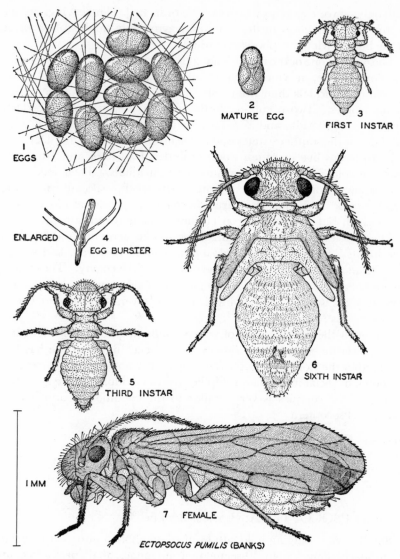

Fig. 196. A winged psocid *Ectopsocus pumilis,* and its life history stages. (After Sommerman)

collected by the beginning student, although they occur abundantly in many habitats. Their food is relatively nonspecific, consisting of fungus mycelium, lichens, dead plant tissue, and dead insects, even of their own species. They live in a wide variety of situations out-of-doors—on clumps of dead leaves, dried standing grass, dead or dying leaves of corn plants, bark of tree trunks, in the leaf cover on top of the ground, on shaded rock outcrops, under fence posts, and in bird and rodent nests. Several species live on moldy or partially moldy foods, bookbindings, and almost anything with available starch or fungus mycelium.

Some of the species are stocky and move slowly, even when disturbed. Many of them are more slender, and a few are quite flat. These usually move with considerable speed, and a few are among the most rapid dodgers to be found among the insects. Studies to date indicate that the entire life span from egg to death of the adult is between 30 and 60 days, of which about half is spent in the adult stage. The eggs are laid on the leaf surface or other spot which the adult frequents. Depending on the species, eggs are deposited singly or in groups up to about 10. After oviposition the female spins strands of silk over the eggs and anchors them to the surface of the support. In some species only a few strands are spun over the eggs; in others a dense web may be spun over each group of eggs. The eggs hatch in a few days, and the nymphs pass through six nymphal stages and become adults in 3 or 4 weeks.

In the more northern states the winter is passed in the egg stage by some species and as nymphs or adults by others. Species inhabiting warm buildings continue to breed throughout the year.

About one hundred and fifty species are known from North America, representing about twelve families and many genera. The group is world-wide in distribution, with an estimated number of species nearing nine hundred.

Economic Status. Several species of psocids cause considerable waste of food and damage to libraries. They consume only small quantities of foodstuffs, because they feed chiefly on mold. At times, however, they become extremely abundant, spread through an entire building, and get into every possible hiding place. In this way they may contaminate otherwise marketable goods to such an extent that quantities of the material must be discarded. Their damage to libraries is more direct. They eat the starch sizing in the bindings of books and along the edges of the pages, defacing titles and necessitating rebinding and repairs. The two most common species are the common booklouse, *Liposcelis divinatorius*, a minute wingless species, fig. 197; and *Trogium pulsatorium*, another small species having the wings reduced to small scales.

References

Chapman, P. J., 1930. Corrodentia of the United States of America: I. Suborder Isotecnomera. *N. Y. Entomol. Soc. J.*, **38**:219–290.
Sommerman, Kathryn M., 1944. Bionomics of *Amapsocus amabilis* (Walsh). *Ann. Entomol. Soc. Amer.*, **37**:359–364.

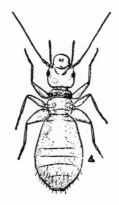

Fig. 197. A wingless booklouse *Liposcelis divinatorius*. (From U.S.D.A., E.R.B.)

Order PHTHIRAPTERA: Chewing and Sucking Lice

Small to medium-sized wingless insects, usually somewhat flattened, figs. 198, 201, that live as ectoparasites on birds and mammals. They have various types of mouthparts; only short three- to five-segmented antennae, sometimes hidden in a recess of the head; reduced or no compound eyes; and no ocelli. The thorax is small, the segments sometimes

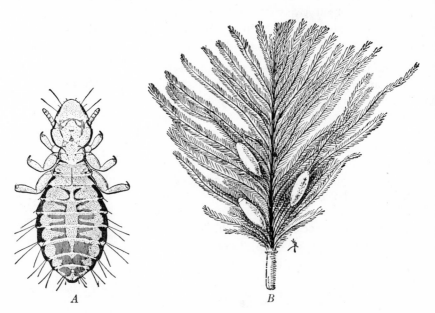

A   B

Fig. 198. The chicken head louse *Cuclotogaster heterographus*. *A*, adult; *B*, eggs on feather. (*A*, from Illinois Nat. Hist. Survey; *B*, from U.S.D.A., E.R.B.)

indistinct; the legs short but stout; and the abdomen has from five to eight distinct segments. Metamorphosis is gradual.

The Phthiraptera contain three distinctive suborders with an interesting evolutionary history. Members of the primitive suborder Mallophaga live on sloughed skin, dried blood at wounds, and other organic material on the body of the host. This suborder has a simple, reduced type of chewing mouthparts. One line of the Mallophaga apparently began to break the skin of the host and feed on exuding blood. Concurrently with this development the mouthparts became reduced to only the labium and mandibles, and an oesophageal pump evolved, used for sucking up this food. The only known representative of this stage is the elephant louse *Haematomyzus*. From a primitive member of this line arose a branch in which a wholly new set of piercing-sucking stylets, fig. 200, evolved in company with the blood-sucking habit. This branch developed into the suborder Anoplura, the sucking lice.

KEY TO SUBORDERS

1. Mandibles sclerotized, toothed, and functional, situated at end of a beaklike projection of the head or on ventral side of head, fig. 199; mouthparts not styliform 2

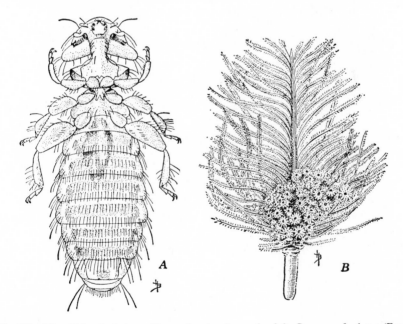

Fig. 199. The chicken body louse *Menacanthus stramineus*. *A*, adult; *B*, eggs on feather. (From U.S.D.A., E.R.B.)

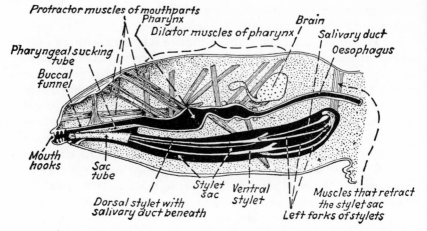

Fig. 200.   Head and mouthparts of *Pediculus humanus*.   (After Metcalf and Flint, *Destructive and useful insects*, by permission of McGraw-Hill Book Co.)

Mandibles apparently absent; mouthparts composed of long stylets retractable into a cavity in the head, fig. 200 . . . .**Anoplura**

2. Front of head produced into a narrow beak longer than the rest of the head.   Contains only *Haematomyzus elephantis*, occurring on Indian elephant

Rhyncophthirina

Front of head not produced into such a beak, figs. 201–203.   On many kinds of mammals and birds . . . .**Mallophaga**

SUBORDER MALLOPHAGA

CHEWING LICE.   The chewing lice average about 3 mm. in length, a few species attaining 10 mm.   They vary considerably in shape and habits, some being long and slender, others short and wide; some are active and rapid of movement, others sedentary and sluggish.   There is little correlation between speed and shape.   Their mouthparts are of the chewing type, but greatly reduced and difficult to interpret without careful study.

There are several hundred species of Mallophaga in North America, comprising about six families and many genera.   Each species occurs on only one species of host, or on a group of closely related species.   The turkey louse, for instance, occurs only on turkeys, but the large poultry louse occurs on many kinds of domestic fowl, such as chickens, turkeys, peacocks, guinea hens, and pigeons.   The small family Trichodectidae occurs only on mammals, and the large family Menoponidae occurs only on birds.

All the Mallophaga live entirely on the host body and have continuous and overlapping generations throughout the year. They feed on scaly skin, bits of feather, hair, clotted blood, and surface debris. The eggs are glued to the hair or feathers of the host and thus kept under incubator conditions. The eggs of various species differ in shape; some are long and simple, as in fig. 198*B*; others are ornamented with tufts of barbs or hair, as in fig. 199*B*.

KEY TO FAMILIES

1. Maxillary palps present; antennae arising from ventral portion of head and usually situated in grooves or cavities, fig. 199 (series **Amblycera**) 2
   Maxillary palps absent; antennae arising from or near lateral margin of head and not situated in grooves, fig. 198 (series **Ischnocera**) 5
2. Tarsus with 1 claw or none. On guinea pigs . . . . **Gyropidae**
   Tarsus with 2 claws. On birds 3
3. Entire head triangular in outline, fig. 199, the posterolateral areas posterior to the eyes considerably expanded laterally; antennae in grooves which are completely open laterally . . . . **Menoponidae**
   Head more elongate and the anterior portion comparatively wide; antennae in nearly circular cavities opening ventrally 4
4. Sides of head with conspicuous swellings anterior to eyes . . . . . . **Laemobothriidae**
   Sides of head straight or nearly so, without such swellings . . . . **Ricinidae**
5. Tarsus with a single claw. On mammals . . . . **Trichodectidae**
   Tarsus with 2 claws. On birds . . . . **Philopteridae**

Economic Status. Many species of Mallophaga infest domestic birds and animals and cause a considerable loss. Poultry are the most important group attacked. Chicken lice, *Menopon gallinae* and *Menacanthus stramineus*, fig. 199, cause loss of weight and reduction of egg laying in chickens, turkeys, and other fowl. The chicken head louse *Cuclotogaster heterographus*, fig. 198, occasionally occurs in outbreak form and causes the death of broods of young chicks. Several other species infest fowl, but the aforementioned, because of their reproductive capacity, are the most common and destructive.

Domestic animals are attacked by various species of the genus *Trichodectes*. Dogs, cats, horses, cattle, sheep, and goats may suffer considerable loss of condition if badly infested with these lice. There is evidence that the biting sheep louse *T. ovis* injures the base of the wool and causes commercial depreciation by lowering the staple length of the sheared product.

SUBORDER ANOPLURA

SUCKING LICE. The North American species represent about twenty genera and one hundred species ranging in length from 2 to 5 mm. All of them occur normally on mammalian hosts and feed on blood which is sucked through a tube formed by an eversible set of fine stylets, fig. 200. Occasionally poultry may have a small infestation of sucking lice, but all cases on record have been accidental colonizations by a common mammalian species. The entire life cycle is spent on the host. The eggs are glued to a hair and soon hatch into nymphs which are very similar to the adults in both appearance and habits. Breeding occurs continuously throughout the year.

KEY TO FAMILIES

1. Body with a dense covering of short, stout spines or of spines and scales; parasitic on marine mammals such as seals, sealions, and walrus. . . . .**Echinophthiriidae**
   Body chiefly with discrete rows of spines or hairs, figs. 201–203, never with scales; on land animals 2
2. Head with no trace of eyes, fig. 201 . . . . **Haematopinidae**
   Head with eyes or eye tubercles present, figs. 202, 203 3
3. Three pairs of spiracles, forming an oblique row on each side, on what appears to be the first abdominal segment (really the fused first three), fig. 202 (*Phthirius*)
   Phthiriidae
   Only one pair of spiracles on first abdominal segment (apparent as well as real), fig. 203 (*Pediculus*) . . . . **Pediculidae**

Economic Status. Sucking lice are a real concern on two counts: (1) losses inflicted on livestock, and (2) their menace to man.

LOSSES TO LIVESTOCK. Horses, cattle, sheep, goats, dogs, and cats are attacked by several species of lice. The loss is due partly to irritation and partly to loss of blood, with resultant poor condition of the animal and failure to gain weight normally. Frequently lice will cause sheep and goats to rub against fences or trees, with heavy damage to the wool. In the main, poorly kept animals are the principal individuals badly attacked, but this is not always the case. An outbreak allowed to go unchecked will usually spread through an entire herd. Most of the lice attacking domestic animals belong to the family Haematopinidae, of which *Haematopinus asini* the horse sucking louse, fig. 201, is a common example.

MENACE TO MAN. Two species of Anoplura are external parasites of man. They are both widespread in distribution and are most abundant under crowded insanitary conditions. They spread from person to person in crowded situations or by clothing and bedding.

CRAB LOUSE, *Phthirius pubis,* FIG. 202. A very small crablike species infesting hairy portions of the body, especially the pubic region. It is

Fig. 201.   The horse sucking louse *Haematopinus asini.*
(From Illinois Nat. Hist. Survey)

seldom found on the head.   This species inflicts painful bites and causes
severe irritation.   It has not been incriminated in the dispersal of any
disease.

BODY LOUSE OR COOTIE, *Pediculus humanus,* FIG. 203.   A larger louse about
4 or 5 mm. long, which occurs on the hairy parts of the body.   There are
two forms of this species, the head louse which occurs chiefly on the head
and glues its eggs to the head hairs, and the body louse which occurs
chiefly on the clothes and reaches to the adjacent body areas to feed.   The

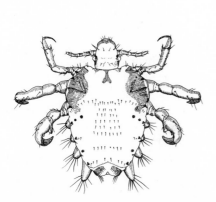

Fig. 202.   The crab louse *Phthirius pubis.*
(Redrawn from Ferris)

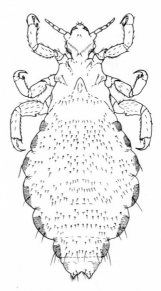

Fig. 203.   The cootie or body
louse *Pediculus humanus.* (Redrawn from Ferris)

body louse glues its eggs to strands of the clothing. Under condition of regular head washing and clothes change cooties are seldom a nuisance. Under insanitary conditions they may develop in tremendous numbers and produce constant irritation.

The cootie has vied with the mosquito in shaping the destiny of history. Cooties transmit typhus fever and trench fever, which until recent years have been the scourges of northern armies, especially in winter. Under insanitary crowded camp or trench conditions, soldiers with heavy clothing provided ideal hosts for cooties. Typhus and trench fever have occurred in outbreak form and with disastrous results throughout many European armies. Napoleon's army in Russia was decimated as much by louse-borne disease as by hunger and exposure. The opposing Russian army suffered fully as much from typhus as the French army. Many claim that the cooties won the campaign, defeating both armies.

In World War II the control of lice by treatment of entire city populations stopped outbreaks of typhus which had reached epidemic proportions. Especially effective results achieved in Naples in 1944 represent one of the most significant modern advances in the annals of preventive medicine.

References

Ewing, H. E., 1929. *A manual of external parasites.* Baltimore: Thomas. Pp. 127–157, figs. 71–85, bibl.
Ferris, G. F., 1934. A summary of the sucking lice (Anoplura). *Entomol. News,* **45:**70–74, 85–88.
1951. The sucking lice. *Pacific Coast Entomol. Soc., Mem.* **1:**1–320.
Riley, W. A., and O. A. Johannsen, 1938. *Medical entomology,* 2nd. ed. New York, McGraw-Hill Book Co. 483 pp.

Order THYSANOPTERA: Thrips*

Small elongate insects, fig. 205, most of them between 2 and 3 mm. long, with six- to nine-segmented antennae, large compound eyes, and compact mouthparts which form a lacerating-sucking cone. Many forms are wingless; others may have short wings or well-developed wings. In the latter case there are two pairs, both very long and narrow, with only one or two veins or none; the front pair often larger, and both with a long fringe of fine hair along at least the hind margin. The legs are stout, the tarsi ending in a blunt tip containing an eversible pad or bladder. Metamorphosis appears to be gradual, but there is present a definite pupa.

The thrips are an extremely interesting group, quite different from any

* In the European literature the name Physopoda is often used for this order.

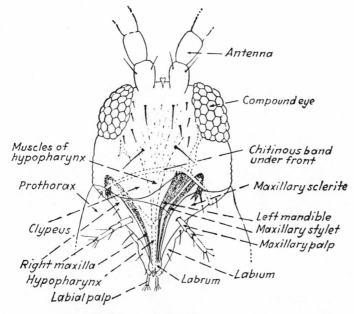

Fig. 204. Mouthparts of the flower thrips. (From Metcalf and Flint, *Destructive and useful insects*, by permission of McGraw-Hill Book Co.)

related forms. They occur commonly in flowers, and they may be found by breaking open almost any blossom and looking around the bases of the stamens or pistils. A large number of diverse forms can be taken by sweeping grasses or sedges in bloom. Many species are destructive to various plants and are found on the leaves of infested hosts. A large number of species, predaceous on mites and small insects, occur under bark of dead trees and in ground cover.

Thrips' mouthparts are of an unusual type, fig. 204. The various parts fit together to form a cone; some of the parts are needle-like stylets, which pierce and lacerate the food tissues; the juices thus released are sucked up into the stomach by a pump in the head capsule which pulls the liquid food through the cone formed by the mouthparts.

The metamorphosis of thrips is as unusual as their morphological features. The early nymphal stages are similar to the adults in structure of legs and mouthparts and in general shape; their feeding habits are also the same as those of the adults. These points are characteristic of insects having gradual metamorphosis. The first two instars have no wing pads, fig. 205,2; the pads appear suddenly in the third instar as fairly large structures, fig. 205,3; in the fourth (last) nymphal instar the

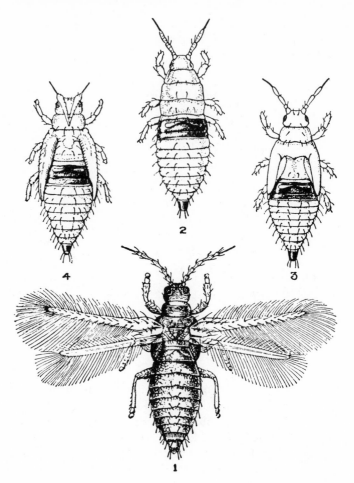

Fig. 205.   The red-banded thrips *Heliothrips rubrocinctus*.   1, adult; 2, nymph or larva; 3, propupa; 4, pupa.   (After U.S.D.A., E.R.B.)

wing pads are greatly enlarged, fig. 205,4.   This fourth instar is quite
unlike the others in habits.   It does no feeding and is completely quiescent, with the antennae held back over the top of the head and pronotum.
Certain thrips have an additional fifth nymphal instar.   In some species
having this, the fourth-instar nymph enters the soil and forms a cocoon,
in which the quiescent fifth stage is passed.   This feature is similar in
so many respects to holometabolous development that the quiescent
stage is called a *pupa*, and the first two instars are called larvae.   The
third-stage form, the active stage with the wing pads, is called a *propupa*.

In some groups this form is not developed, the larvae transforming directly to the quiescent pupae.

The order contains several families represented in North America by about five hundred species, some of them measuring 5 or 6 mm. in length. A large number of species occur in only one species of plant, but a few common species, such as the flower thrips, *Frankliniella tritici,* feed on a great variety of plants and frequent blossoms of almost any species of plant. A few species are predaceous, feeding on red spiders and other mites, and minute insects.

KEY TO COMMON FAMILIES

1. Major anal setae of last segment arising from a ring at apex of segment, which forms an undivided tube in both sexes, fig. 206*F* (suborder TUBULIFERA)

Phlaeothripidae

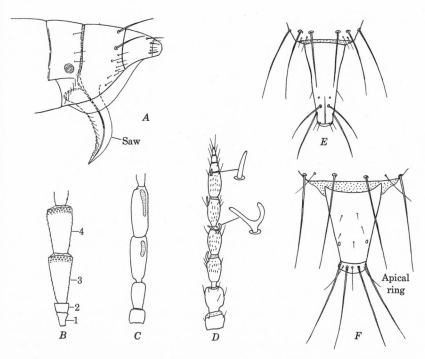

Fig. 206. Structures of Thysanoptera. *A*, apex of abdomen of *Anaphothrips,* Thripidae; *B*, base of antenna of *Heterothrips,* Heterothripidae; *C*, same of *Aeolothrips,* Aeolothripidae; *D*, antenna of *Anaphothrips,* Thripidae; *E*, dorsum of apex of abdomen of *Oxythrips,* Thripidae; *F*, same of *Allothrips,* Phlaeothripidae. (Redrawn from various sources)

Major anal setae of last segment arising from body of segment, fig. 206E; in female, last segment divided ventrally by saw slit, fig. 206A, in male, last segment sometimes tubular (suborder Terebrantia)....................... 2

2. Sensory areas on antennal segments 3 and 4 small and circular or blister-like, never forming conical or fingerlike protuberances, fig. 206B, C; antennae always 9-segmented 3

Sensory areas on segments 3 and 4 conelike or finger-like, projecting from segments, fig. 206D; antennae 8- or 9-segmented; female saw curved down at apex, fig. 206A .... **Thripidae**

3. Apex of third antennal segment having a complete band of small, circular sensoria, fig. 206B; female saw curved down at apex, as in fig. 206A

Heterothripidae

Apex of third antennal segment with one or two ovoid or elongate sensoria, fig. 206C; female saw curved up at apex.................... **Aeolothripidae**

Economic Status.   Several species of this order inflict considerable damage on commercial crops.   The following, all belonging to the family Thripidae, are among the most injurious thrips over the United States as a whole:

The onion or tobacco thrips *Thrips tabaci* is a widespread species varying from lemon yellow to dark brown, which is especially injurious to onions, beans, and tobacco.

The greenhouse thrips *Heliothrips haemorrhoidalis* is a dark species with the body ridged to give it a checked or reticulate surface.   The species is cosmopolitan.   In the temperate region it is chiefly a greenhouse pest, attacking many kinds of hothouse plants.

The pear thrips *Taeniothrips inconsequens* is brown with gray wings.   It attacks pears, plums, and related plants and produces a curious silvery blistered appearance on the injured leaves.

References

Priesner, H., 1949.   Genera Thysanopterorum.   *Soc. Fouad Ier Entomol. Bull.,* 33:31–157.
Watson, J. R., 1923.   Synopsis and catalogue of the Thysanoptera of North America.   *Univ. Florida Agr. Expt. Sta. Bull.,* 168:1–100, bibl.

Order HEMIPTERA: Bugs and Their Allies

A large assemblage of diverse insects, characterized chiefly by (1) piercing-sucking mouthparts that form a beak, (2) gradual metamorphosis, and (3) usually the possession of wings, fig. 207A.   With few exceptions the compound eyes are large, the antennae four- to ten-segmented, the individual segments frequently long, two pairs of wings are present and have relatively simple or reduced venation, and the abdomen has no cerci.   In many families the abdomen has a well-developed sheath and sawlike ovipositor much as in fig. 70.

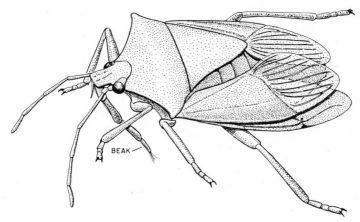

Fig. 207*A.*   A typical stink bug, illustrating beak and wings.   (From Illinois Nat. Hist. Survey)

The order derives its name from the structure of the front wing in many families, figs. 207*A* and 210, in which the basal portion is hard and thick, called the *corium,* and the apex is thinner and transparent, called the *membrane.*   The corium approaches in texture the hard wing cover or elytron of beetles; hence the name "hemelytron" is often applied to this half hard, half soft type of wing.

All but a few of the species comprising the Hemiptera can be placed in two large groups.   In one, the suborder Homoptera, the membranes of the front wings are typically entirely translucent, the beak arises from the posterior portion of the head, and the head has a typical tentorium. In the other, the suborder Heteroptera, the bases of the front wings are thickened and the apices membranous, the beak typically arises from the front end of the head, the cheeks fuse behind the beak to form a gula, and the head lacks a tentorium.   Comparing these two groups with other related orders of insects, it seems probable that the ancestral form of the entire order was a plant feeder and morphologically was much like some of the existing Homoptera, especially small members of the family Cicadidae with their abundant wing venation and primitive type of saw-like ovipositor.

The early homopterous lineage arising from this primeval hemipteran appears to have changed little more than in evolving a slender flagellum. In an early homopterous offshoot, represented by the family Fulgoridae, the antennae became situated beneath the eyes instead of between them. The continuing homopteran line gave rise to two main branches.   In one the pronotum became large and considerably overlapped the mesonotum; this branch includes the cicadas (Cicadidae), spittle bugs (Cercopidae), tree hoppers (Membracidae), and leafhoppers (Cicadellidae).   In the

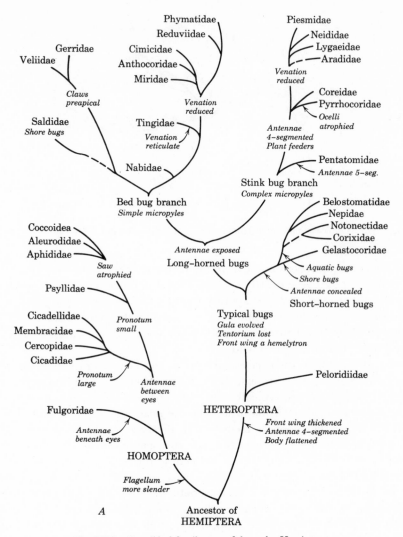

Fig. 207B. Simplified family tree of the order Hemiptera.

other branch the pronotum remained small but the flagellum thickened, and the beak became angled so that it appears to arise from between the front legs; the most primitive family of this branch, the Psyllidae, has a typical sawlike ovipositor, but this structure is lost in the more specialized portion of the branch that includes the aphids (Aphididae), whiteflies (Aleurodidae), and scale insects (Coccoidea).

The suborder Homoptera is still a plant-feeding group. Because of

this uniformity of feeding habits and the lack of morphological specialization, the suborder Homoptera as a whole might be considered little
specialized. In the aphids, whiteflies, and scale insects, however, great
biological specializations evolved, including holometabolous development, complex life histories involving as many as seven different stages or
morphs in the annual cycle of a single species, and seasonal alternation of
hosts. In the other large homopterous branch, the leafhoppers evolved
into such a multitude of species that their number is equaled by very few
other families.

In the heteropterous line, the beak became anterior and the cheeks
fused behind it, forming an extremely strong bridge or gula across the
back of the head below the occipital foramen. This suggests evolutionary
changes associated with a dietary change to a predacious habit. The base
of the forewings became thickened and the antennae became reduced to
four segments. This ancestral heteropteran gave rise to a large number of
families; several have become highly modified for aquatic or amphibious
life and at present these can be placed in the family tree only tentatively.

Some of the aquatic bugs may represent an early offshoot in which the
antennae became very short and ultimately the legs became wide, flat,
and hairy, used as swimming organs. The probable primitive members
of this branch are the toad bugs (Gelastocoridae); truly aquatic forms
include the families Nepidae, Belostomatidae, Corixidae, and Notonectidae. The inclusion of the last two families in this branch is tentative.

From the main stem form of the Heteroptera, two persisting major
branches evolved. In the bedbug branch, the egg micropyles remained
simple; in the stink bug branch, a reticulate subsurface network having a
respiratory function became differentiated in the walls of the raised
micropylar processes.

The family Nabidae, possessing both ocelli and an abundant front-
wing venation, may be little changed from the progenitor of the bed bug
branch. Other lines evidently arising from this progenitor are the
Miridae-Anthocoridae-Cimicidae lineage and the Reduviidae-Phymati-
dae lineage, both with reduced venation. The lace bugs (Tingidae) may
be an offshoot of this branch, as may also be the lineage including the
shore bugs (Saldidae) and water-striders (Veliidae and Gerridae).

The stink bug branch gave rise to two lines: (1) the stink bug lineage
having 5-segmented antennae (Pentatomidae) and (2) the coreid bug
lineage, composed entirely of plant feeders. Most primitive of this coreid
lineage are the families Coreidae and Pyrrhocoridae; families having a
reduction in venation include the Lygaeidae, Neididae, and Piesmidae.
The flat bugs (Aradidae) may possibly belong here.

If indeed the progenitor and stem forms of the Heteroptera were predaceous, then a return to plant feeding occurred independently in the Tingidae, Miridae, perhaps Pentatomidae, and in the progenitor of the entire coreid lineage. Aquatic forms also evolved independently at least twice; in one line the bugs became truly aquatic, evolving efficient paddles used for subsurface swimming; in the other the bugs became amphibious, evolving efficient tarsal structures used in running on the surface of the water.

Reviewing the evolution of the order Hemiptera, it is evident that it arose from an ancestor that combined the characters of the fulgorids, cicadas, and aphids. The line arising from this ancestor that evolved into the typical Heteroptera lost the tentorium, evolved an anteriorly situated beak and a gula behind it, lost all but four antennal segments, became at least somewhat flattened, and the base of the front wing became thickened. Because all of them possess all these specializations, the North American representatives of the Heteroptera give no clues to the order in which these changes occurred. A small family called the Peloridiidae, known from the forests of southern South America, New Zealand, and Australia is unique in being flattened, having somewhat hardened front wings, and having 3-segmented antennae, but having a typical tentorium and lacking a gula, both latter characters being typically homopterous. This combination of characters suggests strongly that the Peloridiidae may be a group of "living fossils" representing a very early offshoot of the heteropterous line. If this is correct, it means that in the heteropterous line the number of antennal segments became reduced, the body became flattened, and the front wings hardened before the beak became anterior, a gula evolved, and the tentorium atrophied.

KEY TO COMMON FAMILIES

1. Hind leg without tarsal claws and having both tarsus and pretarsus flattened and bearing a dense fringe of long hair down each side, fig. 209A; middle tarsus having normal tarsal claws (part of **Heteroptera**). 2
   Hind tarsus having tarsal claws similar to those on middle tarsus; hind leg usually without a long fringe but occasionally having one, fig. 209C. 3
2. Beak forming a triangular striated piece that appears as a ventral sclerite of the head, fig. 209G; middle tarsus having extremely long claws, fig. 209B; front tarsus comblike. . . . . **Corixidae, p. 282**
   Beak cylindrical and rodlike, curving back from the ventral portion of the head, as in fig. 209L; front and middle legs usual in shape. . . . **Notonectidae, p. 282**
3. Beak arising from posterior margin of head, fig. 208L; no gula present behind it (**Homoptera**). 4
   Beak arising from front or venter of head, fig. 209L; the venter of the head posterior to beak forming a sclerotized bridge or gula (**Heteroptera**). 15

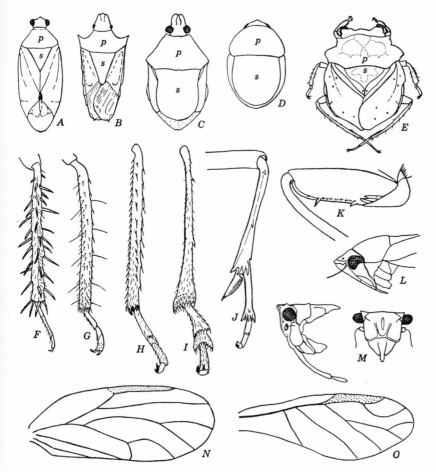

Fig. 208. Diagnostic characters of Hemiptera. *A*, outline of *Lygus*, Miridae; *B*, outline of *Solubia*, Pentatomidae; *C*, outline of *Stiretrus*, Pentatomidae; *D*, outline of *Corimelaena*, Pentatomidae; *E*, *Gelastocoris*, Gelastocoridae; *F*, hind tibia and tarsus of *Pangaeus*, Pentatomidae; *G*, hind tibia and tarsus of *Thyanta*, Pentatomidae; *H*, hind tibia and tarsus of *Aulacizes*, Cicadellidae; *I*, hind tibia and tarsus of *Aphrophora*, Cercopidae; *J*, hind tibia and tarsus of *Stenocranus*, Fulgoridae; *K*, front femur of *Pacarina*, Cicadidae; *L*, head of *Gypona*, Cicadellidae; *M*, head, lateral and anterior views, of *Poblicia*, Fulgoridae; *N*, wing of *Psylla*, Psyllidae; *O*, wing of *Aphis*, Aphididae.

4. Having wings, which are sometimes reduced to short scales 5
   Completely wingless species 13
5. Front femur greatly enlarged in comparison with middle femur, fig. 208*K*; three
   ocelli present . . . . **Cicadidae**, p. 275
   Front femur no larger than middle femur; three, two or no ocelli present 6

6. Antennae arising from sides of head, situated beneath or behind eyes, fig. 208*M*
   Fulgoridae, p. 275
   Antennae arising from front of head between eyes, fig. 208*L* 7
7. Pronotum enlarged dorsally into a large structure which covers most of head and body and may be highly ornamented with spines and processes, fig. 211
   Membracidae, p. 275
   Pronotum much smaller, without dorsal enlargement 8
8. Pronotum forming a broad shield which covers the greater part of the meso-notum, fig. 214; tarsus 3-segmented, fig. 208*H, I* 9
   Pronotum forming a narrow collar which does not extend back over the meso-notum, fig. 217; tarsus 1- or 2-segmented 10
9. Hind tibia bearing a double row of spines down its entire length, its apex usually not enlarged, fig. 208*H* . . . . **Cicadellidae**, p. 276
   Hind tibia with only scattered spines except at apex, which is enlarged and armed with a prominent crown of spines, fig. 208*I* . . . . . . **Cercopidae**, p. 275
10. Having only one pair of wings . . . . male **Coccoidea**, p. 279
    Having two pairs of wings 11
11. Wings milky-opaque, covered with a fine powdery white wax
    Aleurodidae, p. 279
    Wings transparent or patterned, not covered with a waxy secretion 12
12. Front wing with R_s very long, arising before stigma, and Cu branched, fig. 208*N*; abdomen never with cornicles . . . . **Psyllidae**, p. 279
    Front wing with R_s short, arising from some part of the stigma, and Cu un-branched, fig. 208*O*; abdomen in many species having a pair of lateral tubes or cornicles, fig. 217 . . . . **Aphidoidea**, p. 279
13. Eyes large, antennae situated at sides of head below or behind eyes, fig. 208*M*
    Fulgoridae, p. 275
    Either eyes rudimentary or absent, or antennae situated on front of head be-tween eyes, fig. 217 14
14. Tarsus 1-segmented; body covered with a hard shell, waxy secretions, or a detachable scale, fig. 218; abdomen never having cornicles
    Coccoidea, p. 279
    Tarsus 2-segmented; body at most with waxy secretions; abdomen often having a pair of conspicuous cornicles or tubes, fig. 217 . . . . **Aphidoidea**, p. 279
15. Antennae shorter than head, usually recessed in a concavity beneath the eyes or under the lateral margin of the head, as in fig. 209*G* 16
    Antennae at least as long as the head, usually extending free from it, fig. 223, sometimes fitting into a pronotal groove when at rest 18
16. Ocelli present. Small toadlike bugs, fig. 208*E*, found along the margins of lakes and streams . . . . **Gelastocoridae**
    Ocelli absent. Forms living in water, sometimes flying and attracted to lights
    17
17. Tarsi 1-segmented, front tarsus with only a minute claw or none, fig. 209*D*; apex of abdomen with a long or short respiratory tube, fig. 221*B*, each blade of which is concave mesally, the two fitting together to make a hollow tube; hind legs slender and without fringes of long hair . . . . **Nepidae**

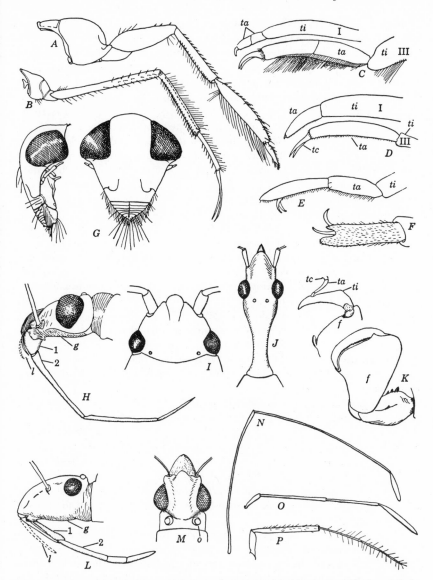

Fig. 209.   Diagnostic characters of Hemiptera.   *A*, and *B*, hind leg and middle leg of *Corixa,* Corixidae; *C*, front and hind tarsi of *Belostoma,* Belostomatidae; *D*, front and hind tarsi of *Nepa,* Nepidae; *E* and *F*, front tarsus, lateral and dorsal aspect of *Gerris,* Gerridae; *G*, head of *Corixa;* *H*, head of *Nabis,* Nabidae; *I*, head of *Lygaeus,* Lygaeidae; *J*, head of *Myodocha,* Lygaeidae; *K*, front leg, inset showing structure of tarsus of *Phymata,* Phymatidae; *L*, head of *Alydus,* Coreidae; *M* and *N*, head and antenna of *Jalysus,* Neididae; *O*, antenna of *Myodocha,* Lygaeidae; *P*, antenna of *Lyctocoris,* Anthocoridae.   *f*, femur; *g*, gula; *l*, labrum; *o*, ocellus; *ta*, tarsus; *tc*, tarsal claws; *ti*, tibia.

Tarsi 2-segmented, front tarsus with a stout curved claw, fig. 209C; apex of abdomen at most with a pair of short flat respiratory filaments; hind tibia and tarsus often flattened, always having fringes of long hair for swimming
....**Belostomatidae, p. 284**

18. Head extremely long and slender, slightly bulbous at apex, where beak arises, the eyes situated at the middle of what appears to be a long neck; rest of body also very slender, fig. 221C.............................**Hydrometridae**
Head much stouter, fig. 209I, or eyes not situated on the neck, fig. 209J..... 19

19. Front leg having femur and tibia chelate, forming a large grasping device, femur swollen and triangular, tibia curved and closing against the end of femur, fig. 209K...**Phymatidae**
Front leg not chelate... 20

20. Claws of front tarsus inserted before apex, fig. 209E, F.................. 21
Claws of front tarsus attached at apex, as in fig. 209C.................. 22

21. Middle pair of legs attached far from front legs, close to hind legs; hind femur very long, fig. 223, beak 4-segmented.................**Gerridae, p. 284**
Middle pair of legs attached about midway between front and hind legs; hind femur only moderately long; beak 3-segmented.................**Veliidae**

22. Scutellum very large, reaching about one-half or more distance from posterior margin of pronotum to end of folded wings, fig. 208B–D; antennae usually 5-segmented. (**Pentatomidae, p. 289**) 23
Scutellum much smaller, reaching about a quarter of the distance from pronotum to tip of body, fig. 208A; antennae usually 4-segmented.......... 26

23. Tibia armed with rows of thick thornlike spines, fig. 208F................ 24
Tibia having series of short even spines, occasionally with a few scattered, very slender hairs, fig. 208G... 25

24. Scutellum triangular and not very large, as in fig. 208B............**Cydninae**
Scutellum large and U-shaped, covering most of abdomen, fig. 208D
Thyreocorinae

25. Scutellum U-shaped and very wide, covering almost all of abdomen, the sides of scutellum curved mesad at extreme base, as in fig. 208D.......**Scutellerinae**
Scutellum V-shaped, fig. 208B, or, if U-shaped, then never larger than in fig. 208C, and slightly contracted just beyond base..........**Pentatominae**

26. Front wing abbreviated, with no membrane, and reaching at most to middle of abdomen, fig. 226 27
Front wing normal, with a large apical membrane, or reaching well beyond middle of abdomen, fig. 210................................... 28

27. Body flat and wide; front wing short, broad, and scalelike, only barely reaching over base of abdomen; sides of pronotum large, round, and flangelike, fig. 226; beak 3-segmented; antennae long and slender......... **Cimicidae, p. 287**
Body narrower, or otherwise different from foregoing, having either a 4-segmented beak, different-shaped wing, or short antennae. A few genera, most of them rare, difficult to key to family, belonging to **Anthocoridae, Miridae, Aradidae, Lygaeidae,** or **Nabidae;** and all nymphs of Heteroptera families listed beyond this point. The wingless species of this group are keyed no farther here.

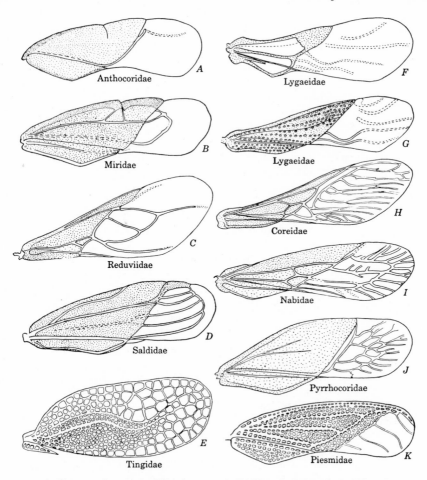

Fig. 210. Elytra or forewings of Hemiptera. *A, Triphleps; B, Lygus; C, Pselliopus; D, Salda; E, Gargaphia; F, Blissus; G, Myodocha; H, Alydus; I, Nabis; J, Euryophthalmus; K, Piesma.*

28. Hemelytra large, covering entire abdomen and reticulate over their entire surface with a net-like pattern, with little or no distinction between corium and membrane, fig. 210*E*.............................. **Tingidae,** p. 285
Hemelytra with a definite apical membrane, fig. 210, all except *E*........ 29
29. Membrane of hemelytron having one or two large basal cells, and none, one, or two short spurlike veins extending distally from these, fig. 210*B, C*...... 30
Membrane of hemelytron having either no closed cells, fig. 210*A*, or at least five or six veins (including the costa) running through the membrane, fig. 210*D, H, I, J*... 31
30. Ocelli prominent, two in number; membrane of hemelytron having a long vein proceeding from top of upper closed cell, fig. 210*C*...... **Reduviidae,** p. 287

A TEXTBOOK OF ENTOMOLOGY

Ocelli absent; membrane with a vein proceeding only from bottom of lower closed cell, or such a vein lacking, fig. 210*B*. . . . . **Miridae,** p. 286
31. Antenna having first two segments stout, last two threadlike, forming a slender terminal filament, fig. 209*P*; ocelli present but small; hemelytral membrane with only one or two weak veins, fig. 210*A*. . . . . **Anthocoridae**
Antenna having one or both of the two apical segments as thick as the first or second, fig. 209*N*, *O* 32
32. Hemelytral corium extending markedly beyond a ridgelike oblique vein near apex of corium, fig. 210*K*; corium entirely reticulate . . . . **Piesmidae**
Hemelytral corium not extending beyond an apical oblique vein, fig. 210*H-J*, or not having such a vein, fig. 210*G*. 33
33. No ocelli present 34
Two ocelli present 35
34. Flat wide warty bugs, fig. 224*A*, *B*; tarsus 2-segmented, the first segment short; hemelytra often small, the periphery of the abdomen extending considerably beyond them . . . . **Aradidae,** p. 288
Stout insects, the body deep; tarsus 3-segmented, the first segment long; hemelytra larger, fig. 210*J*, covering all abdomen except tip and sides near apex
Pyrrhocoridae
35. Hemelytral membrane having four or five large and fairly regular closed cells and no other venation, fig. 210*D*; oval fairly flat bugs found on stream and lake shores . . . . **Saldidae**
Hemelytral membrane having either an irregular network of cells or only one or two small, well-sclerotized ones, fig. 210*F-I*. 36
36. Membrane having a series of about 15 irregular veins, at least on apical portion, fig. 210*H*, *I* 37
Membrane having only five or six veins across it, fig. 210*F*, *G*. 38
37. First segment of beak short and conelike, thicker than the second, fig. 209*H*; front femur thickened, front tibia armed inside with a double row of short black teeth . . . . **Nabidae**
First segment of beak cylindrical and long, similar in general shape to second segment, fig. 209*L*; front femur usually more slender, front tibia never with inner rows of black teeth . . . . **Coreidae,** p. 288
38. Each ocellus situated behind an eye, at the base of a distinct swelling, fig. 209*M*; extremely slender and elongate bugs, with long and slender legs and antennae; last segment of antenna short and oval, forming a small club, fig. 209*N*
Neididae
Ocelli situated closer to or between eyes, and not at the base of a swelling, fig. 209*I*, *J*; chiefly robust short insects or having short legs; antennae either short, fig. 209*O*, or not clubbed . . . . **Lygaeidae,** p. 288

SUBORDER HOMOPTERA

CICADAS, LEAFHOPPERS, APHIDS, SCALE INSECTS, ETC. This suborder contains the cicadas, leafhoppers, aphids, scale insects, and their allies, all plant feeders. The North American fauna is composed of about a dozen families or superfamilies.

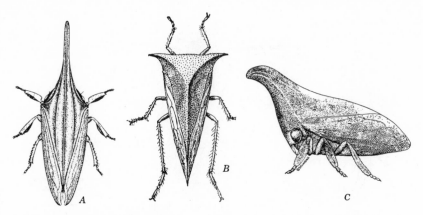

Fig. 211.   Treehoppers, Membracidae.   *A, Campylenchia latipes; B, Ceresa bubalis; C, Enchenopa binotata.*   (*A, C,* from Kansas State College; *B* from U.S.D.A.)

Fig. 212.   A fulgorid *Peregrinus maidis.*   (After Thomas)

The Fulgorid-Cicada Series.   In this series the flagellum of the antenna is needle-like and the tarsus is almost invariably 3-segmented.   In the series are found many forms of bizarre appearance.   Some of the Membracidae, or treehoppers, fig. 211, have the pronotum greatly enlarged and ornamented with ridges, horns, or prongs.   The Fulgoridae are a large family, and many resemble leafhoppers, fig. 212.   Some of the fulgorids have large foliaceous wings, and others, such as our native *Scolops* and the South American lantern fly, or peanut bug, *Lanternaria phosphorea,* have bizarre projections of the head.   Another oddity is the spittle bug family, Cercopidae.   The nymphs of this family produce masses of white froth or spittle-like substance and live hidden beneath it.   Two well-known and abundant families of the series are the Cicadidae (cicadas) and the Cicadellidae (leafhoppers).

CICADIDAE, CICADAS.   These are large insects, many North American species measuring 2 inches or more.   They are distinguished structurally from related families by having three distinct ocelli on the dorsum of the head.   The males have highly developed musical organs, and during warm days and summer evenings they make a shrill noise.   The nymphs

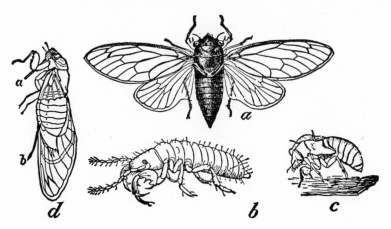

Fig. 213.　The periodical cicada *Magicicada septendecim*.　*a* and *d*, adults; *b*, nymph; *c*, shed nymphal skin.　(From U.S.D.A., E.R.B.)

have enlarged front legs, presumably for digging, and are subterranean, feeding on sap from the roots of deciduous trees.

The nymphal period is long, 2 to 5 years for most species.　The periodic cicadas *Magicicada septendecim* and its immediate relatives, fig. 213, constitute a group of southern species with a nymphal life of 13 years and a group of northern species with a nymphal life of 17 years.　These species have attracted widespread attention because of the periodic nature of their cycles.　In some areas only a single brood occurs, and there the adults appear only every 13 or 17 years.　On these occasions they usually appear in huge swarms, and the ovipositing females may cause serious damage to the twigs and branches of fruit and hardwood trees.

CICADELLIDAE, LEAFHOPPERS.　This family is the largest in the entire order Hemiptera, represented in North America by over twenty-five hundred species.　Leafhoppers are not only numerous in species but also extremely abundant in numbers of individuals.　They are probably collected in general sweeping more commonly than any other insect group.　Most of these are less than 10 mm. long and have long hind tibiae bearing longitudinal rows of spines, but with neither large spurs nor a crown of spines at the tip.　Although a few species are broad or angular, most are slender and nearly parallel-sided, fig. 214.　Female leafhoppers have strong ovipositors which they use to cut slits for eggs in plant stalks (usually herbs) or leaves.　Leafhoppers are often destructive to certain crops, not only by direct damage caused by feeding, but also because they transmit many plant diseases.　The beet leafhopper *Circulifer tenellus* transmits the virus which causes curly top of beets, a most destructive disease

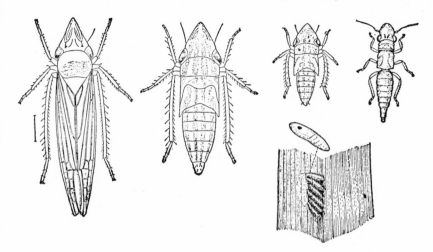

Fig. 214. A leafhopper *Draeculacephala mollipes,* adult, nymphs, and eggs. (From U.S.D.A., E.R.B.)

to the sugar-beet crop; and the plum leafhopper *Macropsis trimaculatus* transmits another destructive virus which causes peach yellows.

The Psyllid-Aphid Series. In this series the antennae are either short and stout or long and threadlike, fig. 217, in at least some stage of the life

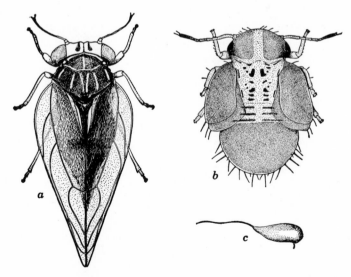

Fig. 215. The pear psylla *Psylla pyricola.* *a,* adult; *b,* nymph; *c,* egg. (From Connecticut Agricultural Experiment Station)

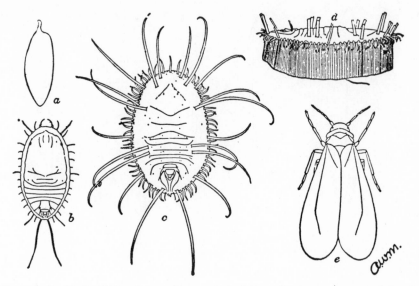

Fig. 216.   A whitefly *Trialeurodes vaporariorum.   a*, egg; *b*, larva, first instar; *c*, puparium, dorsal view; *d*, puparium, lateral view; *e*, adult.   (After Morrill)

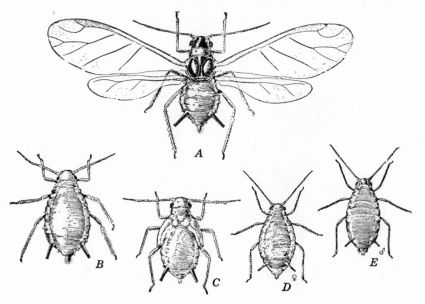

Fig. 217.   The apple aphid *Aphis pomi.   A*, alate viviparous female; *B*, apterous viviparous female; *C*, nymph of alate; *D*, oviparous female; *E*, male.   (From U.S.D.A., E.R.B.)

cycle. The wing venation is greatly reduced, and the tarsi have only one or two segments. Because of the occurrence of diverse body forms within the life cycle of a single species, it is difficult to characterize the families with a brief description.

Two families, the jumping plant lice or Psyllidae, fig. 215, and the whiteflies or Aleurodidae, fig. 216, have a simple life cycle in which adults of both sexes are winged and similar in general appearance. In many Psyllidae and all Aleurodidae the later nymphal instars are flat, inactive or sluggish, and scalelike in appearance. The members of both families are small.

All the other families of the psyllid-aphid series are segregated into two large groups: (1) the aphids and their allies, the superfamily Aphidoidea, and (2) the mealybugs and scale insects, the superfamily Coccoidea. Each group contains several families differentiated chiefly by biological characteristics and including many species of great economic importance.

THE APHIDOIDEA, fig. 217, are characterized by (1) the presence of several veins and a stigmal area in the fore wings of the winged forms; (2) the existence of two-segmented tarsi in most species; and (3) the existence of a complex system of alternating generations including wingless, winged, parthenogenetic, and sexual forms in the life cycle of a single species. This phase is discussed more fully in Chapter 6. The Aphididae, or plant lice, is the most important family in the group. Many species of great economic concern are members of this family, for example, the melon aphid *Aphis gossypii*, a pest of cucurbits and cotton; and the green peach aphid *Myzus persicae*, a pest of many crops and the disseminator of many plant diseases.

THE COCCOIDEA, fig. 218, differ in several important respects from the aphids: (1) The females are always wingless, extremely sluggish or completely fixed in position, and are covered by a waxy secretion or a tough scale, or have a hard integument as in the family Coccidae; (2) the males are small and delicate and have a single pair of wings with only one or two simple veins; and (3) the life cycle is relatively simple.

The family Diaspididae, fig. 218, is one of the most important in the scale insect group. The females are the sedentary, small, scalelike or cushion-like insects found on many species of trees. The actual insect is a delicate oval body hidden beneath the scale, which is a protective covering. The appendages are extremely reduced, the body becoming little more than an egg sac at maturity. As the eggs are gradually discharged, the body shrinks, so that the entire egg mass is laid within the protective covering of the scale. The first-instar nymphs are minute and extremely active. They crawl with rapidity in all directions and thus effect the

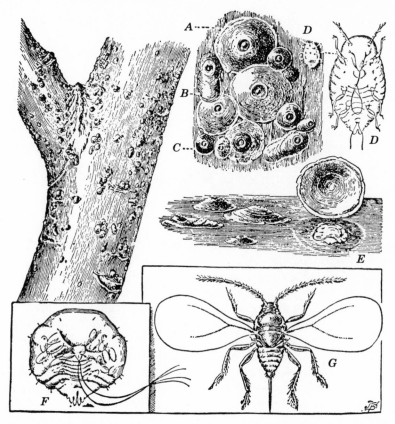

Fig. 218.   The San Jose scale *Aspidiotus perniciosus,* infesting apple.   *A*, scale of adult female; *B*, scale of male; *C*, first-instar young; *D*, same more enlarged; *E*, scale lifted to expose the female body beneath; *F*, body of the female; *G*, adult male.   (From U.S.D.A., E.R.B.)

widespread distribution of these scale insect species.   After the first molt, the nymphs become sedentary, and each forms a scale.   Several species of the family are among the most destructive insects known to commercial agriculture.   The San Jose scale *Aspidiotus perniciosus,* fig. 218, is a persistent pest of deciduous fruit trees and many ornamentals; before advent of oil sprays it threatened to wipe out several of the fruit crops in many areas in the United States.   The cosmopolitan oystershell scale *Lepidosaphes ulmi,* fig. 219, is a common pest of almost all deciduous trees and shrubs in the United States.

The family Eriococcidae or mealy bugs, fig. 220, are another important group, attacking many hosts, especially greenhouse and household

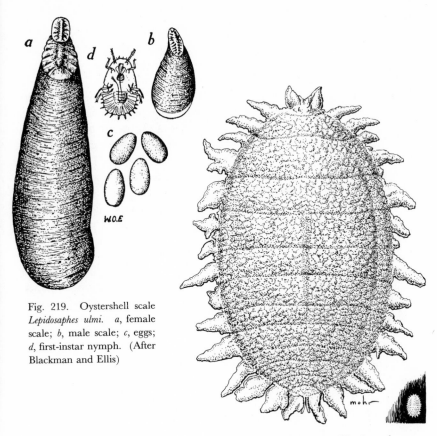

Fig. 219. Oystershell scale *Lepidosaphes ulmi.* *a*, female scale; *b*, male scale; *c*, eggs; *d*, first-instar nymph. (After Blackman and Ellis)

Fig. 220. A mealy bug. The inset shows natural size. (Original by C. O. Mohr)

plants in the more northern areas. The mealy bugs make no scale but secrete waxy filaments which are especially noticeable along the periphery of the body.

SUBORDER HETEROPTERA

BUGS. This suborder contains a wide variety of forms ranging from a few millimeters to a few inches in length, and including terrestrial, semiaquatic and aquatic types. The antennae are four- or five-segmented, and the eyes are well developed except in the ectoparasitic family Polyctenidae.

The nymphs of all forms resemble the adults in general outline, but differ uniformly in having dorsal stink glands on the abdomen.   In this respect they differ also from the nymphs of the Homoptera.

The eggs are usually laid singly or in groups glued to stems or leaves. In some forms the eggs are inserted into plant stems or, rarely, into damp sand.

The suborder contains predaceous species and plant-feeding species; often both types occur in the same family.   The predaceous species feed chiefly on smaller insects.   In certain species of the plant bugs, Miridae, the predaceous habit is only partially developed, and insect blood serves merely to supplement the principal diet of plant juices.   The mixture of food habits in the same families has resulted in some queer anomalies. In the stink bug family Pentatomidae the harlequin bug *Murgantia histrionica* is a serious pest of cabbages; a predatory species *Perillus bioculatus* is one of the most effective natural enemies of the dreaded Colorado potato beetle.

Short-Horned Bugs.   The members of this group have short antennae, usually recessed under the head and not visible from the dorsal aspect, and are either aquatic or shore-inhabiting.   Of the nine families recognized in our fauna, the Corixidae, Notonectidae, and Belastomatidae are the most common.

CORIXIDAE, WATER BOATMEN.   These bugs, fig. 221*A*, are characterized by the short stout labium which looks more like the lower sclerite of the head than like a beak, fig. 209*G*.   The front legs are short, flattened, or scoop-shaped; the hind legs are long, flattened, and fringed with combs of bristles.   Both nymphs and adults are truly aquatic, swimming in the water and incapable of more than clumsy flopping on land.   The fringed hind legs are used for swimming; they swim dorsal side up.   The adults leave the water for dispersal flights and may be observed in swarms over bodies of water.   Sometimes these swarms are attracted to lights.   The eggs are attached to solid supports such as stones, sticks, and shells in the water.   Certain forms, such as *Ramphocorixa*, more often lay their eggs on the body or appendages of the crayfish *Cambarus*, which in some localities may be literally plastered with corixid eggs.

The water boatmen differ from all other Hemiptera in their feeding habits.   They feed in the ooze at the bottom of the water, the stylets of the mouthparts darting in and out in unison like a snake's tongue.   These stylets draw into the pharynx an assortment of diatoms, algae, and minute animal organisms which constitute their food.

NOTONECTIDAE, BACKSWIMMERS.   In form these aquatic bugs superficially resemble the water boatmen, particularly in the long-fringed oar-like hind legs used for swimming.   They are very different, however,

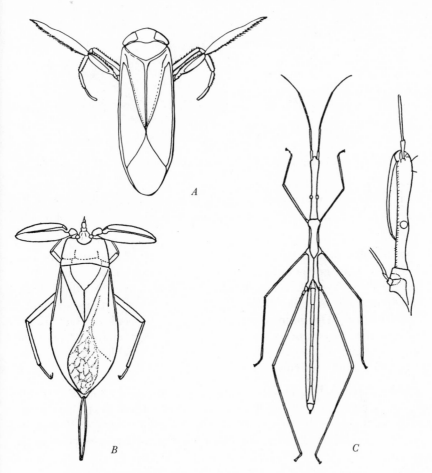

Fig. 221. Heteroptera. *A*, dorsal aspect of *Arctocorixa*, Corixidae; *B*, dorsal aspect of *Nepa*, Nepidae; *C*, ventral and partial lateral aspects of *Hydrometra*, Hydrometridae. (*A* adapted from Hungerford, *B* and *C* from Hemiptera of Connecticut)

in many ways.   Most conspicuous is their habit of always swimming on their backs.   The coloration of the backswimmers is modified to match this change in swimming position.   The ventral side, which is uppermost, is dull brown to match the stream or pond bottom.   The dorsal side, which is hidden from above when the insect is swimming, is usually whitish, creamy, or lightly mottled.   The beak in the Notonectidae is stout and sharp, used to suck the body contents from small aquatic animals such as Crustacea and small insects on which the backswimmers

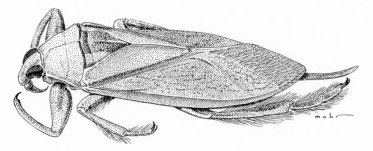

Fig. 222.   A giant water bug *Lethocerus,* Belostomatidae.   (Original by C. O. Mohr)

feed.   Many species of backswimmers deposit their eggs on the surface of objects in the water; others insert their eggs into the stems of aquatic plants.

BELOSTOMATIDAE, GIANT WATER BUGS.   Members of this family are wide and stout, with grasping front legs and crawling and swimming middle and hind legs, fig. 222.   They include some of the largest North American Hemiptera, for example, *Lethocerus americanus,* which attains a length of 3 to 4 inches.   The giant water bugs have strong beaks.   They are predaceous, feeding on insects, snails, small frogs, and small fish.   Commonly they are attracted to lights where they draw considerable attention, owing to their ungainly movements and large size.

Long-Horned Bugs.   In this group the antennae are exposed and elongate, projecting well in front of the head.   Over thirty families are recognized in North America, including a great variety of shapes and habits.   The following nine families afford a good cross section of the group.

BED BUG BRANCH

GERRIDAE, WATER STRIDERS.   These are slender bugs with long legs, fig. 223.   They live on the surface of water, inhabiting chiefly ponds, the margins of lakes, and the more sluggish backwaters and edges of rivers and streams.   The tarsi are fitted with sets of nonwetting hair which allow the bugs to run and stand on the water surface with amazing speed and ease.   All species are predators or scavengers, feeding chiefly on other insects that occur on the water surface.   They lay their eggs in masses attached to aquatic plants or thrust them into submerged stalks.   Several other closely related families live as striders or skaters on the water surface; from these the Gerridae differ in having very long hind femora, which extend considerably beyond the apex of the abdomen.

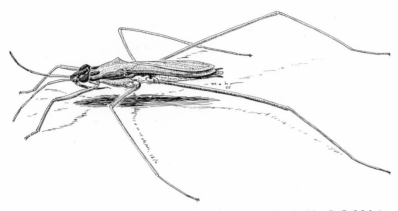

Fig. 223.   A water strider *Gerris rufomaculata,* Gerridae.   (Original by C. O. Mohr)

TINGIDAE, LACE BUGS.   These are small delicate plant-feeding insects, fig. 224*C, D,* usually occurring in large colonies.   The pronotum and hemelytron are wide, reticulate, and lacelike, extending well beyond the sides of the body; in certain genera the pronotum has a large bulbous mesal lobe which extends forward above the head, fig. 224*C.*   The antennae and beak are four-segmented, ocelli are lacking, and the tarsi are two-segmented.   The nymphs differ considerably from the adults in general appearance; some are comparatively smooth and scalelike; others are armed with large numbers of long spines.   The eggs are laid in or on the leaves of the host plant.   The Tingidae are represented in North America by over two hundred species, most of them specific to a single

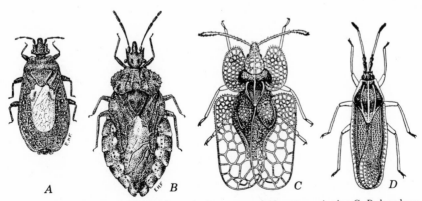

Fig. 224.   Heteroptera.   *A, B,* flat bugs *Aradus acutus* and *Neuroctenus simplex; C, D,* lace bugs *Atheas exiguus* and *Corythuca floridanus.*   (*A, B,* after Froeschner, *C, D,* after Heidemann)

host genus or species. A colony of lace bugs produces a characteristic white-spotted appearance of the leaves that readily betrays the presence of the colony. Examination of alder, oaks, sycamores, hawthorns, apples, birches, and other trees will net many species of lace bugs. Shrubs and herbs also support a considerable fauna.

MIRIDAE, PLANT BUGS. This family is a very large one, containing about fifteen hundred species, over a third of all known North American Heteroptera. The plant bugs, fig. 225, belong to the series of families having a four-segmented beak and no ocelli. With few exceptions they possess fully developed wings; the hemelytron usually has a distinctive sclerite or *cuneus* in the sclerotized portion and one or two simple cells in the membrane, fig. 210*B*.

Most of the species are plant feeders, many attacking only one or a very limited number of host species. Plant-bug feeding causes etiolation or blossom blight of the host and frequently results in marked commercial loss of certain crops. Some of the more destructive economic species are the cotton fleahopper *Psallus seriatus;* the garden fleahopper *Halticus bracteatus,* which damages alfalfa, clover, and garden crops such as beans; and the tarnished plant bug *Lygus lineolaris,* a general feeder and a local pest of many crops. Certain genera, including a few striking ant mimics, are predaceous on aphids and other insects.

The plant bug females insert their eggs into dead herbaceous stalks. Most of the species have only a single generation a year, and the winter

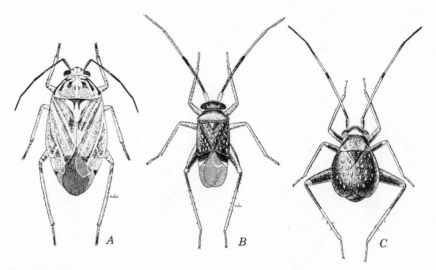

Fig. 225. Miridae, plant bugs. *A*, the tarnished plant bug *Lygus lineolaris; B* and *C*, the garden fleahopper *Halticus bracteatus,* male and female. (From Illinois Nat. Hist. Survey)

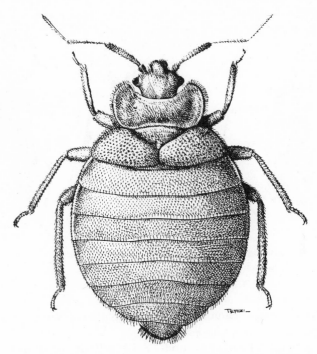

Fig. 226.   The common bed bug *Cimex lectularius*.   (By permission of British Museum (Natural History).)

is passed in the egg stage.   A few species, including the tarnished plant bug, hibernate as adults and deposit their eggs the following spring.

CIMICIDAE, BED BUGS.   This family includes only a few species of wide flat insects that feed on the blood of birds and mammals.   The fore wings or hemelytra are represented only by short scalelike pads; the hind wings are completely atrophied.   They live in bird or mammal nests and in dwellings.   Man is attacked by the common bed bug *Cimex lectularius*, fig. 226, which may become an important pest in living quarters of all kinds.   During the day the bed bugs hide in cracks and crevices of woodwork, furniture, and debris, emerging at night to seek a blood meal.   The female lays up to two hundred cylindrical whitish eggs, depositing them in crevices.

REDUVIIDAE, ASSASSIN BUGS.   These are sluggish predaceous insects, usually medium-sized to large, that feed on other insects.   Most of our species have fully developed wings, fig. 227, and several have wide foliaceous legs.   The nymphs of certain species secrete a sticky substance over the dorsum, on which are carried bits of leaves and debris, providing the

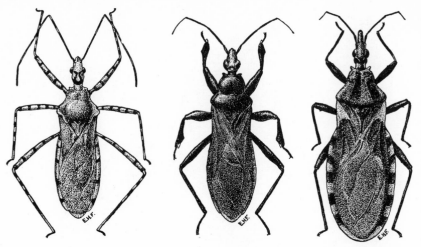

Fig. 227. Three Reduviidae. Left to right, *Pselliopus barberi, Melanolestes picipes,* and *Triatoma sanguisuga.* (Drawings loaned by R. C. Froeschner)

animal with very good camouflage. The eggs are laid singly or in clusters, glued to plants or other supports. Assassin bugs sometimes attack man, inflicting a painful burning wound. All the Reduviidae are terrestrial.

STINK BUG BRANCH

COREIDAE, SQUASH BUGS, COREID BUGS. In general characters the Coreidae are like the Lygaeidae, differing chiefly in having many veins in the hemelytron membrane. Many of these bugs resemble the lygaeid bugs in general shape. Others, such as *Acanthocephala,* bear a striking resemblance to the Reduviidae. Coreid bugs feed on plants. The most widely known is the squash bug *Anasa tristis,* fig. 228C, which attacks squash, cucumbers, and other cucurbit crops. Its eggs are laid in patches on the leaves and stem of its hosts. There are several generations each year, and winter is passed in the adult stage.

ARADIDAE, FLAT BUGS. This family includes a group of moderate-sized species which are the flattest members of the Heteroptera. They live under the bark of dead trees and are thought to feed on fungi. The tarsi are two-segmented, the antennae and beak four-segmented, and ocelli are lacking. The wings are greatly reduced in size and when folded occupy only a small area of the dorsum, fig. 224A, B. This illustration depicts two common species in eastern and central North America.

LYGAEIDAE, CHINCH BUGS, LYGAEID BUGS. Most North American members of this family are fairly small somber-colored or pale forms. A few

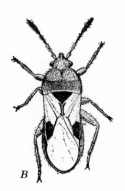

Fig. 228. Heteroptera. *A*, the milkweed bug *Oncopeltus fasciatus* and *B*, the chinch bug *Blissus leucopterus,* Lygaeidae; *C*, the squash bug *Anasa tristis,* Coreidae. (*A* after Froeschner, *B* from Illinois Nat. Hist. Survey, *C* from U.S.D.A.)

genera, such as the milkweed bug *Oncopeltus fasciatus,* fig. 228*A*, are strikingly marked with red and black. The diagnostic family characters include the four-segmented beak and a hemelytron with a few irregular veins crossing the membrane, fig. 210*F*. Most species have distinct ocelli. In North America the most important member of this family is the chinch bug *Blissus leucopterus,* fig. 228*B*, which is one of the major insect pests of corn and small grains in the corn-belt states. The chinch bugs hibernate in ground cover as adults. In early spring they feed on grasses and small grains and lay their eggs on the roots and crown of the food plants. The eggs hatch in about 2 weeks, the nymphs feeding on the same plants and maturing in 6 weeks. By the time this brood matures, the original food crop is almost invariably either mature and becoming dry, or it has been overpopulated and offers little in the way of nourishment. When this happens, the entire brood moves out in search of more succulent food. The exodus takes the form of a mass migration, not by flight but by foot. Mature nymphs and newly emerged adults make up the hurrying mass of insects on the march. The search for a better food supply usually ends in a field of corn, now well established and in prime growth. To protect the corn at this time various types of barriers have been developed to trap and kill these marching hordes. The individuals which reach the corn plants establish themselves and produce the second generation. Both the migrating first generation and the second generation feeding to maturity do extensive damage to the corn crop. When mature, the second generation goes into hibernation until the following spring.

PENTATOMIDAE, STINK BUGS. To this family, figs. 208*B*, *C*, *D*, and 229,

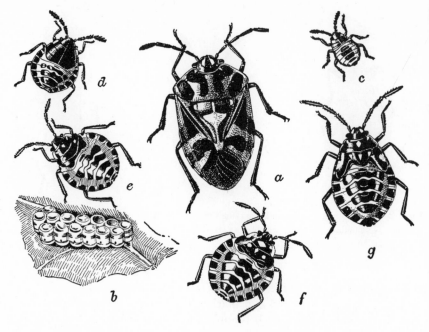

Fig. 229.   Pentatomidae.   The harlequin bug, *Murgantia histrionica*.   *a*, adult; *b*, egg mass; *c*, first stage of nymph; *d*, second stage; *e*, third stage; *f*, fourth stage; *g*, fifth stage.   (From U.S.D.A., E.R.B.)

belong many large or medium-sized bugs, most of them broad, many of them mottled with shades of green, gray, or brown. A few are brightly patterned. Some species are predaceous, feeding on a wide variety of other insects. Others are entirely phytophagous, of which the harlequin bug *Murgantia histrionica,* fig. 229, is a familiar example. This bug feeds on cruciferous plants and often does serious damage to cabbage. Three subfamilies, the Scutellerinae, Cydninae, and Thyreocorinae, are classed as separate families in some works. The identifying characteristics of these subfamilies, given in the key to families, hold fairly well for the nearctic fauna but break down when considered for the world fauna as a whole.

References

Blatchley, W. S., 1926. *Heteroptera or true bugs of eastern North America.* Nature Publishing Co.: Indianapolis, Ind. 1116 pp., illus.
Britton, W. E., 1923. The Hemiptera or sucking insects of Connecticut. *Connecticut State Geol. Nat. Hist. Survey, Bull.,* **34**:807 pp., illus.

Froeschner, R. C., 1941–62. Contributions to a synopsis of the Hemiptera of Missouri, Pts. 1–4. *Am. Midland Naturalist,* **26**:122–146; **27**:591–609; **31**:638–683; **42**:123–188; **67**:208–240.

Torre-Bueno, J. R. de la, 1939–46. A synopsis of the Hemiptera-Heteroptera. *Entomol. Am., n.s.,* **19**:141–310; **21**:4ᵢ–122; **26**:1–141.

Van Duzee, E. P., 1917. Catalogue of the Hemiptera of America north of Mexico. *Univ. Calif. Tech. Bull.,* **2**:1–902.

NEUROPTEROID ORDERS

The ten neuropteroid orders all have a complete metamorphosis and because of this are frequently termed the Holometabola. Living species constitute three distinctive groups: (1) the Hymenoptera, preserving a typical sawlike ovipositor and having a reduction in the longitudinal wing veins; (2) the Coleoptera, lacking a saw but having veinless hard front wings functioning as protective coverings; and (3) the Neuroptera-Mecoptera series having (primitively) an abundant wing venation and a peculiar structure on the coxa called the meron, fig. 230.

Order HYMENOPTERA: Sawflies, Ants, Bees, and Wasps

A large order, including many different body shapes and with a size range from 0.1 mm. in minute parasitic forms to at least 50 mm. in some of the wasps, fig. 231. Integument heavily sclerotized, the pleural sclerites considerably coalesced. Mouthparts are of the chewing type, in many forms modified for lapping or sucking. Wings well developed, reduced, or absent; if well developed, they are transparent, the two pairs similar in texture, and without scales; they have a great range in venation. Legs of primitive forms have the base of the femur set off as an extra trochanter-like segment, fig. 230*A*. Generalized forms, fig. 232, have a considerable reduction and coalescence of veins, but there are a moderate

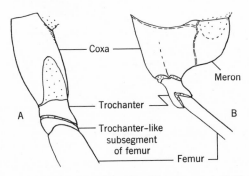

Fig. 230. Lateral view of base of leg of *A, Pamphilius* (Hymenoptera); *B, Chrysopa* (Neuroptera).

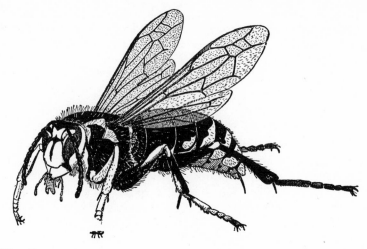

Fig. 231.   The bald-faced hornet *Vespula maculata,* a representative hymenopteran.   (From Illinois Nat. Hist. Survey)

number of crossveins.   Antennae range from 3- to about 60-segmented, and are of many shapes.   Larvae are caterpillar-like or grublike, all having a distinct head and chewing mouthparts, some with thoracic or abdominal legs or both, and others without any legs.

In number of known species, the Hymenoptera is one of the largest orders of insects.   It also contains groups such as the ants, wasps, and bees that are extremely abundant in numbers of individuals and therefore conspicuous and common elements of the fauna.   These forms, however, are highly modified members of the order.

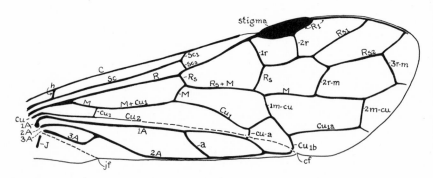

Fig. 232.   Diagram of hymenopterous wing, combining primitive veins of several archaic families.

Of living Hymenoptera, the most primitive are plant-feeding sawflies of the relatively rare families Xyelidae and Pamphiliidae. In these, the larvae have antennae possessing up to seven segments and well-developed legs on each thoracic segment. An early offshoot of this line evolved into the typical sawflies, all leaf feeders, including the common families Argidae, Tenthredinidae, Cimbicidae, and Diprionidae. In all of these and the Xyelidae, the larvae have a pair of legs on each of several abdominal segments; because of this they are often mistaken for caterpillars of Lepidoptera. In these groups, the female cuts slits in leaves with her saw, and lays an egg in each slit. The slit heals over and the egg is then encased in plant tissue.

Another early line arising from presumably a pre-xyelid ancestor evolved into stem borers, comprising the group called the horntails. In these, the saw became narrow, fitted for drilling into harder plant tissue, and the larval legs became greatly reduced. The family Cephidae appears to be the most primitive member of this line; other members include the hard-bodied Siricidae and Xiphydriidae, and the peculiar rare family Orussidae whose larvae are thought to be parasitic on wood-boring beetle larvae.

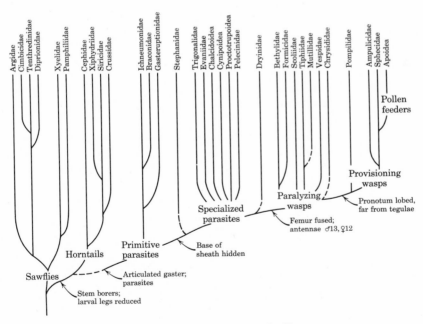

Fig. 233. Simplified family tree of the order Hymenoptera.

The sawflies and horntails constitute the suborder Symphyta, in which the abdomen is joined broadly to the thorax, fig. 234A, B. In the other and much larger portion of the Hymenoptera, the suborder Apocrita, the first abdominal segment is fused solidly with the metathorax and there is a hingelike articulation between the first and second abdominal segments, fig. 234D, the rest of the abdomen becoming a single flexible unit called the gaster. The larvae of the Apocrita are primarily parasites or predators on immature stages of other insects and are legless. The adult saw is slender, adapted to piercing substrates in which the host larvae live, penetrating the host itself, or to stinging.

From which group of Symphyta the Apocrita arose has been, and is, a far from settled question. The Apocrita may well have arisen from the base of the horntail line. It seems plausible that the stem-boring larva of a horntail more primitive than any now living might have overtaken and eaten larvae of other species in its progress through host plant tissue; that this relationship could have become habitual; and that finally it became established as the normal food relationship of the species. The parasitic habit probably originated in this way. That some such change occurred within the horntails, resulting in the parasitic Orussidae, strengthens this supposition regarding the Apocrita. In addition to the hinged gaster, the primeval apocritan also evolved a transverse, flexible suture across the mesothoracic scutum, undoubtedly associated with better flight in an unusually hard insect. The same change occurred also in one line of the horntails; perhaps it first appeared in the horntail-apocritan ancestor.

The earliest known Apocrita are today represented by primitive parasites including the rare Stephanidae and Gasteruptionidae and the abundant Braconidae and Ichneumonidae. In these the base of the saw and sheath are exposed, much as in the sawflies. In a line from some early ancestor of these groups arose a form in which the seventh sternite grew forward, concealed the base of the saw or ovipositor, and became a ventral trough or guide for it. This form gave rise to a large group of specialized parasites, including the Cynipoidea, Chalcidoidea, Proctotrupoidea, and several distinctive but smaller families such as the Pelecinidae, Evaniidae, and Trigonalidae. It is a peculiar circumstance that independently in several lines the basal segment of the femur re-fused with the apical portion and the number of antennal segments decreased. In both the Cynipoidea and Chalcidoidea, plant-feeding forms arose; the other members were typical parasites whose larvae killed their hosts by the simple mechanism of gradually eating their tissues. In all of these parasites, the female does not paralyze the prey.

From some specialized parasite there evolved one having a new kind of

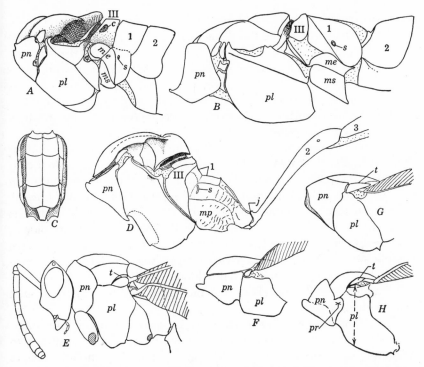

Fig. 234. Diagnostic characters of Hymenoptera. *A*, thorax of *Arge*, Argidae; *B*, thorax of *Janus*, Cephidae; *C*, venter of abdomen of *Chrysis*, Chrysididae; *D*, thorax of *Eremotylus*, Ichneumonidae; *E*, head and thorax of *Chalcis*, Chalcididae; *F*, thorax of *Proctotrupes*, Proctotrupidae; *G*, thorax of *Ancistrocerus*, Vespidae; *H*, thorax of *Sceliphron*, Sphecidae. *c*, cenchrus; *j*, basal articulation of gaster; *me*, metaepimeron; *mp*, metapleuron; *ms*, metaepisternum; *pl*, mesopleuron; *pn*, pronotum; *pr*, pronotal lobe; *s*, first abdominal spiracle; *t*, tegula. III, metanotum. 1, 2, 3, segments of abdomen.

parasitic habit. The female used her ovipositor to sting and immobilize or paralize the host larvae, then laid an egg on it. The prey was thus kept alive while the egg hatched and the larva commenced feeding; in addition, the immobilized prey did not move into a new situation detrimental to the wasp larva. Most of these wasps prey on insect larvae that live in protected situations, such as soil-inhabiting beetle larvae. In the form giving rise to this line of paralyzers, the basal femoral segment had become completely reunited with the apical portion, and the antennae had stabilized at the reduced number of 13 segments in the male, 12 in the female; in a few groups the number went lower. Many lines evolved from the "paralyzer" ancestor. One gave rise to the uncommon wasp family Bethylidae and probably also to the ants (Formicidae); others gave rise to the wasp families Scoliidae, Tiphiidae, Mutillidae, and their relatives; and another gave rise to the social wasps, the yellow jackets or bald-faced hornets (Vespidae). In this latter family, the females evolved the behavior pattern of first making a nest, then either laying an egg in it and provisioning it, or the reverse. The cuckoo wasps (Chrysididae) may be a specialized offshoot of this complex but their relationships are doubtful. The peculiar Dryinidae, primarily parasites of leafhoppers, may be primitive members of the paralyzing wasps.

From the complex of stinging and paralyzing wasps arose one other lineage of note. In this line, as in some Vespidae, the female first constructed a nest (probably excavated in the ground), then hunted and stung the prey, brought it to the nest, and laid an egg on it. In this line the lateral corners of the pronotum became lobate and distant from the tegulae, fig. 234H. These were the primeval solitary wasps (Sphecoidea). In the two living primitive families of this line, the common spider wasps (Pompilidae) and the rare cockroach wasps (Ampulicidae), some members still lay the egg first, then seek provisions. In the more highly evolved and common family Sphecidae, the provisions are normally procured before the egg is laid.

From a branch of the typical sphecoid line arose the bees. The major changes in this evolution were a switch in food from insects or spiders to pollen, and the evolution of branched hairs on the body and appendages associated with gathering pollen. Some investigators believe that the bees may be composed of two branches, each arising from a different wasp group, that evolved pollen-gathering and branched hairs independently.

The evolution of the higher Hymenoptera (the wasps and bees) therefore appears to have occurred chiefly through changes in behavior patterns without drastic accompanying changes in external morphology.

KEY TO SUBORDERS AND COMMON FAMILIES

1. First abdominal segment solidly joined with second, at most a shallow constriction between them, first tergite forming a distinct plate or pair of plates, fig. 234A, B (suborder **Symphyta**, p. 301)............................ 2

   Juncture of first and second abdominal segments constricted to form a ball-and-socket joint, fig. 234D; the first tergite is fused solidly to the thorax, and the remainder of the abdomen forms an articulating unit called the gaster (suborder **Apocrita**, p. 304) 9
2. Antenna 3-segmented, fig. 235A, the third sometimes split longitudinally to form a lyre-shaped prong, fig. 235B.............................**Argidae**

   Antenna at least 6-segmented, the end segment never cleft, fig. 235C–I...... 3
3. Third antennal segment at least as long as combined length of the succeeding 9 segments, the segments beyond the third forming a slender terminal filament, fig. 235E ....**Xyelidae**

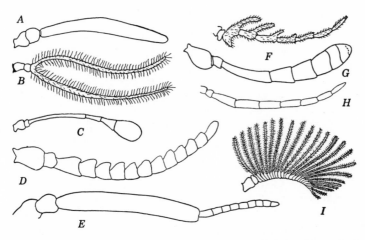

Fig. 235. Antennae of Hymenoptera. *A, Sofus* female, Argidae; *B, Sphacophilus* male, Argidae; *C, Trichiosoma,* Cimbicidae; *D, Augomonoctenus* female, Diprionidae; *E, Pleroneura,* Xyelidae; *F, Cladius* male, Tenthredinidae; *G, Tenthredo,* Tenthredinidae; *H, Pseudodineura,* Tenthredinidae; *I, Monoctenus* male, Diprionidae.

Third antennal segment not longer than the combined length of the next 3 or 4 segments, or antenna clavate, fig. 235*C*. 4

4. Antenna capitate, fig. 235*C*; lateral edge of abdomen sharp and angular; large robust species, fig. 238. . . . .**Cimbicidae**
Antenna pectinate, serrate, filiform, or in a few species as clavate as fig. 235*G*; lateral edge of abdomen round. 5

5. A shallow but distinct constriction between first and second abdominal tergites, and cenchri absent, fig. 234*B*. . . . .**Cephidae**
No constriction between first and second abdominal tergites, and cenchri (*c*) well developed, forming a pair of velvety pads on the metanotum, fig. 234*A*. . . . 6

6. Front tibia having only one apical spur. **Siricidae, p. 304**
Front tibia having two apical spurs. 7

7. Antenna 7- to 9-segmented, fig. 235*F–H*. . . . . **Tenthredinidae, p. 301**
Antenna having 10 or more segments, figs. 235*D, I*. 8

8. Antenna narrow and filiform, proportioned as in fig. 235*H*
Tenthredinidae, p. 301
Antenna serrate in females, fig. 235*D*, pectinate in males, fig. 235*I*
Diprionidae, p. 304

9. Petiole composed of two segments, usually one or both bearing a dorsal hump or node, fig. 248. . . . . **Formicidae, p. 310**
Petiole consisting of only one segment, figs. 249, 250. 10

10. First segment of gaster forming an isolated petiole bearing a dorsal node or projection, fig. 249; includes winged and wingless forms. . . . **Formicidae, p. 310**
First segment of gaster either expanded posteriorly or not bearing a dorsal node
. . . . 11

11. Wings completely atrophied or reduced to small pads 12
    Wings well developed, reaching to or beyond middle of abdomen 13
12. Body fuzzy with dense hair, fig. 246 . . . . **Mutillidae, p.** 308
    Body smooth or with only inconspicuous hair. A few species in each of several
    families of parasitic habit, keyed no further here.
13. Front wing without a stigma (a thickened area along the anterior margin of the
    wing), and with sclerotized venation reduced to a single anterior vein, some-
    times with a "tail" at its tip, fig. 236*D*, sometimes completely atrophied . . . 14
    Front wing either having a definite stigma or having a more extensive venation,
    fig. 236*F* 15
14. Lateral corner of pronotum extending to the tegula, fig. 234*F*. Several families
    of small parasitic wasps, chiefly . . . . **Proctotrupoidea**
    Lateral corner of pronotum not extending to the tegula, fig. 234*E*
                                                           Chalcidoidea, p. 306
15. Pronotum having each posterolateral corner forming a round earlike or epaulet-
    like lobe ending below level of tegula, fig. 234*H* 16

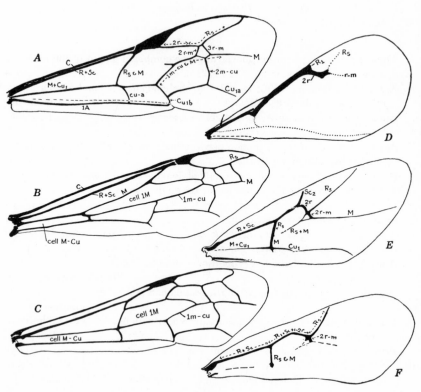

Fig. 236. Wings of Hymenoptera. *A*, Ichneumonidae; *B*, *Vespa*, Vespidae; *C*, *Myzine*, Tiphiidae; *D*, *Tetrastichus*, Chalcidiidae; *E* and *F*, two types of Cynipidae.

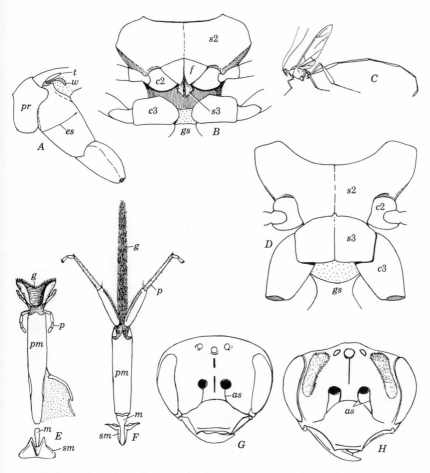

Fig. 237. Parts of wasps and bees. *A*, mesopleuron of *Ceropales,* Pompilidae; *B*, mesosternum of *Myzine,* Tiphiidae; *C*, outline of *Pelecinus,* Pelecinidae; *D*, thoracic sterna of *Scolia,* Scoliidae; *E*, labium of *Colletes,* Colletidae; *F*, same of *Anthidium,* Megachilidae; *G*, head of *Halictus,* Halictidae; *H*, head of *Andrena,* Andrenidae. *as*, antennal sutures; *c*, coxa; *es*, mesopleural suture; *f*, flaplike processes of mesosternum; *g*, glossa; *gs*, base of gaster; *m*, mentum; *p*, palpus; *pm*, prementum; *pr*, pronotum; *s*, sternite; *sm*, submentum; *t*, tegula; *w*, wing base. (*E–H* after Michener)

Pronotum having posterolateral corner truncate or angulate, often abutting tegula, fig. 234*G*, or rounded but not earlike, fig. 237*A* 23

16. Hind basitarsus no wider than succeeding segments, the plantar surface clothed only with dense, short pile; body and appendages without branched hairs, each hair simple, neither branched nor fringed (*Sphecoidea*)

Sphecidae, p. 313

Hind basitarsus slightly wider (in numerous forms many times wider) than succeeding segments, the plantar surface with moderately long and abundant hair; body and appendages having branched or spiral hairs; each hair has many branches or whorls and may appear fringed, fig. 252 (*Apoidea*, p. 314)
.... 17
17. Head having two sutures below each antenna, delimiting a large subantennal sclerite, fig. 237*H* ....**Andrenidae**
Head having only one suture below each antenna, fig. 237*G*............. 18
18. Labium with mentum and submentum virtually absent; front wing with free basal part of *M* (basal vein) usually strongly curved...........**Halictidae**
Labium with mentum and submentum present, fig. 237*E*, *F*; front wing with basal vein straight 19
19. Labium with glossa short and its apex either rounded, truncate, or incised, fig. 237*E* ....**Colletidae**
Labium with glossa elongate and narrow or pointed at apex, fig. 237*F*...... 20
20. Labial palp with segments similar and cylindrical, as in fig. 237*E*....**Melittidae**
Labial palp with first two segments elongate and sheath-like, fig. 237*F*..... 21
21. Front wing with 3 submarginal cells (crossveins 2*r–m* and 3*r–m* both present, as in fig. 236*C*) ....**Apidae**
Front wing with 1 or 2 submarginal cells (one or both of these crossveins lacking) 22
22. Labrum longer than wide and widened at extreme base........**Megachilidae**
Labrum wider than long or narrowed at extreme base..............**Apidae**
23. Front wing having costa and stem of radius running very close together, obliterating the costal cell, fig. 236*A*, and antenna with more than 16 segments 24
Either front wing with an open costal cell between radius and costa, fig. 236*B*, *C*, or antenna with no more than 14 segments......................... 25
24. Front wing having crossvein 2*m–cu*, fig. 236*A*.........**Ichneumonidae**, p. 304
Front wing lacking crossvein 2*m–cu*...................**Braconidae**, p. 305
25. Gaster with only three apparent dorsal segments, the tergites heavily sclerotized; the three large sternites each divided longitudinally into a pair of concave armored plates, fig. 234*C*. Robust, hard, shining metallic wasps capable of curling up into a ball. Cuckoo wasps.....................**Chrysididae**
Gaster with at least five apparent dorsal segments..................... 26
26. Front wings having no definite stigma, but instead a clear, triangular area bounded posteriorly by a vein, fig. 236*E*, *F*, but without an anterior vein
Cynipoidea, p. 307
Front wings having a thickened stigma, or an anterior vein, fig. 236*B*, *C*.... 27
27. Gaster extremely elongate and slender, fig. 237*C*; head and thorax black and shining. Females only (males are rare) of *Pelecinus polyturator*....**Pelecinidae**
Gaster much shorter and thicker; texture and color of various sorts........ 28
28. Corner of pronotum ending below level of wing and not in vicinity of tegula; mesopleuron almost always having a straight, fine transverse suture at about its midpoint, fig. 237*A*, although in some species it is difficult to detect
Pompilidae, p. 313

Corner of pronotum ending at or above level of wing and usually abutting against the tegula, fig. 234*G*; mesopleuron with a crooked suture or none. . 29

29. Front wing having cell 1*M* longer than cell *M–Cu*, fig. 236*B*; either wings pleated lengthwise when folded or antennae clublike . . . . **Vespidae**, p. 309

 Front wing having cell 1*M* shorter than cell *M–Cu*, fig. 236*C*, or former cell open due to atrophy of 1st *m–cu* 30

30. Metasternum large and rectangular, fused with the mesosternum, the two forming a large, level plate overlying the bases of the four posterior coxae; hind coxae wide apart, fig. 237*D* . . . . **Scoliidae**, p. 308

 Metasternum small and triangular, or inconspicuous, not fused with mesosternum; hind coxae close together, fig. 237*B* 31

31. Hind margin of mesosternum produced into a pair of triangular plates partially overlapping the bases of the middle coxae, fig. 237*B* . . . . . . **Tiphiidae**, p. 308

 Hind margin of mesosternum without such lobes 32

32. Second segment of gaster large and bulbous compared with third segment, fig. 246; body often conspicuously hairy . . . . **Mutillidae**, p. 308

 Second segment of gaster not markedly wider than third segment; body never conspicuously hairy . . . . **Tiphiidae**, p. 308

SUBORDER SYMPHYTA

SAWFLIES AND HORNTAILS. The Symphyta, with the exception of the small parasitic family Orussidae, are a plant-feeding group. The larvae either feed externally on foliage or mine in leaves, leaf petioles, or stems. The adults of many groups feed on the pubescence of the host plant, cropping it by means of their sickle-shaped mandibles as a cow does grass; in other groups they may be predaceous on smaller insects or feed on nectar and pollen. The group is a large one; the North American forms represent twelve families and include in their host selection a great diversity of plant groups.

Distinguishing features of sawfly larvae are: a distinct head, with simple chewing mouthparts; antennae slender or platelike with one to seven segments; eyes with only a single lens; and abdominal legs (when present) without hooks or crochets.

The adults of most sawflies are compact and fairly robust. Of the leaf-feeding families, the largest common species is *Cimbex americana*, fig. 238, in which the antennae are capitate; the males and females are differently colored. The females of all but a few species have a well-developed saw used to cut egg slits in leaves or petioles.

The Tenthredinidae is the largest family, characterized chiefly by the simple 9- to 16-segmented antennae. Most of the species are external leaf feeders, and among them are several of economic importance, such as the rose-slug *Endelomyia aethiops;* the imported currant worm *Nematus ribesi*, fig. 239; and the larch sawfly *Pristiphora erichsoni.* In certain species

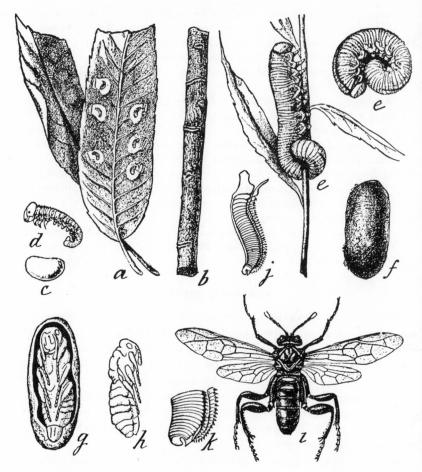

Fig. 238. A large sawfly *Cimbex americana*. *a*, willow leaves showing location of eggs; *b*, twig showing incisions made by adult; *c*, egg; *d*, newly hatched larva; *e, e*, mature larvae; *f*, cocoon; *g*, open cocoon showing pupa; *h*, pupa, side view; *i*, mature sawfly; *j, k*, saw of female. (After Riley)

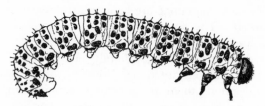

Fig. 239. The feeding stage larva of the imported currantworm *Nematus ribesi*. (From Connecticut Agricultural Experiment Station)

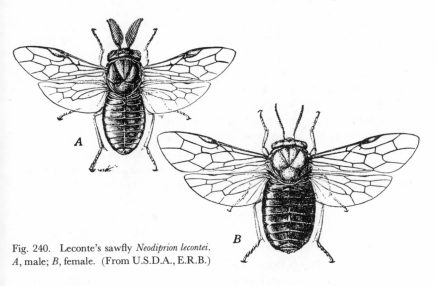

Fig. 240. Leconte's sawfly *Neodiprion lecontei*. *A*, male; *B*, female. (From U.S.D.A., E.R.B.)

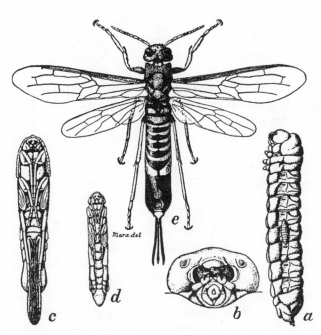

Fig. 241. A horntail wasp *Tremex columba*. *a*, larva; *b*, larval head, ventral aspect; *c*, *d*, female and male pupa; *e*, adult. Note small parasite larva attached to horntail larva. (After Riley)

the larvae mine in the leaf tissue, for example, *Heterarthrus nemorata,* one of the birch leaf miners. Species of other genera, including *Euura,* produce true galls.

The Diprionidae is another economically important family. In this family the antennae are at least 13-segmented, serrate in the female and pectinate in the male, fig. 240. All the species are stout and more or less drab in color. The larvae are caterpillar-like and external feeders on coniferous needles. Many species are among the worst defoliators of spruce and pine forests. Of special note are the ravages caused to spruce in northeastern America by the European spruce sawfly *Diprion hercyniae.* Especially injurious to young pines is the red-headed pine sawfly *Neodiprion lecontei,* fig. 240.

The Siricidae contain some of the largest members of the suborder. They are elongate, sometimes attaining a body length of 40 mm. The larvae bore in tree trunks and are cylindrical and almost legless. Both adults and larvae have a horny spikelike projection at the posterior end of the body, the character which gives them the name "horntails." *Tremex columba,* fig. 241, is a common species attacking maple, elm, beech, oak, and some other deciduous trees.

SUBORDER APOCRITA

ANTS, BEES, AND WASPS. In general the Apocrita are more graceful, active, and more rapid of movement than the Symphyta. The larvae are chiefly internal or external parasites, or are fed by the adults, or make plant galls. They are legless, have a distinct exposed head capsule bearing greatly reduced mouthparts and antennae, and frequently exhibit hypermetamorphosis.

In the Apocrita the first segment belonging to the abdomen is fused solidly with the thorax, so that what appears to be the abdomen really does not have its anterior segment. This body region that appears to be the abdomen is termed the *gaster.*

PRIMITIVE PARASITES

ICHNEUMONIDAE, THE ICHNEUMON WASPS. Usually slender wasps, having long and many-segmented antennae, fig. 242, and having subcosta fused with the stem of radius in the front wings. All members of this family are parasites on insects or spiders. Their favorite hosts are the larvae of Lepidoptera, for example, *Glypta rufiscutellaris,* a parasite of the oriental peach moth. In addition, a number of ichneumon wasps parasitize the larvae of Coleoptera, Hymenoptera, and Diptera, and a few other insects. The adult ichneumon wasp female deposits its eggs on or inside the body of the host. If the eggs are laid on the epidermis of the host, the newly

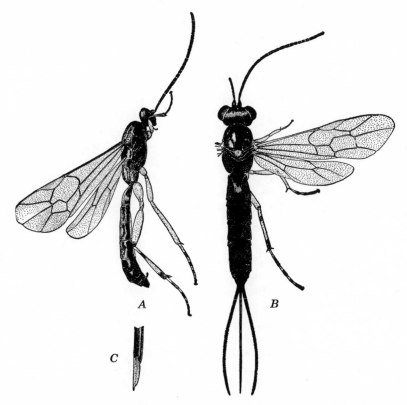

Fig. 242. An ichneumon wasp *Glypta rufiscutellaris*. *A*, male; *B*, female; *C*, tip of ovipositor. (From Connecticut Agricultural Experiment Station)

hatched larvae may bore into the body. The larvae develop into legless grubs which either attach to the outside of the host or develop within the body of the host. When mature, the larvae spin pupal cocoons near the host; the grubs may pupate within the host or leave it to spin cocoons. Ichneumon wasps have a wide range in size. Many of the small forms only a few millimeters long parasitize small moth larvae.

BRACONIDAE. This is a large family closely related to the ichneumon wasps. The species average smaller than the ichneumon wasps, and many braconids have reduced wing venation. A number of species are important as parasites of economic pests. One of these, *Apanteles melanoscelus*, fig. 147, has been imported for biological control of gypsy-moth larvae. This small parasite exhibits the interesting hypermetamorphosis prevalent among most of the parasitic families of Hymenoptera. In fig. 147

are illustrated the different shapes of the larva in various stages of development; the anal vesicle (*a*) may be used to identify the posterior end of the larva.

SPECIALIZED PARASITES

SUPERFAMILY CHALCIDOIDEA, THE CHALCID WASPS. Small wasps, sometimes less than a millimeter in length, with a greatly reduced wing venation, fig. 243, and usually with elbowed antennae. These wasps are largely internal parasites, especially of larval Lepidoptera and of the larvae of other parasitic Hymenoptera, which they attack within the body of the primary host. These parasites of parasites are called hyperparasites. The few chalcids which are not parasitic develop in various seeds, or in plant stems, especially grasses. To this nonparasitic group belongs the clover-seed chalcid *Bruchophagus gibbus,* whose larva develops in the seeds of clover and alfalfa; the wheat jointworm *Harmolita tritici,* whose larva bores in the stems of wheat; and the wheat straw-worm *Harmolita grandis,* fig. 243. Locally and sporadically the wheat jointworm causes serious damage to the crop.

Of unique interest is the specialized life history of certain tiny chalcid flies belonging to the family Agaontidae. These develop in the seeds of figs. The males are wingless and live only within the fig fruit in which they develop, fertilizing the females even before the latter emerge from the fig seed. The females are winged and fly from flower to flower in search of suitable seeds for oviposition, carrying the pollen on their bodies and pollinating each flower visited. This is the only method by which figs are pollinated. Many commercial varieties of figs do not require pollination to develop their fruits, but the fruit of the choice Smyrna fig will not develop without pollination. In order to grow these in North

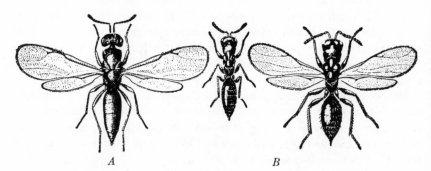

A   *B*

Fig. 243. *A*, the wheat jointworm *Harmolita tritici,* and *B*, wingless and winged forms of the wheat straw-worm *Harmolita grandis.* (From U.S.D.A., E.R.B.)

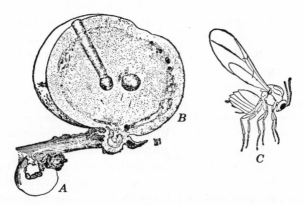

Fig. 244. A cynipid, *C*, and its gall; *A*, immature gall, and *B*, a section of mature gall showing cells and exits. (From Essig, *College entomology*, by permission of The Macmillan Co.)

America it was necessary to introduce the European fig chalcid *Blastophaga psenes* to effect pollination.

SUPERFAMILY CYNIPOIDEA, THE GALL WASPS. These are small wasps, most of them characterized by the large triangular cell near the apex of the front wing, fig. 236*E*, and by the deep but bilaterally compressed abdomen. Many groups of the superfamily are parasitic on dipterous larvae, aphids, and other insects, but the best-known group produces galls on plants. As a matter of fact, the gall wasps themselves are seldom seen, but every naturalist is familiar with some of the many different types of galls that are produced on the leaves, stems, or roots of oak, roses, and other plants by the larvae of these insects. One of these is shown in fig. 244. There are hundreds of species of gall wasps, nearly every species producing a different type of gall. Some species that live on oaks have an alternation of generations, with one generation producing a gall on the roots and the alternate generation making a gall on the leaves or twigs.

PARALYZING WASPS

These insects are of special interest because social life has been developed independently in two groups. An account of the development of the social habit is given in Chapter 6. Most of the families, however, are solitary, and parasitic or predatory in habit.

Most of the nonsocial wasps secrete a remarkable substance that is discharged with their sting. When injected into their prey, this secretion typically causes complete motor paralysis without death resulting. Such a paralysis is used by wasps that lay an egg on the prey or provision a nest with prey. The induced paralysis has a triple advantage: It keeps the prey edible until the wasp larva hatches and begins feeding, it insures that the prey will not move away from the legless wasp larva, and it prevents the attraction of scavenger insects to the odor of dead insects.

SCOLIIDAE, THE SCOLIID OR DIGGER WASPS. These fairly large insects have wings in both sexes, and most of the species are black or are banded or spotted with black and yellow, such as *Scolia dubia*, fig. 245. The female wasps dig through the soil in search of their prey, white grubs (larvae of the beetle family Scarabeidae). When the female encounters a suitable host larva, she stings it, thereby paralyzing it; digs a crude cell around it; lays an egg on the doomed larva; and then moves on in search of another victim. The egg soon hatches into a legless grub which attaches to the paralyzed beetle larva and begins eating it. Within a period of about 2 weeks, the wasp larva has consumed the host and is full grown. It then spins a cocoon in the earthen cell and usually passes the winter in this stage. The next spring or summer the larva pupates, and later the adult chews its way out of the cocoon and digs to the surface.

Closely related to the scoliid wasps are many families of somewhat similar habits. Species of the family Mutillidae are parasites of wasps and bees. In many mutillids the females are wingless and have a close resemblance to ants. The Mutillidae females, however, lack the "node" on the petiole of the gaster, fig. 246, and in addition are covered with dense velvety or silky pile. From this latter character the family has received the name velvet ants. These velvet ants have a powerful sting and use it freely if interfered with.

Many oriental species of *Tiphia*, of the related family Tiphiidae, have

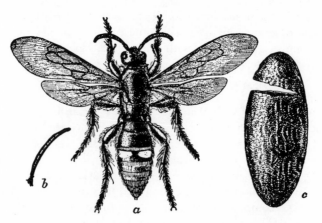

Fig. 245. A digger wasp *Scolia dubia*. *a*, female wasp; *b*, antenna of male; *c*, cocoon showing escape opening. (From U.S.D.A., E.R.B.)

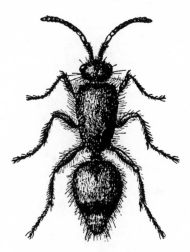

Fig. 246.   A mutillid wasp female *Dasy-mutilla bioculata.*   (After Washburn)

been brought to the United States and propagated for parasitizing grubs of the Japanese beetle, and a few have shown definite promise of assisting in the control of the beetle.

VESPIDAE, YELLOW JACKETS AND HORNETS, WASPS.   This family contains species varying from 10 to about 30 mm. in length, many of them having elaborate yellow and black or white and black markings, fig. 231.   In all subfamilies except the Masarinae the wings in repose are folded longitudinally like a fan.   In the Masarinae the antennae are capitate.   Most of the Vespidae are solitary in habit.   The adults make a burrow in wood or soil or construct a pottery container for the abode of the grub.   The nest is usually stocked by the adult with paralyzed caterpillars or with pollen and honey.   Certain of the Vespidae are social in habit.   By masticating wood fibers with an oral secretion, they produce a paper which they fashion into a platelike or baglike nest.   The most familiar of these are the platelike nests of *Polistes,* fig. 247, which are made up of a single horizontal comb of larval cells.   These are commonly found hanging from eaves of buildings and in similar sheltered places.   Colonies of *Polistes* rarely have over a few dozen members.   The largest colonies of vespids found in North America are made by the bald-faced hornet *Vespula maculata.*   These colonial nests, oval in shape and with an opening at the bottom, are most often attached to tree branches.   Each contains several layers of larval cells, or combs, arranged one above the other.   The workers in a colony forage for insect prey, such as flies and caterpillars. These are crushed and mangled by the wasps and fed to the maggot-like larvae in the cells of the nest.

In temperate regions of North America the colonies die out at the end

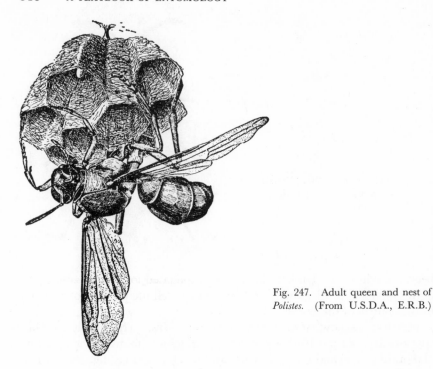

Fig. 247. Adult queen and nest of
Polistes. (From U.S.D.A., E.R.B.)

of autumn. In late summer a brood of males and females is produced.
With cold weather, the workers and males die; the autumn brood of females, by this time fertilized, hibernate in rotten logs or stumps. These
females emerge the following spring and begin new colonies.

FORMICIDAE, ANTS. In these insects the first segment of the gaster forms
a petiole or stalk and bears a dorsal projection or node, figs. 248, 249.
This structure differentiates ants from other antlike wasps. In addition
to the normal males and females, ant species usually have a third form,
the nonreproductive workers, which are always wingless. These workers
are the ants we usually see scurrying about. They perform most of the
work of the colony, such as building the nest, excavating the subterranean
chambers, and gathering food for the colony.

A typical colony starts with the swarming flights of the winged reproductive males and females. At periodic intervals (frequently once a year)
large numbers of winged males and females are produced in an established colony. When weather conditions are favorable, the sexual forms
leave the nest as a swarm, embark on their nuptial flight, and mate in
flight. The male dies soon after mating. The fertilized female seeks a
suitable nest site in the ground, an old log, or other situation, bites off

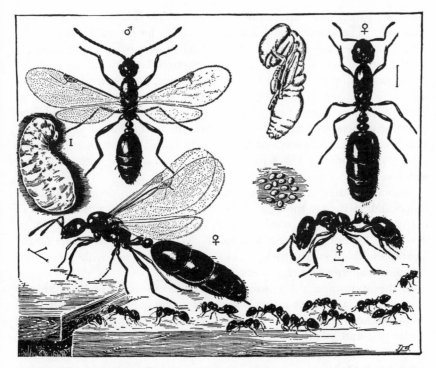

Fig. 248. The little black ant *Monomorium minimum,* showing several stages and activities. (After Marlatt, U.S.D.A., E.R.B.)

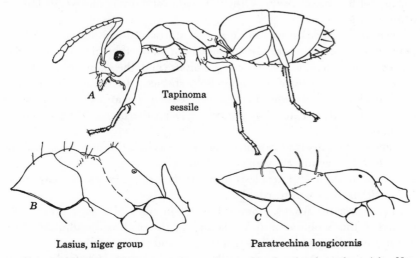

Fig. 249. Lateral aspect of three ant genera to illustrate the dorsal node on the petiole. Note that in *A* and *C* this node is small but distinct.

her wings, and seals up a small hollow, which becomes the first nest chamber. The female remains in this chamber for several weeks, during which time the eggs are laid and the grublike larvae are fed to maturity by the female. The food is produced apparently from the histolysis of the wing muscles and the fat body of the female. It is extruded from her mouth as a secretion. The mature larvae pupate and soon emerge as small workers. These break out of the nest chamber, seek food, and henceforth keep the female, or queen, and the next brood of workers provided with food. Subsequent broods help with the task of keeping the colony provisioned. The female continues to lay eggs, without further fertilization, for several years.

Colonies of many species contain only a few dozen or a few hundred individuals, whereas those of other species may attain a population of many thousands. The small colonies are usually situated under stones, in stumps, logs, or in galleries in the soil. Many of the large colonies build large mounds of earth, sticks, and debris, interspersed with a complex system of galleries and chambers.

In the main, ants are omnivorous, feeding on living or dead animal matter (especially other insects), vegetable substances such as fungi, and sweet exudates or secretions of plants, such as honeydew, nectar, wound discharges, and glandular products. Certain insects such as aphids and some scale insects produce honeydew or other secretions; the ants tend these insects with great care and "harvest" the sweet substances produced. Some ant species are extremely savage and live almost entirely by the capture of live insect prey. Others live almost entirely on sweets, fats, or dead insects.

Several ant species invade houses or stores and are among the most persistent domestic pests. In the northern states the thief ant *Solenopsis molesta,* Pharaoh ant *Monomorium pharaonis,* and the odorous house ant *Tapinoma sessile* are common household species. In the southern states the introduced Argentine ant *Iridomyrmex humilis* is an exceedingly common household pest and has almost replaced the native ant population in many localities.

Some of the ants feed primarily on plant seeds. Of these ants, various species of the genus *Pogonomyrmex,* known as "agricultural ants," have become abundant and are destructive in the grain and grass areas of the Great Plains and westward.

OTHER VESPOID WASPS. Several small families of wasps are closely related to the Scoliidae and Vespidae. Many of these families are rare or their members are small and secretive, and seldom collected. One is frequently seen: the green, metallic cuckoo wasps or Chrysididae that lay their eggs in the burrows of other kinds of wasps, and whose larvae feed on the provisions stored by the host wasp.

PROVISIONING WASPS AND BEES

These wasps not only sting their prey, but usually carry the prey to a previously prepared cell, then lay an egg on the provisions when enough have been stored. The most primitive of these appear to be the spider wasps or Pompilidae (sometimes called Psammocharidae) that provision their nests with spiders. The bees evolved from the wasps.

In these wasps and the bees, each corner of the pronotum ends in a round lobe that is situated below the level of, and does not touch, the tegulae. This lobe is only moderately indicated in the Pompilidae, fig. 237*A*, but well defined in the bees and other wasps, fig. 234*H*.

SUPERFAMILY SPHECOIDEA, THE SOLITARY WASPS. In these insects the hairs are simple and undivided, in contrast to those of the bees, and the hind tibiae and tarsi are not modified for pollen gathering. The group is a large one and includes a great diversity of sizes, shapes, and colors. As currently defined it comprises the rare family Ampulicidae and the heterogeneous and large family Sphecidae. The habits of all Sphecidae are essentially the same. The female wasp makes a mud nest or a nest excavated in pith, wood, or soil and provisions it with a particular kind

Fig. 250.   The yellow and black mud dauber wasp, *Sceliphron servillei*, that builds series of mud cells on stones and walls and provisions them with spiders.   (From Essig, *Insects of Western North America*, by permission of The Macmillan Co.)

of paralyzed prey. An egg is deposited in each stocked compartment of the nest, this egg hatching into a legless grub which feeds on the provender stored for it by the parent. In temperate climates the larva overwinters in its cocoon, pupating the following year in early summer.

The various wasp groups are usually specific in the prey they choose for provisioning the larval cells. The Pemphredoninae (often only 2 or 3 mm. long) capture aphids, the Sphecinae usually use caterpillars, the Trypoxylinae use spiders, and so on. One of the most interesting and showy species in the central and eastern states is the cicada killer *Sphecius speciosus.* This is a large black and yellow species which attains a length of 40 or 50 mm. It commonly captures and paralyzes the common cicada *Tibicen linnei* and carries it to a burrow in the ground, provisioning each burrow with one cicada. When attacked by the wasp, the cicada makes a loud piercing noise, but this subsides in a moment as the cicada is stung and becomes paralyzed.

The most familiar of the solitary wasps are the thread-waisted mud daubers, especially species of *Sceliphron,* fig. 250. These mud daubers build a mud nest of several cells. The nests are common under bridges, eaves of houses, or other sheltered places. The wasps gather the mud at the edge of near-by pools or puddles, flying back and forth from water's edge to nest with mouthfuls of "plaster." The cells are provisioned with spiders.

SUPERFAMILY APOIDEA, THE BEES. This group, fig. 251, includes all the native and domestic bees, comprising the six families keyed on p. 300. They are morphologically very similar to the Sphecidae, differing in having branched body hairs, fig. 252, which give them a fuzzy or velvety appearance. Typically they have the hind legs or the venter of the

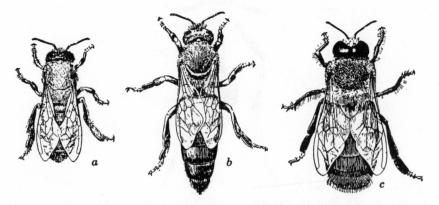

Fig. 251. The honey bee. *a,* worker; *b,* queen; *c,* drone. (From U.S.D.A., E.R.B.)

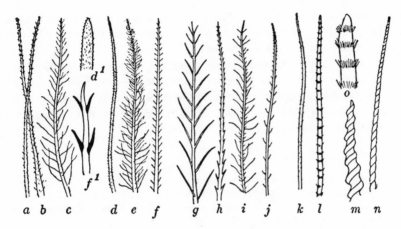

a b c   d e f   g h i j   k l   m n

Fig. 252. Hairs of bees. (After J. B. Smith)

abdomen provided with areas of fairly long hairs used in collecting pollen; in the males of some genera and in both sexes of parasitic genera these structures are lacking. Most of our 3000 species are solitary in habit, making cells in burrows or cavities as do many solitary wasps. The bees provision the nest cells with honey and pollen, which constitutes the food of the larvae. Certain genera of bees are "parasitic," laying their eggs in the cells or nests of other bees. The intruder larva matures faster than the host larva and eats up the stored food. These "parasitic" bees could be called "cowbird bees."

Of our native fauna only the bumblebees (the genus *Bombus*, family Apidae) have developed social living. In this group fertilized females overwinter in log cavities or ground cover. They emerge in spring, find a protected site in a hole in the ground or in a deserted mouse nest, and begin a colony. A brood of eggs and young is raised, the mother, or queen, feeding the larvae on honey. This brood is composed of sterile females or workers that take over the task of gathering food for succeeding larvae. Toward fall no more workers are produced, but instead a swarm of males and functional females. The males die soon after mating, and the workers also die with the approach of winter. The new brood of fertilized queens disperses for hibernation, and the entire colony is disbanded.

The habits of the domestic honey bee *Apis mellifera* are much more specialized than those of the bumblebee. In the first place the colonies do not die out during the winter; their members live during this period on honey stored up throughout the summer. Individual queens have lost the ability to forage for themselves; hence they cannot start a new colony alone but must be accompanied by some workers from the parent colony. The phenomenon of "swarming" is the colonization flight in which the old queens set out to form a new nest.

The honey bee is of considerable importance in that it affords a large cash return from the sale of honey. But the greatest role of the bees, including both our native species and the honey bee, is the pollination of a great variety of wild and domestic plants, including most of our commercial fruits and legume crops.

References

Michener, C. D., 1944. Comparative external morphology, phylogeny, and a classification of the bees. *Am. Museum Nat. Hist. Bull.,* **82**:151–326.
Muesebeck, C. F. W., ET AL., 1951. Hymenoptera of America north of Mexico, synoptic catalogue. *USDA Agr. Monograph,* **2**:1–1420.
Ross, H. H., 1937. A generic classification of the Nearctic sawflies. *Illinois Biol. Monograph,* **15**(2):1–173.
Viereck, ET AL., 1916. The Hymenoptera of Connecticut. *Connecticut State Geol. Nat. Hist. Survey Bull.,* **22**:824 pp. 10 pls., 15 figs.

Order COLEOPTERA: Beetles and Weevils

The adults usually have two pairs of wings: The first pair is veinless, hard, and shell-like, and folds together over the back to make a stout wing cover; the second pair, used for flight, is membranous, usually veined, and in repose folds up under the wing covers or elytra, fig. 253. The body is normally hard and compact. The mouthparts are of the chewing type; the antennae are well developed, usually 10- to 14-segmented; the compound eyes are usually conspicuous; and the legs are heavily sclerotized. The larvae, fig. 261, normally have distinct head capsules, chewing mouthparts, antennae, and thoracic legs, but no

Fig. 253. A stag beetle *Copris minutus.* (From Illinois Nat. Hist. Survey)

abdominal legs. The pupae, fig. 270, have the adult appendages folded against but not fused with the body. The adults vary in size from less than a millimeter long to several inches (up to 100 mm.). The shape and coloring varies just as much, but the most brilliantly colored, weird-shaped, and gigantic species occur in the tropical regions of the world.

Most kinds of beetles are either plant feeders or are predaceous on other insects. Usually both adults and larvae of the same species have similar food habits; that is, both forms will be phytophagous, or both will be predaceous, although the larva may not feed on either the same species of plant or the same part of the plant. Thus the June beetles, which are phytophagous, utilize different parts of plants for food during their development. The adults feed on the foliage of forest trees, but the larvae, known as white grubs, feed on the roots of trees, shrubs, herbs, or grasses. The predaceous beetles are active hunters, stalking their prey. Certain groups have more specialized food habits. Some are endoparasites of other insects or feed on insect egg masses.

The order as a whole is terrestrial. Certain families, however, are aquatic, both larvae and adults living in water. The larvae usually leave the water to pupate, making an earthen cell in nearby soil. As is the case with the land forms, the aquatic group includes both herbivorous and predaceous species, the latter predominating.

The great majority of beetles have a single generation per year and a simple life cycle similar to the following. The oval or round eggs are laid in spring or early summer and hatch in 1 or 2 weeks. The larvae are voracious feeders, usually attaining full growth during the summer and pupating in the soil. The adults emerge in a few weeks, feeding and maturing throughout the remainder of the summer and autumn. With the advent of cold weather they hibernate. The following spring these adults emerge and lay eggs, and the cycle begins again. The old adults usually die soon after egg laying is completed.

There are many deviations from this biological pattern. For example, some ladybird beetles have continuous and overlapping generations throughout the warmer seasons. In other groups, notably many species whose larvae live in soil or rotten wood, the winter is passed in the larval stage, and the adults occur for only a limited span during spring or summer.

Since they first differentiated from their somewhat neuropteroid ancestors over 300 million years ago, the beetles have evolved into more species than any other group of living things. Probably about 200,000 species have persisted to the present; of these about 25,000 occur in North America, representing about 150 families. By adding together primitive characters of living beetles, it is apparent that all of them arose

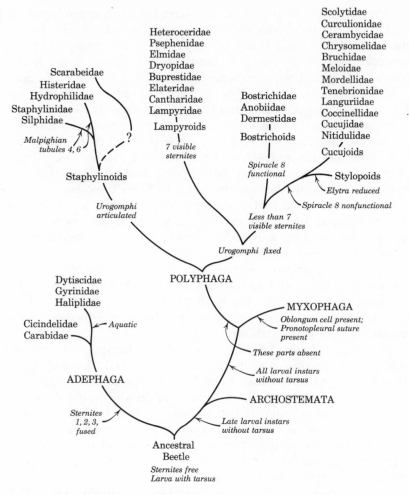

Fig. 254. Simplified family tree of the order Coleoptera (compiled chiefly from Crowson).

from an ancestral beetle in which the abdominal sternites of the adults were free and distinct and the larval legs possessed a tarsal segment. From this ancestor two lines arose. In one line the larval tarsi persisted but the first three abdominal sternites fused and the hind coxae became immovably attached to and divided the first abdominal sternite, fig. 255, 1. This line gave rise to the suborder Adephaga, including a primitive small family Rhysodidae, found in rotten logs, the terrestrial predators related to the ground beetles (Carabidae), and the predaceous water beetles (Dytiscidae and others).

In the other line arising from the primeval coleopteran, the abdominal segments remained separate but the larval tarsi were lost. The suborder Archostemata seems to be the most primitive existing member of this line, because the tarsal segment of the larval legs is still present in early instars and lost only in later instars. This suborder includes only a few small, rare families including the Cupedidae and Micromalthidae. In a line arising from an ancestor having this condition the larval tarsus was lost completely. This line evolved into the suborder Polyphaga, containing the great bulk of living beetles. In many branches of the Polyphaga, certain families have become extremely specialized and quite different morphologically from more primitive members of their own branch. Certain minute fungus beetles, for example, have lost all their hind-wing venation, most of their tarsal segments, and undergone peculiar changes of body sclerites associated with minute size. In describing the various branches it is therefore impossible to do more than indicate the characters of primitive members of each branch.

The most primitive members of the line that gave rise to the Polyphaga are a few rare, small, beetles (Hydroscaphidae and Sphaeriidae) possessing the crossvein setting off the oblongum cell in the hind wing and the suture between the pronotum and propleuron. These are considered a separate suborder, the Myxophaga. The oblongum cell and pronotopleural suture were lost in the line that evolved into the suborder Polyphaga. The ancestor of this suborder gave rise to two branches. In the more primitive (the staphylinoid branch), the urogomphi (cerci) of the larvae remained articulated; in the other the urogomphi fused solidly with their parent segment.

The staphylinoid branch contained two major lineages, one having four malpighian tubules, including the rove beetles (Staphylinidae), carrion beetles (Silphidae) and their allies, the other having six malpighian tubules, including the Histeridae and Hydrophilidae. The affinities of the scarab beetles (Scarabeidae) are difficult to determine; it has been suggested that they are an unusual offshoot of the staphylinoid branch.

The other branch of the typical Polyphaga gave rise in turn to two subsequent branches. In the lampyroid branch the adult body of the more primitive families retained at least seven visible abdominal sternites. This branch evolved into an amazing assortment of families, including the soldier beetles (Cantharidae) and fireflies (Lampyridae), the click beetles (Elateridae) and flat-headed borers (Buprestidae), many water beetles (Psephenidae and allies), and many others.

In the sister branch of the lampyroids, the abdomen has at most six visible sternites. This branch gave rise to the bostrichoids, in which all

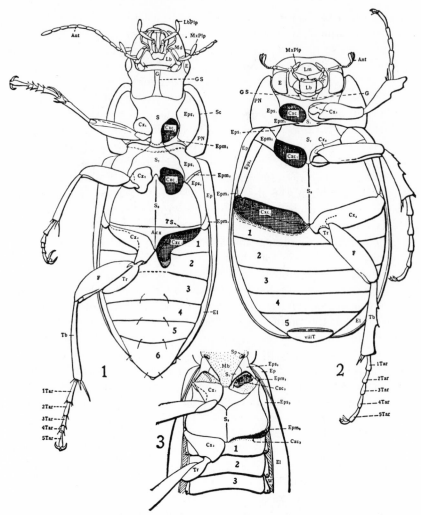

Fig. 255.   (1) Ventral view of a ground beetle *Harpalus* sp., Carabidae.   Left legs are removed.
(2) Ventral view of a May beetle *Phyllophaga* sp., Scarabaeidae.   Right legs are removed.
(3) Ventral view of part of thorax and abdomen of a soldier beetle *Chauliognathus pennsylvanicus*,
Cantharidae.   *Acx*, antecoxal piece; *Ant*, antenna; *Cx*$_{1-3}$, 1st, 2d, and 3d coxae; *Cxc*$_{1-3}$, 1st,
2d, and 3d coxal cavities; *E*, eye; *El*, elytron; *Ep*, epipleuron; *Epm*$_{1-3}$, epimera of the pro-,
meso-, and metathorax; *Eps*$_{1-3}$, episterna of pro-, meso-, and metathorax; *F*, femur; *G*, gula;
GS, gular suture; *Lb*, labium; *LbPlp*, labial palpus; *Lm*, labrum; *Mb*, membrane; *Md*, mandible; *MxPlp*, maxillary palpus; *PN*, pronotum; *S*$_1$, *S*$_2$, *S*$_3$ pro-, meso-, and metasterna;
Sc, suture separating pronotum from episternum; *Sp*, spiracle; 1 *Tar* to 5 *Tar*, the five tarsal
segments; *Tb*, tibia; *Tr*, trochanter; *TS*, transverse suture; 1 to 6, abdominal sternites; viii*T*,
8th tergite.   (After Matheson, *Entomology for introductory courses*, by permission of Comstock
Publishing Co.)

the spiracles in the adult remained functional, and the cucujoids, in which the spiracle on the adult eighth abdominal segment became non-functional. The bostrichoids evolved into the larder beetles and powder-post beetles (Dermestidae, Anobiidae, and Bostrichidae) and their allies. The cucujoids evolved into many families diverse in both habits and appearance. More common forms include the ladybird beetles (Coccinellidae), the darkling beetles (Tenebrionidae), and the principal group of plant-feeding beetles (Chrysomelidae, Scolytidae, and allies).

The structures of the stylopids (superfamily Stylopoidea) are so highly modified that the exact affinities of the group are obscure. In the adult the spiracles of the eighth abdominal segment are non-functional. On this and other tenuous evidence it is probable that the group arose either as a sister group of the cucujoids or from some ancestor among the cucujoids.

KEY TO COMMON FAMILIES

1. Front of head produced into a definite beak, fig. 278, that may be long or short, the antennae arising from side of beak; palpi vestigial .. **Curculionidae,** p. 344
   Front of head not produced into a beak; if slightly so, antennae arising between the eyes, or maxillary palpi prominent, fig. 255 2
2. Middle and hind legs very wide and flat, almost paper-thin, fitted for swimming, the basitarsus large and triangular, the next two produced laterally to form long swimming "fingers"; front legs tubular, fitted for grasping, fig. 256*L*; each eye completely divided, one part on dorsum of head, the other on ventral aspect of head, fig. 256*K* . . . . **Gyrinidae,** p. 329
   Middle and hind legs having some or most segments robust and not flattened, occasionally furnished with rows of long hairs and fitted for swimming in this fashion; eye seldom divided, and then only by a continuation of a head flange or by an antennal base, fig. 258*K* 3
3. Maxillary palpi longer than antennae and slender, resembling antennae, fig. 256*Q*; chiefly aquatic . . . . **Hydrophilidae**
   Maxillary palpi shorter than antennae, and not antenna-like; antennae various 4
4. Antenna with each of last 3 to 7 segments enlarged on one side to form an eccentric plate or lamella, fig. 256*A*, *C*, each lamella situated nearly at a right angle to long axis of antenna 5
   Antenna not lamellate, but frequently having the end segments fairly evenly enlarged to form a club, fig. 256*E*, or most of the segments with anterior projections giving a saw tooth, fig. 256*G*, *J*, or pectinate outline 6
5. Elytra short and squarely truncate, exposing three full tergites of abdomen, these tergites heavily sclerotized and hard, fig. 259*B* . . . . **Silphidae**
   Elytra longer, usually rounded at apex, usually covering entire dorsum of abdomen but in a few species exposing one or two tergites, fig. 262

Scarabeidae, p. 331

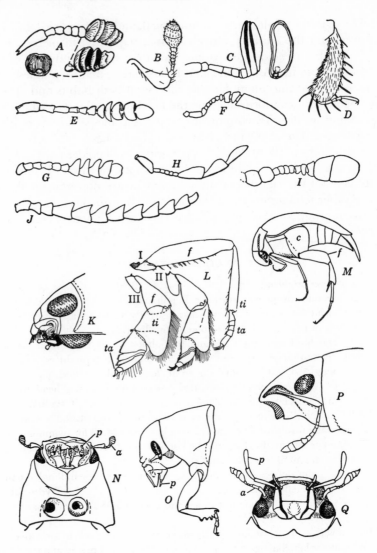

Fig. 256. Diagnostic characters of Coleoptera. *A*, antenna of *Nicrophorus,* Silphidae, insets showing concave end segments and an end view of one; *B*, antenna of *Scolytus,* Scolytidae; *C*, antenna of *Thyce,* Scarabeidae; *D*, tibia and tarsus of *Heterocerus,* Heteroceridae; *E*, antenna of *Silpha,* Silphidae; *F*, antenna of *Attagenus,* Dermestidae; *G*, antenna of *Languria,* Languriidae; *H*, antenna of *Stegobium,* Anobiidae; *I*, antenna of *Anthrenus,* Dermestidae; *J*, antenna of *Melanotus,* Elateridae; *K*, head of *Dineutes,* Gyrinidae; *L*, legs of *Dineutes,* Gyrinidae; *M*, profile of *Mordellistena,* Mordellidae; *N* and *O*, head and prothorax of *Dendroctonus,* Scolytidae; *P*, head of a short-snouted weevil, *Curculionidae; Q*, head of *Tropisternus,* Hydrophilidae. *a*, antenna; *c*, coxal plate; *f*, femur; *p*, maxillary palpus; *ta*, tarsus, *ti*, tibia.

6. Elytra short, exposing five or more sclerotized abdominal tergites, fig. 261*A* . . 7
Elytra covering all or most of abdomen, never more than two or three sclerotized tergites visible from above; occasionally the abdomen of a female of this group may be extremely distended with eggs, and a third or fourth segment may project from beneath the elytra, but all except the apical two are soft and semimembranous 8

7. Elytra truncate, parallel-sided, and abutting evenly down the meson, fig. 261; abdomen hard and regular in outline . . . . **Staphylinidae,** p. 330
Elytra ovate, overlapping considerably at base; abdomen flabby and shrinking irregularly when the specimen dries . . . . **Meloidae,** p. 338

8. Hind tarsus 4-segmented, front and usually middle tarsi 5-segmented 9
Either hind tarsus having 3 or 5 segments, or all tarsi having the same number of segments 11

9. Body narrow and deep, bilaterally compressed, and with the hind coxa forming a large plate that appears as a major sclerite in the side of the thorax, fig. 256*M*; small beetles often found abundantly in flowers . . . . . **Mordellidae**
Body wider than deep, often flattened, hind coxae no larger than in fig. 255, 2 10

10. Each front coxal cavity closed posteriorly by a projection of the pleuron which meets the apex of the sternum, fig. 258*D*; lateral edge of pronotum forming a sharp flange or delineated by a ridge or carina (compare fig. 258*H*)
                                              Tenebrionidae, p. 341
Front coxal cavities open posteriorly, the posteromesal corner of propleuron not extending mesad of outer portion of coxa, fig. 258*C*; lateral edge of pronotum rounding inconspicuously into pleural region . . . . **Meloidae,** p. 338

11. Head not retracted within prothorax 12
Head retracted within prothorax, so that only anterior portion protrudes, fig. 258*H* 14

12. Ventral portion of head having a large convex gular region with a single gular suture down the middle; and the palpi very short or indistinct, fig. 256*N*; antennae always elbowed, with a long first segment, and sometimes ending in a flat club, fig. 256*B* 13
Ventral portion of head having either a small gular area or two gular sutures, and the maxillary palpi usually much longer, fig. 255, 1; antennae not elbowed or the first segment shorter in proportion to the remainder 14

13. Antenna long, fig. 256*P*, usually ending in a small cylindrical, elliptic club; side of head having a deep groove for reception of antenna
                                               Curculionidae, p. 344
Antenna short, fig. 256*N, O*, ending in a large flat club or comb, fig. 256*B*; side of head without antennal groove . . . . **Scolytidae,** p. 345

14. Hind and middle tarsi either 3- or 4-segmented, or fourth segment very small in comparison with the third, fig. 257*b, c* 15
Hind and middle tarsi 5-segmented, the fourth as large as or as thick as the third, fig. 257*a* 27

15. All tarsi having the third segment enlarged, fig. 257*b, c*, deeply bilobed or channeled dorsally; the fourth segment extremely small, either sunken into the

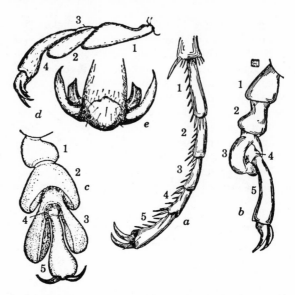

Fig. 257. Tarsi of Coleoptera. *a, Harpalus,* Carabidae; *b, Leptinotarsa,* Chrysomelidae; *c, Chelymorpha,* Chrysomelidae; *d, Epilachna,* Coccinellidae; *e,* toothed tarsal claws. (From Matheson, *Entomology for introductory courses,* by permission of the Comstock Publishing Co.)

cleft of the third or arising from dorsum of the base of the third, fig. 257*b*; fifth segment large and normal; frequently the fourth segment cannot be seen, or it appears as a minute subdivision of the base of the fifth segment, in which case each tarsus appears 4-segmented............................ 16
Either tarsi only 3-segmented, or third segment not enlarged, channeled, or bilobed 42
16. Antennae about as long as or longer than the body, fig. 277
Many **Cerambycidae**, p. 343
Antennae distinctly shorter than body............................. 17
17. Last tergite (pygidium) exposed, almost completely visible beyond or below end of elytra, fig. 258*E, F*.. 18
Last tergite almost entirely or completely covered by elytra............. 21
18. Hind tibia little, if any, longer than basitarsus, and having a pair of long apical spurs, fig. 258*P* .... **Bruchidae**, p. 343
Hind tibia much longer than basitarsus, frequently with only short spurs or none
.... 19
19. Elytra short, each only about twice as long as wide; stocky short species, the pygidium oblique, and at a definite angle to dorsal contour of the body, fig. 258*F* 20
Elytra long, each four times or more as long as wide; elongate species, the pygidium nearly horizontal, following dorsal contour of the body, fig. 258*E*
some **Cerambycidae**, p. 343

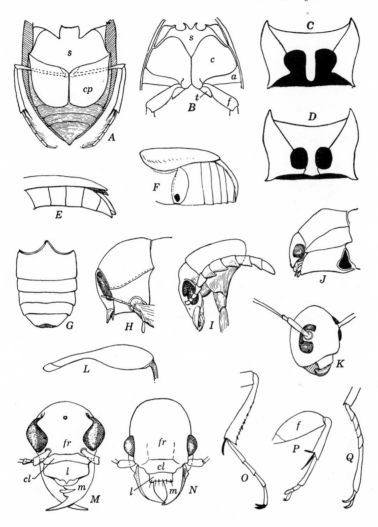

Fig. 258.   Diagnostic characters of Coleoptera.   *A*, venter of *Haliplus,* Haliplidae; *B*, venter of *Dytiscus,* Dytiscidae; *C*, prosternum showing open coxal cavities, diagrammatic; *D*, prosternum showing closed coxal cavities, diagrammatic; *E*, abdomen of *Leptura,* Cerambycidae; *F*, abdomen of *Bruchus,* Bruchidae; *G*, venter of *Criocerus,* Chrysomelidae; *H*, head and prothorax of *Cryptocephalus,* Chrysomelidae; *I*, head and prothorax of *Acanthoscelides,* Bruchidae; *J*, head and prothorax of *Helichus,* Dryopidae; *K*, head of *Saperda,* Cerambycidae; *L*, hind femur of *Neoclytus,* Cerambycidae; *M*, head of *Cicindela,* Cicindelidae; *N*, head of a Carabidae; *O*, hind leg of *Amphicerus,* Bostrichidae; *P*, hind leg of *Amblycerus,* Bruchidae; *Q*, hind tarsus of *Psephenus,* Psephenidae.   *a*, abdomen; *c*, coxa; *cl*, clypeus; *cp*, coxal process; *f*, femur; *fr*, frons; *l*, labrum; *m*, mandible; *s*, sternum; *t*, trochanter.

20. Anterior part of prothorax forming a cylinder against which the flat head fits like a lid; eyes oval, fitting against margin of prothorax, fig. 258*H*
      some **Chrysomelidae**, p. 341
   Anterior part of prothorax narrow, head projecting freely beyond it; eyes incised and V-shaped, head constricted behind them, fig. 258*I*
      most **Bruchidae**, p. 343
21. Last 3 to 5 antennal segments enlarged to form a large loose club, fig. 256*G*; elongate, smooth, and highly polished species . . . . **Languriidae**
   Antenna of uniform thickness throughout, or widening gradually to form a club, or forming a round compact club; form various 22
22. Hind femur greatly enlarged, short, and oval in outline; flea beetles and their relatives . . . . **Chrysomelidae**, p. 341
   Hind femur elongate or more parallel-sided, or constricted at base and enlarged only at apex, fig. 258*L* 23
23. First abdominal sternite very long, its mesal length nearly equal to that of the following four sternites, combined, fig. 258*G*; several genera of varied form (e.g., **Donacia, Crioceris**) . . . . **Chrysomelidae**, p. 341
   First abdominal sternite considerably shorter in proportion to following segments 24
24. Either antenna as long as or longer than body, or hind femur having basal portion slender and apical portion enlarged and clavate, fig. 258*L*
      Cerambycidae, p. 343
   Antenna shorter than body, and hind femur without a basal stalklike portion
   25
25. Tibia having a pair of well-developed tibial spurs at apex
      Cerambycidae, p. 343
   Tibia having either no tibial spurs or very minute ones 26
26. Mesal margin of eye deeply incised, with the antenna situated in the incision, fig. 258*K*, or the eye completely divided, the antenna situated between the two parts . . . . **Cerambycidae**, p. 343
   Either eye not incised, or antenna not at all in incision of eye
      Chrysomelidae, p. 341
27. Hind coxa having a wide long ventral plate which covers coxa, trochanter, and most of the femur, fig. 258*A*; small to medium-sized stout aquatic beetles
      Haliplidae
   Hind coxa at most having only a small outer plate, which does not cover trochanter or femur 28
28. Apices of hind coxae forming a double-knobbed process; base of each coxa extremely large and platelike, appearing as a dominant sclerite of the sternum, fig. 258*B*; hind legs fringed for swimming . . . . **Dytiscidae**
   Hind coxa neither with such a knobbed apex nor with such a platelike base, fig. 255, 1, 2 29
29. First abdominal sternite completely divided by the hind coxal cavities, the sternite appearing as a pair of triangular sclerites, one on each side of the coxae; first three abdominal sternites immovably united, the separating

sutures appearing partly as extremely fine lines, fig. 255, 1 30
Either first abdominal sternite not completely or not at all cut into by the coxal
cavities, or the first three sternites with separating sutures extremely well
developed across the entire width of the segment, as are those between the
more apical segments, fig. 255, 2 31

30. Clypeus fairly narrow, the antennal sockets wider apart than the width of the
clypeus, fig. 258*N* . . . . **Carabidae**, p. 328
Clypeus much wider, fig. 258*M*, the antennal sockets situated closer together
than the width of the clypeus . . . .**Cicindelidae**

31. Abdomen having 6 or more exposed sternites 32
Abdomen having not more than 5 exposed sternites, fig. 255, 2 35

32. Last 3 to 5 antennal segments enlarged to form a club, fig. 256*E* . . . . .**Silphidae**
Antenna the same width throughout, often beadlike or serrate 33

33. Tarsus slender and smooth, segments 1 to 4 very short and ringlike, segment 5
long, about equal in length to first 4 combined, fig. 258*Q*; aquatic forms
Psephenidae
Tarsus stout and densely setose, segment 5 no longer than segment 1, some of the
basal segments elongate, as in fig. 257*a* 34

34. Prothorax with broad anterior and lateral margins forming a hood that covers
most or all the head from above, fig. 264; head partially retracted, viewed
from side . . . . **Lampyridae**, p. 333
Prothorax not having wide margins, head not retracted in pronotum, hanging
down or projecting forward freely . . . .**Cantharidae**

35. Antenna elbowed and capitate, the club appearing as one round segment, sometimes having faint cross sutures; very hard shining black beetles having short
stout legs, the tibiae expanded and spurred for digging, much as in fig. 255, 2
Histeridae
Antenna not elbowed, and either not thickened toward tip, or the enlarged portion composed of 2 or 3 well-separated segments or a single elongate segment;
legs or body shape different from above 36

36. Antennae elongate and serrate throughout, fig. 256*J* 37
Antennae short, filiform, or clavate 38

37. Pronotum having a sharp projection at each posterolateral corner, fig. 265
Elateridae, p. 334
Pronotum having posterolateral corners rounded and not pointed, fig. 266
Buprestidae, p. 335

38. Prosternum produced anteriorly to form a long concave shelf under the head,
but the latter largely retracted into the opening of the prothorax, fig. 258*J*
. . . . 39
Prosternum not so produced, much shorter than pronotum and proportioned
more as in fig. 258*I*; head usually held downward against chest 40

39. Front coxae round, situated some distance from lateral edge of pronotum;
antennae elongate or definitely clavate . . . .**Elmidae**
Base of front coxae triangular, the point of the triangle extending nearly to
lateral edge of prosternum, fig. 258*J*; antennae short, the flagellum of equal

thickness throughout, with wide flat segments.......... **Dryopidae,** p. 335
40. Antennae ending in a distinct club composed of 1 to 3 segments, fig. 256*F*, I
    Dermestidae, p. 337
    Last 3 segments of antenna greatly enlarged but well separated to form a chain,
    fig. 256*H* 41
41. Tibiae without spurs; small convex forms including drugstore and cigarette
    beetles, infesting dried food products........................ **Anobiidae**
    Tibiae with distinct spurs, fig. 258*O*; includes many subcylindrical forms orna-
    mented with horny processes and ridged areas; the powder post beetles
    Bostrichidae
42. Front and middle tarsi 4-segmented, hind tarsus 3-segmented......**Cucujidae**
    All tarsi having the same number of segments........................ 43
43. Tarsi 4-segmented, all segments well marked, and second and third of about
    equal width 44
    Tarsi 3-segmented, or third segment minute and hidden at the base of an
    enlarged second segment, fig. 257*d*............................. 45
44. Tibiae dilated and armed with a series of stout spurs, fitted for digging, fig. 256*D*;
    molelike beetles covered with dense pile, inhabiting wet mud or sand banks
    Heteroceridae
    Tibiae either not dilated or not armed with a series of spurs; flat beetles found
    under bark or in stored grain and feed........................ **Cucujidae**
45. Second segment of tarsus dilated, with a large ventral pad, fig. 257*d*; almost
    hemispherical beetles, usually polished and strikingly patterned
    Coccinellidae, p. 337
    Second segment of tarsus not much if at all wider than first segment; beetles
    often somewhat angular in outline, and flat; frequenting flowers, and tree sap
    Nitidulidae

SUBORDER ADEPHAGA

This suborder contains eight families, most of them predaceous, feeding on other insects. Two common families, the Cicindelidae (tiger
beetles) and Carabidae, are terrestrial. Three other common families,
the Gyrinidae, Haliplidae, and Dytiscidae, are aquatic, both adults and
larvae living in the water.

CARABIDAE, GROUND BEETLES. These are active terrestrial species with
long slender antennae of even thickness, long elytra, and long legs suitable
for running, fig. 259*A*. They vary in size from 1 mm. in length to large
metallic-colored species 35 mm. long. The mouthparts are well developed, and the mandibles are long, strong, and sharp. The adults are
almost entirely nocturnal, hiding during the day in logs or cavities and
under ground cover or stones. They come out at night to forage for prey.
A wide variety of animals, including snails, worms, and adult and immature insects of many kinds, make up their food. The larvae are slender,
with strong mouthparts, well-developed legs, and a pair of terminal

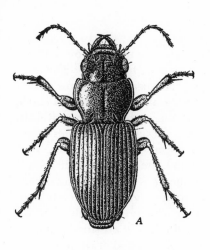

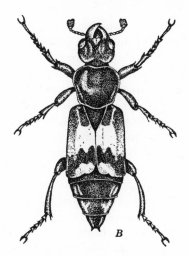

Fig. 259. Beetles. *A*, ground beetle *Pterostichus substriatus; B*, carrion beetle *Nicrophorus marginatus.* (From Kansas State College)

urogomphi (cercus-like organs). They are predaceous, feeding on other insects. They are usually found in the soil or under ground cover, where they hunt their prey. The family is a large one, embracing over two thousand nearctic species.

GYRINIDAE, WHIRLIGIG BEETLES. Aquatic beetles, the adults hard, convex, shining, and dark. The middle and hind legs are broad, fringed

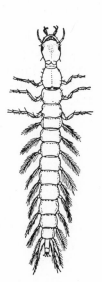

Fig. 260. Larva of whirligig beetle. (Redrawn from Böving and Craighead)

with hair, and used for swimming.   The head has two peculiarities: (1) short antennae with an earlike expansion of the third segment, and (2) eyes that are completely divided into an upper and a lower half, so that the beetle appears to have two pairs of eyes, fig. 256K.   The adults are common in lakes and streams.   They congregate in swarms of thousands, and zigzag along the surface, moving at high speed and leaving silvery crisscrossing wakes which are a familiar sight to every naturalist. During these surface gyrations, the divided eyes give the beetles the opportunity of seeing into the air with the dorsal eyes and into the water with the ventral pair.   The larvae, fig. 260, are slender, white-bodied, and elongate.   Each segment of the abdomen has a pair of long tracheal gills, and at the end of the abdomen are a pair of hooks.   Both adults and larvae are predaceous, the adults feeding chiefly on small organisms falling or alighting on the water surface, the larvae on small organisms they find in sheltered places on the bottom of the pond or stream.

SUBORDERS ARCHOSTEMATA AND MYXOPHAGA

In North America, the Archostemata are represented by two small, old-log-inhabiting families, Cupedidae and Micromalthidae.   The latter includes only a single rare, small species (adults 1.8 to 2.5 mm.) which is remarkable for its paedogenetic larvae.

Of the three families of Myxophaga, two occur in North America. The minute (1.3 mm.) Hydroscaphidae are aquatic and the larger (2–7 mm.) Scaphidiidae live in ground cover and old logs.

SUBORDER POLYPHAGA

This includes the great bulk of the beetles, containing over one hundred and thirty families, and embracing forms diverse in appearance and life history.   Some families, such as the Chrysomelidae and Coccinellidae, are abundant and include species of considerable economic importance. Many families, however, are rare or seldom seen except as a result of specialized collecting and have no known economic importance.   The families mentioned below are only a few of the many beetle families in the suborder Polyphaga.   Concerning families not discussed here, students will find a great deal of information on life history and identification in the books listed at the end of the section on Coleoptera.

STAPHYLINOID SERIES

STAPHYLINIDAE, ROVE BEETLES.   Slender elongate beetles, fig. 261A, with short truncate elytra beneath which the pair of flying wings are folded.   The antennae are fairly long, either filiform or slightly enlarged at the tip.   The larvae are also elongate and look much like small cara-

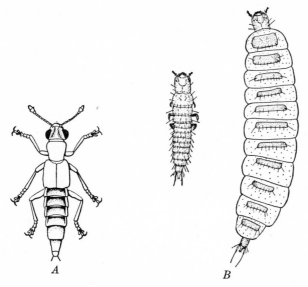

Fig. 261. Rove beetles. *A, Stenus* adult; *B*, first-instar larva of *Aleochara curtula*, left, before feeding; right, after feeding. (*A*, after Sanderson; *B*, modified from Kemner)

bid larvae. Both adults and larvae are scavengers or predators, with the exception of a few species that have parasitic larvae. The adult rove beetles are found on flowers, in ground cover, under bark, in rotting organic material, in ant and termite nests, and in many other situations. The larvae are more secretive and occur chiefly in humid places. The species that are predaceous feed on mites, small insects, insect and mite eggs, and especially on small dipterous larvae. Certain species are valuable factors in the natural control of pests. An interesting example is the genus *Baryodma*, whose larva is a parasite of the cabbage-maggot pupa. The larva of *Baryodma* is unusual in that during the first instar it is a slender free-running form with relatively large tergal plates. It searches in the soil for a fly puparium and enters it by boring an opening. In the puparium the beetle larva feeds on the fly pupa and grows amazingly, becoming greatly distended and grublike before molting. Later instars are always grublike. The same phenomenon occurs in some species of *Aleochara*, fig. 261*B*.

SCARABEIDAE, THE LAMELLICORN BEETLES, OR SCARABS. This is one of the largest families of beetles, characterized most conspicuously by the lamellicorn antennae, in which the apical segments are leaflike and appressed in repose. The scarabs vary greatly in size and shape; most of them are stout and very hard shelled; the larvae are sluggish, stout,

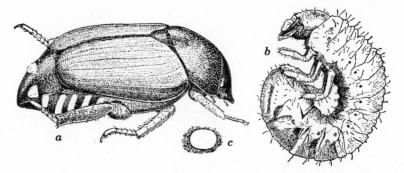

Fig. 262. The Japanese beetle *Popillia japonica*. (From Connecticut Agricultural Experiment Station)

usually white, and with a characteristic curved outline, fig. 262; a number of groups of scarabs are scavengers, and feed on dung, rotting hides, or fungi. Of unusual interest are *Canthon* and certain other genera; the adult fashions a ball of dung, rolls it away, and buries it. Eggs are laid on this ball, and the developing larvae utilize it as food. The remainder of the scarabs are phytophagous, many species of great economic importance. The most publicized member of the family is the Japanese beetle *Popillia japonica*, fig. 262; the larvae feed on grass roots and are especially destructive to lawns and golf courses, and the adults defoliate fruit and shade trees. Another group of destructive species are the June beetles, fig. 145, members of the genus *Phyllophaga;* the adults defoliate deciduous trees, and the larvae, known as white grubs, eat the roots of various grass crops, including corn, small grains, and pasture plants.

Fig. 263. *Dynastes tityus*, Illinois' largest beetle, and a shrew *Sorex longirostris*, Illinois' smallest mammal. (Drawing loaned by C. O. Mohr)

In the tropics there are many species of large scarabs, often brilliantly colored or ornamented with spines or projections on the head or pronotum. In the United States the largest species is *Dynastes tityus* the rhinoceros beetle, which is larger than a shrew, fig. 263. The larva lives in rotten wood.

LAMPYROID SERIES

LAMPYRIDAE, FIREFLIES. These are moderate-sized soft-bodied beetles, having serrate antennae and having the margins of the pronotum projecting like a flange or shelf, which partially covers the head. The elytra are relatively soft. A common eastern species is *Photurus pennsylvanicus*, fig. 264. The adults occur in summer and fly actively on warm nights, almost invariably following a dipping up-and-down course only a few feet above the ground. As they start their "upstroke," each individual flashes a bright light, a mating signal. When swarms of adults are on the wing, the entire countryside is lighted up with these tiny dots of light. From this comes the name firefly. In certain genera the females are wingless and grublike; these and the larvae are also luminous and are called glowworms. Both larvae and adults are predaceous, although some adults may feed partly on plant material or not at all.

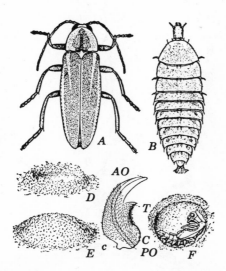

Fig. 264. A firefly *Photurus pennsylvanicus*. *A*, adult male; *B*, mature larva; *C*, left mandible of larva to show mandibular canal. *AO*, opening of canal; *PO*, opening of canal to mouth; *c*, condyle; *T*, tooth on mandible; *D*, beginning of pupal chamber; *E*, completed pupal chamber; *F*, pupa in chamber. (From Matheson, after Hess)

ELATERIDAE, CLICK BEETLES OR WIREWORMS. The adults are trim and hard bodied, fig. 265, having five-segmented tarsi, a large pronotum with sharp posterior corners, serrate antennae, and a long stout sharp process projecting backwards from the prosternum. If placed on their backs, these beetles can spring several inches into the air, at the same time making a loud click, usually alighting right side up. This leap is engineered by using the ventral process of the prosternum as a sort of spring release when the body is tensed. The adults are frequently encountered during spring and summer. These beetles and their acrobatics are well known to youngsters in rural districts.

The larvae, called wireworms, fig. 265, are wormlike and hard-bodied and live in soil or rotten wood. The soil-inhabiting larvae feed on roots of grasses and related plants and are extremely destructive to many of the grain crops. Especially injurious are species of the genera *Melanotus, Agriotes,* and *Monocrepidius.* Several hundred species occur in North America. Of unusual interest to the collector of immature insects are

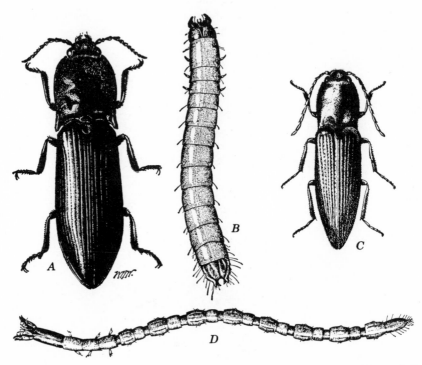

Fig. 265. Elateridae. *A, B,* the dry-land wireworm *Ludius inflatus,* adult and larva; *C, D,* the sand wireworm *Horistonotus uhlerii,* adult and larva. (From U.S.D.A., E.R.B.)

the odd wormlike larvae of the group to which *Horistonotus* belongs, fig. 265*D*.

BUPRESTIDAE, METALLIC OR FLATHEADED WOOD BORERS. These beetles, fig. 266, resemble the click beetles in that the adults have serrate antennae, five-segmented tarsi, and hard bodies, but they differ in having the first two sternites of the abdomen fused and in the coppery or bright metallic coloring of the body and elytra. They are usually more robust and have a shorter pronotum. The larvae of the larger species, fig. 266, bore in wood, attacking live trees or newly felled or killed trees, and feeding either beneath the bark or into the solid wood. These larvae are legless and elongate, having the thorax expanded and flattened. They attack a wide variety of trees, including deciduous and coniferous species. The flatheaded apple tree borer *Chrysobothris femorata* is often an orchard pest of importance. A few genera of small buprestids have leaf-mining larvae, which are more cylindrical than the wood borers and have minute legs.

DRYOPIDAE AND THEIR ALLIES, THE DRYOPID BEETLES. In this group of aquatic beetles both the adults and larvae live in the water. Both are sluggish, crawling over stones or submerged wood, and feeding on surface encrustments. The adult is clothed with fine hair which holds a film of air when under water. The insect uses this film as a means of gas exchange with the surrounding water. The larvae respire by means of tracheal gills situated on various parts of the body, many retractile within ventral pouches. As with other aquatic Coleoptera, the mature larvae leave the water and pupate in damp soil. Most of the dryopids live in cold rapid streams and are frequently found in large numbers in siftings from gravel bars. The adults leave the water periodically for mating or dispersal flights.

Fig. 266. Buprestidae. Left, the flatheaded apple tree borer, *Chrysobothris femorata;* center, the Pacific flatheaded borer, *C. mali;* and, right, larva in burrow. (From U.S.D.A., E.R.B.)

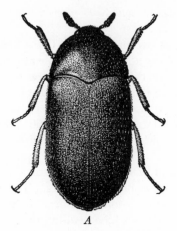

A

Fig. 267.   The black carpet beetle *Attagenus piceus.* *A*, adult; *B*, larva. (From Connecticut Agricultural Experiment Station)

B

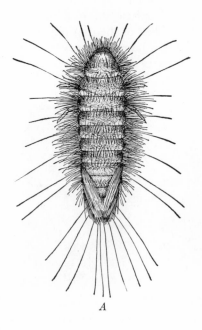

A

B

Fig. 268.   The carpet beetle *Anthrenus scrophulariae.* *A*, larva; *B*, adult. (From Connecticut Agricultural Experiment Station)

BOSTRICHOID SERIES

DERMESTIDAE, CARPET BEETLES. Convex oval beetles, figs. 267, 268, having short clubbed antennae, five-segmented tarsi, and abdomens with only five sclerotized sternites. The larvae are elongate or oval, clothed distinctively with large tufts or bands of long barbed hair. They feed on dried animal products, including fur, skins, and dried meat. The last is attacked readily by species of the genus *Dermestes,* especially *D. lardarius* the larder beetle. Some of the smaller species, particularly the carpet beetle *Anthrenus scrophulariae,* fig. 268, and the black carpet beetle *Attagenus piceus,* fig. 267, attack furs, carpets, and upholstery, in fact, anything made from animal hair. These pests are so widespread that they are a constant menace to household goods and many stored materials. In nature the species feed on dead insects or animal carcasses. The adults feed on the same material as the larvae, but during dispersal flights they feed on pollen and at this time are often found on garden flowers.

CUCUJOID SERIES

COCCINELLIDAE, LADYBIRD BEETLES. Moderately small round convex shining beetles, fig. 269, sometimes prettily patterned with red, yellow, black, or blue markings. The antennae are short and clavate. The tarsi are four-segmented but appear to be three-segmented; the third segment is extremely minute, situated between the padlike second segment and the large end segment bearing the claws. The larvae either are warty creatures or are covered with a waxy secretion and have extremely short antennae but long legs. Two categories of ladybird beetles are of economic importance. Species of the first feed on aphids and scale insects and function as effective means of natural control against some of these pests. The best known of these predators is the vedalia *Rodolia*

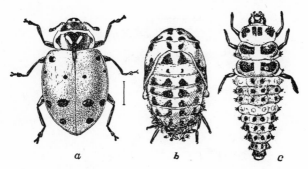

Fig. 269. A ladybird beetle *Hippodamia convergens.* *a,* adult; *b,* pupa; *c,* larva. (From U.S.D.A., E.R.B.)

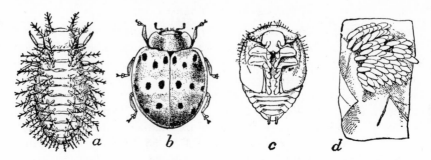

Fig. 270.   The Mexican bean beetle *Epilachna varivestis*.   *a*, larva; *b*, adult; *c*, pupa; *d*, eggs.
(From U.S.D.A., E.R.B.)

cardinalis, a native of Australia, which was introduced into California for
the control of the cottony-cushion scale.   The vedalia has been very
effective in this capacity.   A common and widespread native species is
Hippodamia convergens, fig. 269.   The second important category of ladybird beetles includes plant-feeding species.   In North America the
Mexican bean beetle, *Epilachna varivestis*, fig. 270, is one of the most
destructive defoliators of beans in many central and southern areas.
Both adults and larvae feed on the foliage.

MELOIDAE, BLISTER BEETLES.   The adults, fig. 271, are moderately large
beetles with relatively soft bodies and elytra, long simple antennae, and a
prominent round or oval head that is well set off from the thorax.   The
tarsi of the front and middle legs are five-segmented, but the tarsi of the
hind legs are four-segmented.   This characteristic marks off a group of
some twenty beetle families sometimes referred to as the *Heteromera*.   The
blister beetles contain an oil, cantharidin, which is a powerful skin irritant

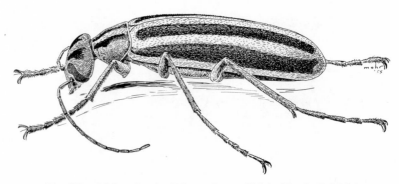

Fig. 271.   A blister beetle, *Epicauta vittata*.   (Original by C. O. Mohr)

and causes the formation of large water blisters on human skin. A sufficient amount of cantharidin to cause irritation is picked up by just handling the live adults.

In most subfamilies the larvae feed on bee larvae or as inquilines on provisions of bee nests. In one subfamily, containing the common and widespread genera *Epicauta* and *Henous*, the larvae feed on grasshopper egg pods. In all of them the life history is complex. The eggs are laid in masses of 50 to 300 about an inch deep in the soil. The first instar is an active, slender, well-sclerotized form called a *triungulin*. In the bee inquiline groups the triungulins crawl into flowers, attach to bees, and are thus carried to bee nests, where the triungulins dismount and begin feeding. In the grasshopper egg predators the triungulins simply search out the grasshopper egg pods.

The development of the larvae is peculiar in many ways. This is well illustrated by that found in *Epicauta pennsylvanica*, fig. 272. The triungulin seeks out a grasshopper egg mass, digs down to it, punctures an egg, and starts feeding on its contents. In a few days, when the triungulin is swollen and fully fed, the first molt occurs. The grublike thin-skinned second instar is relatively inactive and continues feeding on the grasshopper egg mass.

The third, fourth, and fifth instars follow in rapid succession and are similar to the second. The fifth instar, when full grown, leaves the food mass, burrows a few inches farther into the soil, and makes an earthen cell in which it molts to form the sixth-instar larva. This sixth instar, called the coarctate form, is unique among beetles. It is nonfeeding, heavily sclerotized, oval in shape, and rigid. Only humplike legs are present. Usually the winter is passed in this stage. The next summer the coarctate larva molts to form a seventh instar much like the fifth; this larva gives rise to the pupa, which transforms in a few weeks to the adult. The coarctate larva is extremely resistant to desiccation, and it provides a margin of safety to the species in drought years. For, if conditions are too dry during the summer after hibernation, the coarctate larva will not molt, but will "lay over" an additional year or even 2 years, if necessary, when less arid conditions prevail and the normal life cycle can be resumed.

Meloid adults are leaf feeders and cause appreciable damage to potatoes, tomatoes, squash, certain legumes, and other crops. The three-striped blister beetle *Epicauta vittata*, fig. 271, is a colorful representative often injurious to these plants. One of the species frequently feeding on potatoes, *Meloe angusticollis*, is unusual among beetles in that the elytra overlap at the base. Another short-winged meloid is the squash blister beetle *Henous confertus*.

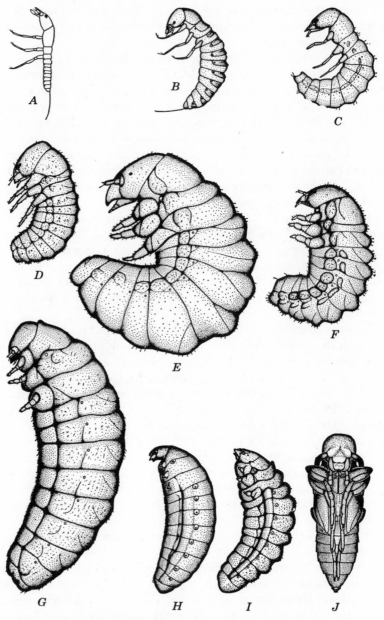

Fig. 272. Immature stages of the black blister beetle *Epicauta pennsylvanica*. *A*, unfed first instar; *B*, fully fed first instar; *C*, *D*, *E*, second, third, and fourth instars; *F*, newly molted fifth instar; *G*, gorged fifth instar; *H*, sixth instar; *I*, seventh instar; *J*, pupa. *A–E*, ×17; *F*, *G*, ×9; *H-J*, ×5. (After Horsfall)

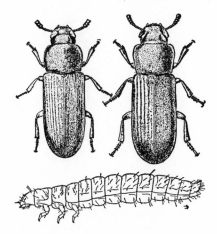

Fig. 273. The rust-red flour beetle *Tribolium castaneum*, and the confused flour beetle *Tribolium confusum*. Below, larva of *T. castaneum*. (From U.S.D.A., E.R.B.)

TENEBRIONIDAE, THE DARKLING BEETLES. Hard-shelled beetles, normally dark in color, and oval or parallel-sided in outline, fig. 273; tarsi of the front and middle pairs of legs are five-segmented, those of the posterior legs four-segmented; antennae moderately long, usually filiform or clavate. The larvae feed chiefly on dead plant material and fungi, especially bracket types, or mycelium in rotten wood; a few are predaceous, and a few attack stored grain. The larvae are elongate and cylindrical, with stout legs. Those of the western genus *Eleodes* feed on plant roots and are called false wireworms because of their resemblance to true wireworms (Elateridae larvae). Another western false wireworm is *Embaphion muricatum*, whose larva is destructive to wheat. Some of the species attacking stored grains and prepared foods are widespread and cause large commercial loss. Among this group are the mealworms *Tenebrio* sp., which are relatively large, the adults reaching a length of 15 mm. and the larvae 25 or 30 mm.; and the confused flour beetle *Tribolium confusum*, and several related species of *Tribolium*, which are very small, the adults being only 3 or 4 mm. in length.

CHRYSOMELIDAE, THE LEAF BEETLES. These comprise a large family, the species small to moderate in size, usually oval, stout, or wide bodied, and having filiform fairly long antennae, fig. 274. The most outstanding characteristics are found in the tarsi, fig. 257*b*, *c*. These appear four-segmented; the third segment is enlarged to form a large kidney-shaped pad; the last segment, really the fifth, is long and slender and appears to be attached within the median incision of the third; actually the fourth segment is an extremely reduced ring at the base of the fifth, but it is so small that it is seldom seen without one's first making a special preparation of the leg. The larvae of the leaf beetles are varied, but most of them

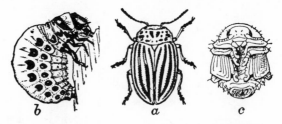

Fig. 274. The Colorado potato beetle *Leptinotarsa decemlineata*. *a*, adult; *b*, larva; *c*, pupa. (From U.S.D.A., E.R.B.)

are stout grubs with short legs and antennae; a number bear spines and processes; and some of the leaf-mining species are long and flat. The eggs are laid in soil or under bark or deposited on stems or leaves.

With few exceptions adult Chrysomelidae feed on plant foliage, and their larvae on roots or leaves. Many attack commercial crops, and the family includes a large number of important economic species. The Colorado potato beetle *Leptinotarsa decemlineata*, fig. 274, is one of the most destructive insects attacking potato; both larvae and adults feed on the foliage. The asparagus beetle *Crioceris asparagi* is a common showy species wherever asparagus is grown; both larvae and adults feed on the foliage; the eggs are black and stuck by one end into the heads of the plants. The larvae of many species, such as *Acalymma vittata*, fig. 275, are known as rootworms. The adults of many small species jump like fleas and for this reason are called flea beetles. Of these, the genera *Phyllotreta* and *Epitrix* contain several species whose adults eat holes in leaves, and the elongate larvae eat roots of cabbage, turnips, potatoes, cucumbers, and other plants. An important species is the tobacco flea

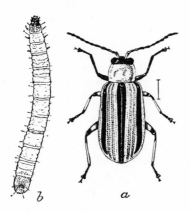

Fig. 275. The striped cucumber beetle *Acalymma vittata*. *a*, adult; *b*, larva. (From U.S.D.A., E.R.B.)

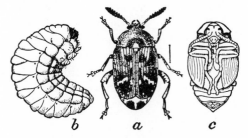

Fig. 276.   The pea weevil *Bruchus pisorum*.   *a*, adult; *b*, larva; *c*, pupa.   (From U.S.D.A., E.R.B.)

beetle *Epitrix hirtipennis*.   Not all flea beetles have root-feeding larvae; those of the genus *Haltica*, for instance, are leaf feeders like the adults.

BRUCHIDAE, BEAN AND PEA WEEVILS.   This is a most interesting family closely related to the Chrysomelidae.   The adults are short and stout; the larvae are grublike and almost legless, living inside legume seeds. Several species are pests of considerable importance in various kinds of stored peas and beans; a common example is the pea weevil *Bruchus pisorum*, fig. 276.

CERAMBYCIDAE, LONGHORN BEETLES.   Elongate beetles, many of them attractively colored, having long legs and unusually long antennae, fig. 277.   In other characters the longhorns are almost identical with the

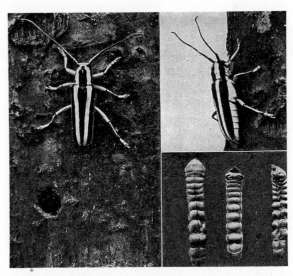

Fig. 277.   The roundheaded apple tree borer *Saperda candida;* larvae, adults, and exit holes, natural size.   (After Rumsey and Brooks)

Chrysomelidae, including the curious tarsi with the enlarged third segment and the reduced fourth. The larvae of the longhorns are cylindrical and elongate, with a round head, and either no legs or minute ones; they are known as roundheaded borers. Most of them bore either in the cambium layer or through the heartwood of trees; a few bore in the roots and lower stems of succulent herbs such as milkweeds and ragweeds, or in the stems of shrubs such as willow and raspberries. Certain species are of considerable economic importance.

The roundheaded apple tree borer *Saperda candida*, fig. 277, is a brown and white striped species that is locally a serious pest of apple, the larvae boring through the trunk and making extensive tunnels. The locust borer *Cyllene robiniae* has a handsome adult with a geometric yellow pattern on black; its larvae bore in young black locust trees and weaken them so that wind breaks them easily; many black locust plantings, established for soil-erosion purposes, have been entirely destroyed in this manner.

Most of the longhorn species are pests of forest trees, attacking both deciduous and coniferous species. Under improper or careless lumbering conditions or unusual weather, various longhorn species may become abundant enough to cause considerable loss to commercial stands of trees.

CURCULIONIDAE, TYPICAL WEEVILS. Head with a definite beak, sometimes elongate and curved, fig. 278; antennae usually elbowed and clubbed; body, elytra, and legs very hard, forming a solid well-armored exterior. The larvae are legless grubs, usually having dark head capsules and white bodies.

The Curculionidae is a large family, containing over two thousand nearctic species and many extremely important economic species. The larvae feed on plant material in a variety of ways; they include root feeders, stem borers, leaf feeders, and those that feed in fruits such as hazelnuts, acorns, cherries, and plums; in rotten wood; or in stored grain. The adults usually feed on the leaves or fruiting bodies of the plant species which serves as host for the larvae.

The boll weevil *Anthonomus grandis*, fig. 278, is one of the most serious cotton pests in the United States and is high on the list of insects causing excessive commercial damage. The adults attack the plants early in the season, feeding on the leaves and in the flowers and young bolls. The females deposit their eggs in feeding holes in the bolls, one egg to each hole. The larvae feed inside the boll, thus making a cell in which they pupate. Adults emerge in a few weeks, cause additional damage by their feeding, and then go into hibernation at the onset of winter. The boll weevil is a native of tropical America and entered Texas from Mexico

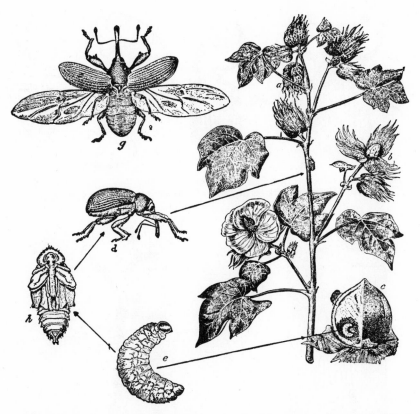

Fig. 278. The boll weevil *Anthonomus grandis*. On the right a cotton plant attacked by the boll weevil, showing at *a*, a hanging dry infested square; at *b*, a flared square with weevil punctures; at *c*, a cotton boll sectioned to show attacking weevil and larva in its cell; *g*, adult female with wings spread as in flight; *d*, adult from the side; *h*, pupa ventral view and *e*, larva. (From Metcalf and Flint, *Destructive and useful insects*, by permission of McGraw-Hill Book Co.)

about 1890. Since that time it has spread gradually throughout most of the cotton-growing region of the United States.

The plum curculio *Conotrachelus nenuphar* attacks plums, cherries, and related fruits; the larvae feed in the body of the fruit. This species is frequently a serious pest.

Of especial importance to stored grains are the granary weevil *Sitophilus granarius* and the rice weevil *Sitophilus oryza;* the larvae of both species live and feed inside grain kernels, and the adults feed either in the old larval burrows or promiscuously on the grain.

SCOLYTIDAE, THE BARK BEETLES. The species of this family feed chiefly

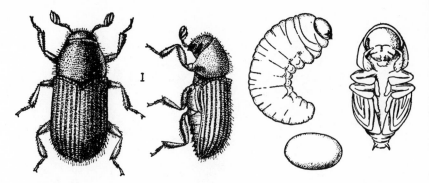

Fig. 279.   The peach bark beetle *Phloeotribus liminaris.*   (From U.S.D.A., E.R.B.)

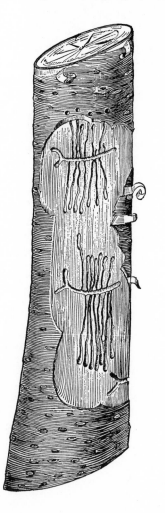

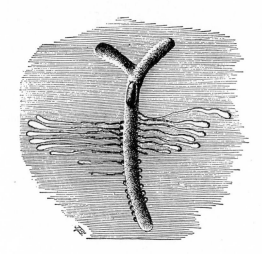

Fig. 280.   Brood chambers and larval galleries of peach tree bark beetle.   At right is shown a brood chamber with egg pockets and early larval galleries.   (From U.S.D.A., E.R.B.)

in trees, either alive or dead. The beetles have only an indistinct snout and are almost cylindrical in body shape, fig. 279. The larvae are legless grubs with typical weevil characteristics. Both adults and larvae feed in galleries, fig. 280. In some species these galleries are in the cambium layer of the tree and if sufficiently numerous result in girdling the tree. Many species of this family attack commercial timber, especially pine, and sometimes cause the premature death of large stands of it.

STYLOPOID SERIES

The stylopoids or twisted-wing insects, fig. 281, comprise the super-family Stylopoidea (sometimes called the order or suborder Strepsiptera) and are a remarkable group of Coleoptera. They include two families, the Stylopidae and Mengeidae, both seen only rarely by collectors.

STYLOPIDAE. The male is winged and free flying, more like a fly than a beetle, its elytra reduced to twisted finger-like organs, the hind wings large and fan-shaped, and the metathorax greatly enlarged. The anten-nae are short and have at least some leaflike segments; the eyes are berry-shaped, with each ommatidium protruding and distinct. The female is extremely degenerate, being only a sac enveloped by the last larval skin; it remains embedded in the host, with only an anterior por-tion, called a cephalothorax, projecting on the outside of the host integument.

The female is viviparous; the first-instar larvae develop inside her body and escape to the outside by crawling through a slit in the exposed cepha-lothorax. These larvae, called triungulinids because of their similarity to meloid triungulins, are curious active creatures, having three pairs of legs, well-developed sclerites, reduced mouthparts and ocelli, several well-developed eye facets, and one or two pairs of long terminal filaments. The larvae can run and jump with great agility. Each attaches and burrows into an individual host and soon molts into a legless grub. The grub lies quiescent in the host body and absorbs its food by diffusion from the host bloodstream. Thus there is no actual destruction of tissues like that caused by the usual type of insect parasitic larva which macerates and ingests its food by mouth. When full grown, the stylopid larva pushes its anterior end between the abdominal sclerites of the host so that the head and thoracic region is exposed, forming a round or flattened structure. If the larva is a male, it transforms within the larval skin to a typical beetle pupa; when mature, the adult breaks through the exposed larval skin and escapes to the outside. If the larva is a female, it has no definite pupal stage, molting directly into the saclike adult female which remains within the larval skin. The eggs develop and hatch within the body of the mother, until she is merely a sac of eggs or young. The number of progeny

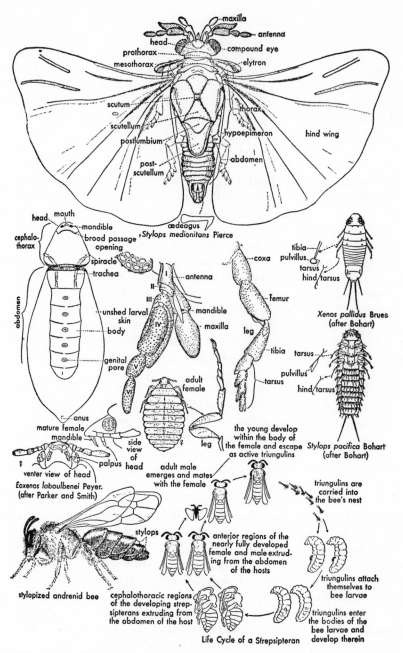

Fig. 281. Stylopoidea, important characteristics and life cycle. (From Essig, *College entomology,* by permission of The Macmillan Co.)

of one of these females is enormous; counts of young range from twenty-five hundred to seven thousand per female.

The North American species of stylopids attack chiefly adults of various bees and wasps. In other parts of the world different groups of stylopids attack Hemiptera (Pentatomidae, Cicadellidae, Fulgoridae, and certain allied families) and Cursoria (Mantidae).

The closely allied family Mengeidae is similar in most respects to the Stylopidae, but has a free-living larviform female. The host of the only nearctic genus is unknown, but European species parasitize silverfish.

The Stylopidae and their close allies are regarded by some authors as comprising a separate order, the Strepsiptera. Most workers agree, however, that this group originated with the beetle complex. The triungulinid larva, reduced wing venation, and parasitic habit suggest a relationship near the Meloidae.

References

Blatchley, W. S., 1910. *The Coleoptera or beetles known to occur in Indiana.* Indianapolis, Ind.; Nature Publishing Co. 1386 pp., 590 figs.

Blatchley, W. S., and C. W. Leng, 1916. *Rhyncophora or weevils of northeastern America.* Indianapolis, Ind.: Nature Publishing Co. 682 pp., 155 figs.

Bohart, R. M., 1941. A revision of the Strepsiptera. *Univ. Calif. Publs. Entomol.,* 7:91–160.

Bradley, J. C., 1930. *Manual of the genera of beetles of America, north of Mexico.* Ithaca, N. Y.: Daw, Illston & Co. 360 pp.

Clausen, Curtis P., 1940. *Entomophagous insects: Coleoptera.* New York: McGraw-Hill Book Co., pp. 522–584.

Crowson, R. A., 1955. *The natural classification of the families of Coleoptera.* London, Nathaniel Lloyd & Co., Ltd. 187 pp.

Edwards, J. G., 1949. *Coleoptera or beetles east of the Great Plains.* Ann Arbor, Mich., Edwards Bros. 181 pp.

Hatch, M. H., 1953. The beetles of the Pacific Northwest. *Univ. Wash. Publs. Biol.,* 16:1–340.

Order RAPHIDIODEA: Snakeflies

Large insects, with two pairs of transparent net-veined wings, similar in many features to the Megaloptera but distinguished by the long serpentine neck, fig. 282. They have long antennae, chewing mouthparts, large eyes, and two pairs of very similar wings. The female has a conspicuous terminal ovipositor. The larvae are terrestrial. They have segmented antennae, faceted eyes, well-developed thoracic legs, but no processes or appendages on the abdomen. The entire known world fauna of five genera and about a hundred species is divided into two families, the Raphidiidae (having ocelli) and the Inocelliidae (lacking ocelli). Two genera occur in North America, both confined to the West, *Agulla*

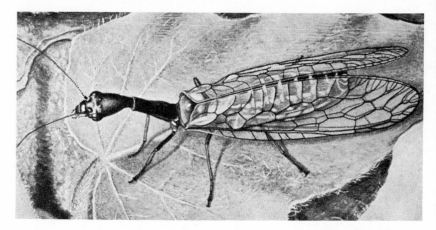

Fig. 282.   A European snakefly *Raphidia ratzeburgi*.   (From Essig, *College entomology*, by permission of The Macmillan Co.)

(Raphidiidae) and *Inocellia* (Inocelliidae).   The larvae occur under loose bark of conifers and are predaceous on other insects.   When mature, the larvae do not spin cocoons but form an oval retreat in a sheltered position, and here the pupal stage is passed.   The adults are also predaceous. They are occasionally swept from foliage and are indeed strange-appearing creatures to find in the net.

References

Carpenter, F. M., 1936.   Revision of the nearctic Raphidiodea (recent and fossil).   *Am. Acad. Arts. Sci. Proc.*, **71**:89–157, 13 figs., 2 pls., bibl.

Order MEGALOPTERA: Dobsonflies and Alderflies

Large insects having aquatic larvae and terrestrial adults and pupae. The adults, fig. 283, have long antennae, chewing-type mouthparts, large eyes, and two pairs of wings.   The wings are similar in texture and venation and have all the major veins plus some additional terminal branches and a large number of crossveins.   The pronotum is large and wide, the abdomen without projecting cerci.   The larvae, fig. 283, have strong biting mouthparts; distinct, segmented antennae; large eyespots each composed of a group of about six facets; well-developed thoracic legs, and paired abdominal processes or gills. The apex of the abdomen has a long mesal process in Sialidae, and a pair of stout hooked larvapods in Corydalidae.

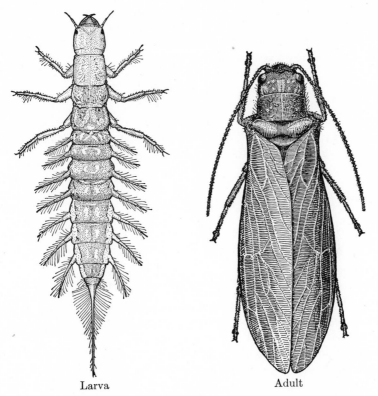

Larva   Adult

Fig. 283.   An alderfly *Sialis* sp., larva and adult.   (From Illinois Nat. Hist. Survey)

KEY TO FAMILIES

Head with no ocelli; tarsi with fourth segment dilated and bilobed, fig. 283
   Alderflies. . . . . **Sialidae**
Head with three ocelli; tarsi with all segments cylindrical.   Dobsonflies
                                                             Corydalidae

Both families together are represented in North America by only five genera and less than fifty species.   The adults range in color from black to mottled or yellow; in one genus (*Nigronia*) the wings are banded with black and white.

The larvae are aquatic, occurring in both lakes and streams.   They are predaceous on small aquatic animals.   The mature larvae of *Corydalus* may attain a length of 80 mm.   They are ferocious larvae, highly prized for bait by fishermen, and called hellgrammites.   The smaller species of the order mature in a year and have an annual life cycle.   The hellgrammites require 2 or 3 years to reach full growth.

When mature, the larvae leave the water and make a pupal cell in damp earth or rotten wood near by. Here the larvae transform to pupae. Megalopteran pupae are active if irritated and capable of considerable locomotion. The pupal stage usually lasts about 2 weeks.

The adults are good fliers, but not agile compared to some of the flies and moths. Some *Corydalus* adults may have a wing span of 5 inches and are among our largest North American insects. The females lay their eggs in large clusters of several hundred each on stones and other objects overhanging the water. These hatch soon after deposition, and the minute larvae fall or twist their way into the water.

References

Ross, H. H., 1937. Nearctic alderflies of the genus Sialis. *Illinois Nat. Hist. Survey. Bull.,* 21(3):57–78, 63 figs.; bibl.

Weele, H. W. van der, 1910. Megaloptera. *Coll. Zool. Selys Longchamps, Brussels,* fasc., 5(115):1–93, 70 figs., 4 pls.

Order NEUROPTERA: Lacewings, Mantispids

The adults are minute to large insects, usually with two pairs of clear wings having many veins and crossveins, with chewing-type mouthparts, long and multisegmented antennae, and large eyes, fig. 284A. The larvae are varied: Most of them are terrestrial and predaceous; one family (Sisyridae) is aquatic, and the larvae feed in fresh-water sponges. All the larvae have thoracic legs, but no abdominal ones, well-developed heads, and mandibulate mouthparts.

KEY TO FAMILIES

1. Front legs with apical segments enlarged for grasping, fig. 286 . . . . . **Mantispidae**
   Front legs with apical segments slender, same as other legs, fig. 284 2
2. Wings with very few veins or crossveins, fig. 285B. Minute insects covered with a waxy bloom and gray in appearance . . . . **Coniopterygidae**
   Wings with veins and crossveins numerous, fig. 285C–G. Larger insects never covered with waxy bloom 3
3. Front wings with a regular, fencelike series of 12 or more crossveins (gradate veins) between R_1 and R_s, fig. 285E 4
   Front wings either with only 1–5 well-separated crossveins between R_1 and R_s, fig. 285C, G, or R_1 and stem of R_s fused, fig. 285D 6
4. Antennae long and slender, fig. 284, tapering to apex . . . . **Chrysopidae**
   Antennae either short and clavate or knobbed at apex 5
5. Antennae short, gradually thickened towards apex . . . . **Myrmeleontidae**
   Antennae long, knobbed at apex . . . . **Ascalaphidae**
6. Front wings with 2 or more branches of R_s arising from fused R_1 and R_s, fig. 285D
   Hemerobiidae

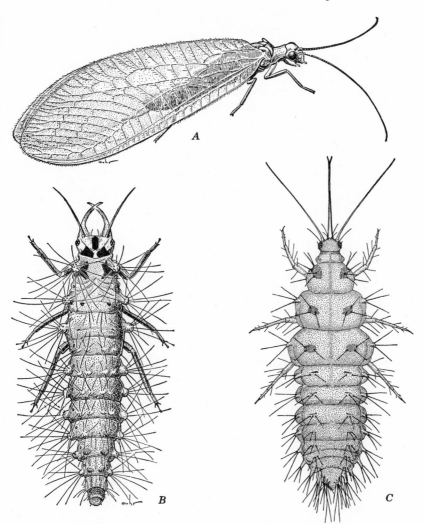

Fig. 284. Neuroptera. *A*, *B*, adult and larva of a lacewing *Chrysopa* sp.; *C*, larva of a spongefeeder *Sisyra* sp. (*A*, *B*, from Illinois Nat. Hist. Survey; *C*, after Townsend)

Front wings with all branches of R_s arising from a separate R_s stem, fig. 285*C*, *F*, *G*
. . . . 7

7. Front wings with almost all costal crossveins forked, fig. 285*F*, *G*. 8
Front wings with few or no costal crossveins forked and with apical margin evenly
rounded, as in fig. 285*A*, *C*. 9

8. Front wings only slightly incised and with recurrent costal vein, fig. 285*G*
Polystoechotidae

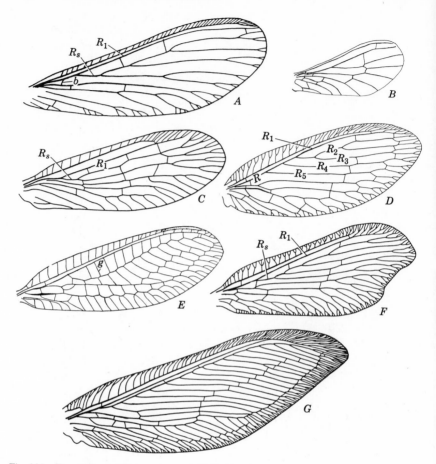

Fig. 285. Front wings of Neuroptera. *A, Nallachius,* Dilaridae; *B, Semidalis,* Coniopteryidae; *C, Climacia,* Sisyridae; *D, Hemerobius,* Hemerobiidae; *E, Chrysopa,* Chrysopidae; *F, Lomamyia,* Berothidae; *G, Polystoechotes,* Polystoechotidae. *b,* basal vein; *g,* gradate vein. (From various sources)

Front wings markedly incised and with no recurrent costal vein, fig. 285*F*
                                                     Berothidae
9. Front wings with S_c and R_1 not fused before apex; basal vein (*b*) present; fig. 285*A*
                                                     Dilaridae
Front wings with S_c and R_1 fused some distance before apex; basal vein absent; fig. 285*C* . . . . **Sisyridae**

SPONGE FEEDERS. The spongeflies are a small family comprising the Sisyridae. The adults look like typical lacewings, but the larvae are robust creatures that live in and eat fresh-water sponges. Their mouthparts form a long beak which sticks out in front of the larva, fig. 284*C*. Only a few species are known to occur in the nearctic region.

SEDENTARY PREDATORS. The mantispids, fig. 286, are another small family, Mantispidae. The adults have a striking resemblance to praying mantids. The front legs are greatly enlarged and fitted for grasping insect prey and are attached at the anterior end of the very long pronotum. The larvae feed on egg sacs of spiders or contents of wasp nests. The first-instar larvae are slender and active, and hunt for a suitable food reservoir. Once this is found, the larvae enter a parasitoid stage, and succeeding instars are grublike and have degenerate legs.

ACTIVE PREDATORS. The larvae of all the other families are active predators. The adults have transparent and abundantly veined wings, which give them the name "lacewings," fig. 284*A*. Most of these insects are relatively slow on the wing. The eggs are laid either attached directly to foliage or at the end of a long hairlike stalk, which is attached to a leaf, fig. 134*N*. This latter method is used only by the Chrysopidae. The

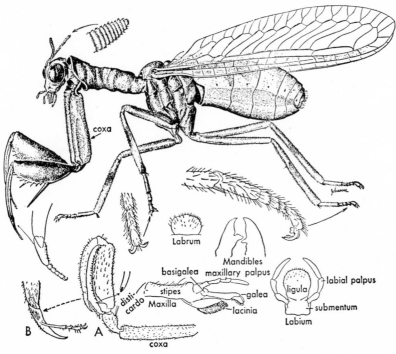

Fig. 286. A mantispid *Mantispa brunnea*. (From Essig, *College entomology*, by permission of The Macmillan Co.)

larvae are active, but sluggish and soft-bodied, and frequently bear warts, tubercles, and long hair. The mouthparts are modified for sucking body juices from the prey. The mandibles and ends of the maxillae are long, bladelike, and sickle-shaped, and a maxillary blade fits beneath each mandible; each of these opposing pieces has a groove, the two fitting together to form a canal from near the tip of the mandible into the mouth opening. The two mandibular-maxillary blades are thrust into the body of the prey from opposite sides, and its body juices are sucked out through the canals.

The larvae of Chrysopidae, fig. 284B and Hemerobiidae crawl freely about on plants and feed on aphids, other small insects, and insect eggs. Their frequent attacks on aphids have earned them the name "aphid-lions." When full grown, the larvae spin a woolly ovoid cocoon under a leaf or in some sheltered spot, and pupation ensues. The larvae of Myrmeleontidae live in sandy soil and dig cone-shaped pits that trap ants and other prey which fall into them. These larvae, called "antlions," differ from the aphidlions only in being more robust. The antlion digs the pit by throwing out sand from the center by upward jerks of the head, using the long mandibles as shovels. The pitfalls may be an inch deep, with sides sloping as much as the texture of the loose sand will allow. The antlion stays in the soil with its head just below the bottom of the crater, constantly in wait for unwary prey. These curious larvae are known to most people by the name "doodlebug." When mature, the larva forms a cocoon in the soil and pupates in it.

References

Carpenter, F. M., 1940. A revision of nearctic Hemerobiidae, Berothidae, Sisyridae, Poly-stoechotidae, and Dilaridae. *Am. Acad. Arts and Sci. Proc.,* 74(7):193–280, 75 figs., 3 pls.
Smith, Roger C., 1934. Notes on the Neuroptera and Mecoptera of Kansas, with keys for the identification of species. *J. Kansas Entomol. Soc.,* 7(4):120–145, 11 figs.

Order MECOPTERA: Scorpionflies

The adults, ranging in size from small to medium, either have two pairs of large net-veined wings, fig. 287, or have the wings short or aborted. The antennae are long; eyes large; and legs slender, in some families long and spindly. The mouthparts are of the chewing type, often situated at the end of a snoutlike elongation of the head. The larvae are grublike or caterpillar-like, always with thoracic legs and in some groups having abdominal larvapods also, fig. 288.

The adults are omnivorous, feeding chiefly on small insects, but supplementing their diet with nectar, pollen, petals, fruits, and mosses. The

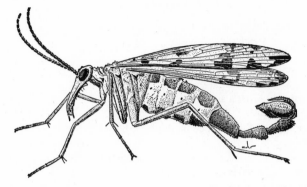

Fig. 287.   A male scorpionfly *Panorpa chelata.*   (From Illinois Nat. Hist. Survey)

winged forms are active fliers.   The males of the Panorpidae, fig. 287, have a large bulbous genital capsule which resembles to some extent the abdomen of a scorpion, and from this the order derives its name "scorpionflies."

The eggs are ovoid and are laid in or on the ground, either singly or in clusters of one hundred or more.   The larvae live in moss, rotten wood, or the rich mud and humus around seepage areas in densely wooded situations.   Their food consists of various types of organic matter.   Pupation occurs in the soil.   There is only one generation per year.

In the small species found in the genus *Boreus* (family Boreidae) the adults have small short wings and, since they mature in winter or early spring, are often found running about on the snow.   The robust, grublike larvae live in the rotting bases of moss mats.   Because they differ from other Mecoptera in many details of both adults and larvae, the Boreidae are considered a separate order Neomecoptera by some authors.   The structure of the mouthparts, however, shows conclusively that the Boreidae are a specialized offshoot related to the families Panorpodidae and Panorpidae, and are not a distinctive order.

Fig. 288.   Larva of *Apterobittacus apterus.*   (Redrawn from Applegarth)

KEY TO FAMILIES

1. Ocelli lacking; wings always well developed, oval, less than 3 times as long as wide, with very reticulate venation. Contains only the rare eastern *Merope tuber*
  Meropeidae
  Ocelli present; if well-developed, wings more than 3 times as long as wide, with only moderately reticulate venation, fig. 287 2
2. Tarsus raptorial, with a single claw . . . . **Bittacidae**
  Tarsus with 2 claws 3
3. Wings, whether well developed or short, with a well-defined and complete set of veins 4
  Wings reduced to small oval pads or short, hard, tapered structures, devoid of venation. Sole genus, *Boreus* . . . . **Boreidae**
4. Head produced into a ventral beak as long as in fig. 287. Sole N. Am. genus, *Panorpa* . . . . **Panorpidae**
  Head short, without a ventral beaklike prolongation. Sole N. Am. genus, *Brachypanorpa* . . . . **Panorpodidae**

References

Byers, G. W., 1954.   Notes on North American Mecoptera.   *Ann. Entomol. Soc. Amer.,* **47**:484–510.
Carpenter, F. M., 1931.   A revision of the nearctic Mecoptera.   *Museum Comp. Zool. Harvard Coll. Bull.,* **76**:206–277.

Order SIPHONAPTERA: Fleas

All adults are wingless, fig. 289, and have long stout spiny legs and short clubbed antennae which in repose fit into a depression along the side of the head. The mouthparts are fitted for piercing skin and sucking

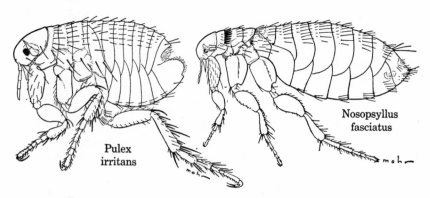

Fig. 289. Two common fleas. (From Illinois Nat. Hist. Survey)

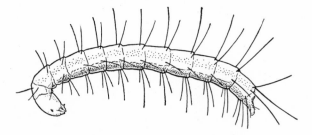

Fig. 290.   Larva of a flea.   (From Illinois Nat. Hist. Survey)

blood and consist of a beak, a pair of palps, and a pair of short bladelike maxillae.   These insects are small, most of them 2 to 4 mm., but a few species attain a length of 6 to 8 mm.   The larvae are slender and worm-like, fig. 290.   They have round heads, no legs, and long hairs on each body segment.   The segments of thorax and abdomen are similar in appearance.   The mouthparts are minute and inconspicuous, of the chewing type.   The pupae are formed in a cakelike cocoon made of earth or grass.

Fleas feed on the blood of mammals or birds and are found on the body of the host or in the nest or runways of the host.   The eggs are minute, whitish, and oval.   They are dropped by the female either on the host or in the nest.   The eggs have no adhesive, so that, if laid on the host animal, they slip through the hair and fall into the litter of the nest or retreat, where they soon hatch.   The larvae live in the soil or debris in the nest and feed on debris or grass.   When full grown, the larvae spin an irregular cocoon in the nest of the host and pupate.   The eggs, larvae, and pupae are seldom seen except by diligent search.

The adult fleas of most species are extremely active.   They slip through hair or feathers with great ease; and in many species the body has combs of spines which further aid this progress.   In some species the adults stay on the body of the host almost all the time; in others the adults stay in the nest and get on the host only for feeding periods.   For this reason it is necessary to collect both from the bodies and in the nests of the hosts to be sure of finding all the species connected with them.

The order Siphonaptera is not a large one.   The North American fauna includes about sixty genera and over two hundred species.   The great percentage of these occur with native mammals, especially various kinds of mice, shrews, ground and tree squirrels, gophers, and rabbits. Bears, beavers, coyotes, and many others support a flea fauna.   A few fleas prey on birds and occasionally are a pest of domestic fowl.

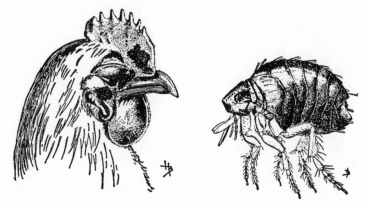

Fig. 291. The sticktight flea *Echidnophaga gallinacea*. Left, infested head of rooster (dark patches are clusters of fleas); right, adult female. (From U.S.D.A., E.R.B.)

KEY TO COMMON FAMILIES

1. Thorax extremely short, all three segments combined shorter than the first abdominal tergite, fig. 291.....................**Tungidae** (*Hectopsyllidae*)
   Thorax considerably longer than first abdominal tergite, fig. 289.......... 2
2. Abdominal tergites with only a single row of setae, fig. 289, *Pulex;* when telescoped, abdomen therefore appears to have a series of similar rows of setae
   Pulicidae
   Abdominal tergites with a double row of setae, the basal row composed of short setae, the apical row of long setae, fig. 289, *Nosopsyllus;* when telescoped the abdomen appears to have an alternation of short-spined rows and long-spined rows of setae 3
3. Head without black teeth, armed only with setae and the genal process, fig. 289, *Nosopsyllus* ....**Ceratophyllidae** (*Dolichopsyllidae*)
   Head with at least one, and usually with a comb of long black teeth (actually flat, black, tooth-like setae) along ventral margin or on genae, like those on prothorax of *Nosopsyllus* 4
4. Head with two black teeth situated at tip. Bat fleas..........**Ischnopsyllidae**
   Head with either more than two teeth, forming a comb, or with teeth situated some distance from tip. Not on bats..................**Hystrichopsyllidae**

PULICIDAE.   Many species of this family are of especial importance to man.   The dog and cat fleas, *Ctenocephalides canis* and *C. felis* and the human flea *Pulex irritans* attack man and invade dwellings, causing great discomfort and inconvenience.   The bites inflicted by the fleas are painful, usually cause hard itching swellings, and may be the source of secondary infections through scratching.   Most important of the fleas are certain rat species, particularly *Xenopsylla cheopis* (and *Nosopsyllus fasciatus* of the Ceratophyllidae) which transmit the dread bubonic plague from rats to man. In their role as disseminators of this disease, which many times has spread like wildfire through crowded cities in many parts of the world, fleas have been responsible for millions of human deaths. At present, public health organizations all over the globe keep close watch on the rat-flea-plague focal points, in an effort to break up new incipient outbreaks of the disease, through control of the rats and fleas.

Two species of Tungidae, the sticktight and chigoe fleas, have unusual habits in contrast with other fleas. The females of both species are minute and attach themselves firmly to the host, feeding more or less continuously. The sticktight flea *Echidnophaga gallinacea,* fig. 291, attacks domestic fowl, attaching to the face and wattles. The chigoe *Tunga penetrans* attacks man, especially on the feet. It burrows beneath the skin and forms a painful swelling out of which protrudes only the end of the flea's abdomen. Both species drop eggs to the ground and have typical flea metamorphosis.

References

Ewing, H. E., and Irving Fox, 1943. The fleas of North America. *USDA Misc. Publ.,* **500,** 128 pp.

Fox, Irving, 1940. *Fleas of eastern United States.* Ames, Iowa: Collegiate Press. 191 pp., 166 figs.

Holland, G. P., 1949. The Siphonaptera of Canada. *Can. Dept. Agric. Tech. Bull.,* **70:**1–306.

——— 1964. Evolution, classification, and host relationships of Siphonaptera. *Ann. Rev. Ent.,* **9:**123–146.

Hubbard, C. A., 1947. *Fleas of western North America.* Ames, Iowa: Iowa State College Press. 532 pp., illus.

Order DIPTERA: Flies

Typical adults, fig. 292, have a single (front) pair of membranous wings, rarely scaled. The wings have few crossveins and a moderate number of

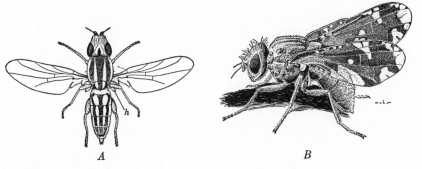

A   *B*

Fig. 292. Typical flies. *A, Meromyza americana; B, Euarosta solidaginis; h,* haltere. (From Illinois Nat. Hist. Survey)

veins. The hind wings are represented only by a pair of slender knobbed balancing organs, called *halteres*. Mouthparts are of various types; in some groups they are modified for piercing and sucking, in other groups for rasping and lapping. The body form is diverse. In a few groups the adults are completely apterous. The eyes are usually large; the antennae vary from 3- to 40-segmented. These are holometabolous insects with legless larvae, usually either with a distinct mandibulate head, fig. 303, or with an internal sclerotized skeleton attached to a pair of hooklike mandibles. The pupa is either free or formed within the skin of the third instar larva.

The order Diptera is a large one, including over fifteen thousand North American species. The food and habitat of the adults are usually very different from those of the larvae.

Adults of many families feed chiefly on nectar and plant sap or on free liquids associated with rotting organic matter. Certain groups, such as mosquitoes and horse flies, feed on animal blood; these have mouthparts highly modified for piercing and sucking. In a few groups, for example, the bot flies, the mouthparts are so vestigial that it is doubtful if the adults take any nourishment.

As a group, fly larvae are moisture-loving, the great majority living in water, in rotting flesh, inside the bodies of other animals, in decaying fruit or other moist organic material, or inside living plant tissue. A few live in relatively dry soil or move about exposed to the air, but these are the exceptions rather than the rule.

For the most part, fly eggs are simple, ovoid or elongate, and are normally laid singly, in, on, or near the larval food. Some, such as those of *Drosophila*, have lateral or polar floats which prevent them from sinking into semiliquid food and drowning. Eggs of certain mosquitoes are sufficiently well protected against the elements to withstand months of alternate drying, wetting, and freezing. In some groups, such as the flesh flies and some parasitic flies, the eggs may hatch just before leaving the body of the female and are deposited as minute larvae. This habit is carried to extreme development in the sheep-tick groups (Pupipara) where the larvae hatch and grow to their full size in the body of the female. Of unusual interest in the order is the paedogenesis exhibited by some species of midges (see p. 192).

Phylogeny. Living Diptera exhibit an extraordinary range of differences in the larval head and its parts and in the adult antennae, wing venation, mouthparts, and other structures. In the group of families embracing the fungus gnats and craneflies, the condition of these structures approaches the condition found in more primitive insects such as the

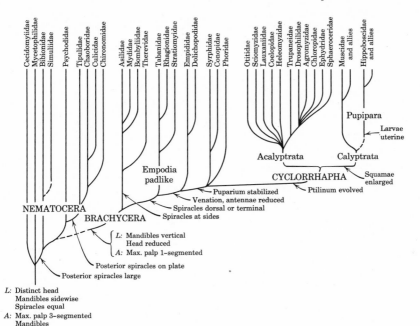

Fig. 293.   Simplified family tree of the order Diptera, *A*, adult; *L*, larva.

Mecoptera. These fly families are therefore considered the primitive end of the evolutionary scale in the Diptera and designated the suborder Nematocera.

By adding together these primitive conditions we obtain the following partial reconstruction of the ancestral dipteran. In the larva the head formed a distinct, typical, sclerotized capsule, the mandibles moved sidewise as in a grasshopper, and the abdomen bore lateral functional spiracles on the first eight segments. In the adult, the antennae were long and composed of many freely-articulating segments, and vein Cu_{1b} extended as a free vein to the wing margin.

All the Nematocera possess the capsule-like larval head, sidewise moving larval mandibles, and primitive adult antennae. Only in the Cecidomyiidae (gall midges) and Mycetophilidae (fungus gnats) are all the larval abdominal spiracles about equal, indicating that these two families represent two of the most archaic lines in the order. In the larvae of the Bibionidae (March fly) and the Psychodidae (moth fly) lineages, the posterior abdominal spiracles became much larger, the others smaller, and respiratory gaseous exchange in the larva was effected chiefly or entirely by the posterior pair of spiracles. Probably from the base of the psychodid line arose the Tipulidae (crane fly) and Culicidae (mosquito) branch, in which anterior larval spiracles were either extremely minute and functionless or atrophied, and the entire respiratory exchange occurred through the posterior spiracles.

These spiracular adaptations were associated first with larval life in shallow water, mud, or extremely wet organic media. The larva fed on submerged organic material, but kept its posterior spiracles above the water surface membrane. This larval feeding pattern is the basis for practically all later evolutionary developments of the Diptera. In mosquitoes the larvae evolved into swimmers able to live in water of some depth. In the higher Diptera, the larvae of various families exploited other essentially aquatic media, including the internal tissues of other insects, dung, and semiliquid organic material such as decaying mushrooms and rotting flesh.

In two common families of Nematocera, the Simuliidae (black flies) and Chironomidae (midges), the larvae became completely aquatic and lost their posterior spiracles. The exact relationship of these families is not known. In the accompanying tree, these families are placed tentatively primarily on similarities to other families with respect to wing venation.

From the base of either the psychodid or tipulid line evolved a dipteron that changed in many respects. The larval mandibles became hooklike and moved up and down, in and out, in a shredding motion; the larval head lost much of its sclerotization; the adult maxillary palps were reduced from 3- to 1-segmented; and in the wings, veins Cu_1 and $1A$ typically met at the wing margin. This marked the beginning of the suborder Brachycera. In the most primitive lineage, including the Asilidae (robber flies) and their allies, the posterior larval spiracles are at the sides of the segment, much as in the nematoceran Bibionidae. The larvae of the asiloid branch are predaceous or parasitic on other insects. In the more specialized lineage, the posterior larval spiracles were close together and situated in a dorsal or terminal depression or cavity, and the segments posterior to them became reduced to small ventral lobes situated under the eighth segment. This ancestor gave rise to two branches, (1) the Tabanidae (horse fly) branch in which the adult tarsal empodia became pulvillus-like, and (2) the Empididae (smoke fly) branch in which the empodia remained slender but the wing venation became reduced. During the evolution of the Brachycera the adult antennae became progressively solidified and reduced. By the time the empidid line arose, the antennae consisted of only three segments, the last one bearing a long style or arista, fig. 296H.

In the families discussed up to this point, the pupa is typically free, although in a few isolated families (e.g., Cecidomyiidae and Stratiomyidae) the pupa forms inside the old larval skin, which hardens to form a protective covering or puparium. The emerging adult makes a longitudinal dorsal slit in the puparium, through which it emerges. From the base of the empidid branch arose a form in which the formation of the puparium became fixed. This form gave rise to a number of families, including the Syrphidae (flower flies) and their allies, that preserve this habit.

From the same puparium-forming ancestor arose another lineage in which the emerging adult pushed off the entire anterior end or cap of the puparium, producing a circular or elliptic anterior opening. On the front of the head evolved an invaginated bulb or ptilinum that became everted at eclosion and pushed off the cap of the puparium. The larvae lost all the external sclerotization of the head, becoming typical fly maggots. This was the ancestral form of the suborder Cyclorrhapha, whose descendants became the most abundant and diverse groups of flies. In these, vein Cu_{1a} became bent backward at its base; the same change also occurred in some of the more primitive flies.

Details of the phylogeny of the Cyclorrhapha are poorly understood. This abundant group of flies seems to represent two large branches. The more primitive one, the series Acalyptrata, has many families that probably resemble the progenitor of the Cyclorrhapha closely in basic characters. Examples are the Otitidae (picture-winged flies) and Sciomyzidae (marsh flies). In one lineage of the Acalyptrata, vein Sc became reduced or atrophied. An early member of this lineage may be the Trupaneidae (fruit flies); more specialized members include the family Drosophilidae and its allies.

In the Calyptrata branch, the connecting membranes at the bases of the wing became enlarged, forming distinct lobes. Early stages in this development are seen in the Anthomyiidae and Fanniidae; more advanced stages occur in the Tachinidae and Sarcophagidae.

In one line of Cyclorrhapha, comprising the series Pupipara, the larvae develop in a type of uterine pouch inside the mother fly. The exact origin of the Pupipara is not clear but they probably arose as an early branch of the Calyptrata.

One pair of structures of great evolutionary interest are the adult mandibles. The Simuliidae, Psychodidae, Ceratopogonidae, Culicidae, Tabanidae, and Rhagionidae all have adult mandibles. They are blade-like, have typical mandibular attachments and muscles, and in shape are suggestive of the adult mandibles of Siphonaptera and some Mecoptera. The number of families in which mandibles still occur in the adults indicates strongly that the primeval dipteran had these structures, and that

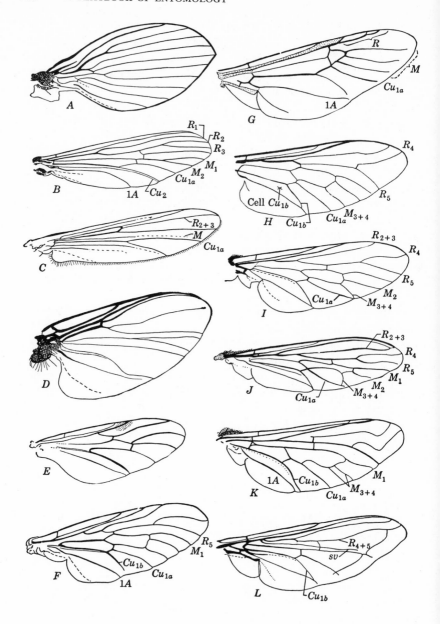

Fig. 294.   Wings of Diptera.   *A, Psychoda,* Psychodidae; *B, Aedes,* Culicidae; *C, Chironomus,* Chironomidae; *D, Simulium,* Simuliidae; *E, Bibio,* Bibionidae; *F, Rhagio,* Rhagionidae; *G, Stratiomys,* Stratiomyidae; *H, Tabanus,* Tabanidae; *I, Thereva,* Therevidae; *J, Asilus,* Asilidae; *K, Anthrax,* Bombyliidae; *L, Scaeva,* Syrphidae; *sv,* spurious vein.

they have become atrophied in many branches of the order. They are not present in the lineage that arose from the base of the Tabanidae line and later evolved into the Cyclorrhapha.

KEY TO COMMON FAMILIES

1. Abdomen only indistinctly segmented; the coxae of the 2 legs of each segment far apart. Adults living as parasites on birds, mammals, or bees 2
   Abdomen having distinct segments; the 2 legs of each segment held fairly closely together, sometimes the coxae almost contiguous. Not living as ectoparasites in the adult stage 4
2. Mesonotum short, resembling the abdominal segments; minute (1.5 mm. long), wingless, parasitic on honeybees. Includes only *Braula caeca*
   Braulidae, p. 390
   Mesonotum different in appearance from abdominal segments, fig. 316. 3
3. Palpi long and slender, forming a sheath for the mouthparts. Living on birds and mammals with the exception of bats . . . . **Hippoboscidae,** p. 390
   Palpi broader than long, not encasing the mouthparts. Parasitic on bats
   Streblidae
4. Antenna having more than 3 segments, fig. 297*F–H* not counting a style or arista, borne by the third 5
   Antenna having 3 segments or less; usually the third bears a style or arista, fig. 297*I*, 297*J*. 21
5. Small mothlike flies, never longer than 5mm., having body and wings densely clothed with hair or scales; wings having about 10 longitudinal veins, and having crossveins only at extreme base, fig. 294*A*; aquatic or semiaquatic moth flies . . . . **Psychodidae**
   Appearance not mothlike, or venation of a different type 6
6. Mesonotum having a distinct, V-shaped suture, fig. 297*A*; elongate species having long legs, fig. 301 . . . . **Tipulidae,** p. 377
   Mesonotum with suture transverse, indistinct, or not present, fig. 297*B, C*. . . . 7
7. Antenna having 6 or more well-marked ringlike or beadlike segments, fig. 297*F*
   8
   Antenna having only 4 or 5 segments, fig. 297*G*, or the terminal segments sometimes indistinctly subdivided, fig. 297*H*. 17
8. Wing having cell Cu_{1b} either open at apex, fig. 294*F*, or lost due to atrophy of veins 9
   Wing having cell Cu_{1b} entirely closed at apex by fusion of veins Cu_{1b} and 1*A*, fig. 294*H* 17
9. Wing having both R_{2+3} and M_{1+2} branched, and the venation fairly parallel, fig. 294*B* 10
   Wing having either R_{2+3} or M_{1+2} unbranched, the venation frequently markedly divergent, fig. 294*E* 11
10. Mouthparts forming a long slender beak, fig. 303 . . . . **Culicidae,** p. 377
    Mouthparts not forming a long beak. Mosquitolike midges . . . . . **Chaoboridae**
11. Ocelli present 12

Ocelli absent 14

12. Anterior margin of wing only slightly more heavily sclerotized than apical and posterior margins .... **Cecidomyiidae,** p. 374
Anterior margin of wing having a sclerotized thickening that stops abruptly at or just beyond juncture with R_{4+5}............................... 13

13. Antennae inserted below level of eyes, fig. 297L; front femur often enlarged; the March flies ....**Bibionidae**
Antennae inserted on a level with middle of eyes; coxae often greatly elongated; the fungus gnats ....**Mycetophilidae**

14. Wing having 2 or 3 strong parallel veins near anterior margin and a group of 6 or 8 oblique very weak veins running from anterior region to or near posterior margin of wing, fig. 294D; antennae short, 12-segmented, the last 10 annular and closely knit, fig. 299................. **Simuliidae,** p. 376
Wing having a different venation, either more veins equally sclerotized, or most of them longitudinal in general course rather than oblique, fig. 294C; antennae usually elongate, with well-separated segments.................. 15

15. Anterior margin of wing only slightly more heavily sclerotized than apical and posterior margins .... **Cecidomyiidae,** p. 374
Anterior margin of wing having a sclerotized thickening that stops abruptly at or just beyond juncture with R_{4+5}, fig. 294C....................... 16

16. Postnotum very large, projecting some distance posterior to scutellum, fig. 297C; slender elongate flies............................ **Chironomidae,** p. 378
Postnotum smaller, scarcely projecting at all from beneath the scutellum, fig. 297B; small stouter flies, the punkies or "no-see-ums".. **Ceratopogonidae**

17. Tarsus having 3 whitish pulvillar pads, fig. 297D; the middle one (the empodium) is sometimes dorsad of the lateral pulvilli, which are sometimes small 18
Tarsus at most having 2 pulvillar pads, the empodium reduced to a seta, fig. 297E; the 2 pulvilli may be reduced, in which case the tarsus lacks pads 19

18. Wing with branches of R_s close to front margin, forming a group of narrow cells along it, fig. 294G; tibia without apical spurs; soldier flies..... **Stratiomyidae**
Wing with branches of R_s not crowded to front margin, R_4 and R_5 diverging to form a triangular cell embracing apex of wing, fig. 294H.. **Tabanidae,** p. 381

19. Antenna having third segment elongate, fourth clavate, without arista or style, fig. 297G; very large species........................ **Mydaidae,** p. 381
Antenna having apical segment not clavate........................ 20

20. Top of head sunken to form a deep excavation between eyes, fig. 297M; wing with M_2 present and M_{3+4} having its base almost or entirely free from Cu_{1a}, the latter 2 veins sometimes fused for a short distance at base and/or apex, fig. 294J .... **Asilidae,** p. 381
Top of head without a deep excavation; wing having M_2 atrophied, and M_{3+4} fused near base for a considerable distance with Cu_{1a}, fig. 294K; hover flies
Bombyliidae

21. Wing at most having S_c and stem of R sclerotized, remaining venation consisting of 3 or 4 weak veins arranged as in fig. 295A; antenna composed of 1 segment

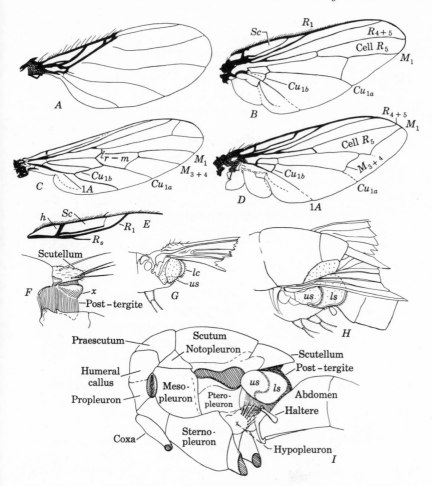

Fig. 295. Diagnostic parts of Diptera. *A*, wing of *Megaselia*, Phoridae; *B*, wing of *Hylemya*, Anthomyiidae; *C*, wing of *Empis*, Empididae; *D*, wing of *Sarcophaga*, Sarcophagidae; *E*, costal region of *Rhagoletis*, Tephritidae; *F*, scutellum of *Zenillia*, Tachinidae; *G*, squamae of *Anthomyia*, Anthomyiidae; *H*, squamae of *Musca*, Muscidae; *I*, thorax, diagrammatic, of higher Diptera. *ls*, lower squama; *us*, upper squama; *x*, secondary convexity on scutellum.

and its arista . . . . **Phoridae,** p. 384
Wing having a more extensive venation, figs. 295*B–D*; antenna may have 1 to 3
   segments 22
22. Tarsus having 3 pulvillar pads, fig. 297*D*, 2 lateral and 1 mesal 23
   Tarsus having only 2 (lateral) pulvillar pads, fig. 297*E*, or none 24
23. Antenna having distinct but faint annular lines on third segment, fig. 297*I*;
   wing having R_s and its branches forming a series of narrow cells along anterior

margin, fig. 294*G* .... Stratiomyidae
Antenna without annulations on third segment; wing having some branches
of R_s ending at or below apex of wing, fig. 294*F*; predaceous snipe flies
Rhagionidae
24. Wing having R_s with 3 branches, fig. 294*I–K* 25
    Wing having R_s with only 1 or 2 branches, fig. 295*B, D* 28
25. Top of head sunken to form a deep excavation between eyes, fig. 297*M*
    Asilidae, p. 381
    Top of head flat or convex, confluent in outline with top of eyes 26
26. Wing having M_{3+4} not fused at base with Cu_{1a} but frequently fusing with Cu_{1a}
    near margin; M_2 present, fig. 294*I*. Moderate-sized species similar to
    Asilidae in habits .... Therevidae
    Wing having M_2 atrophied and M_{3+4} fused at base with Cu_{1a}, fig. 294*K* 27
27. Wing having Cu_{1b} reaching wing margin or fusing with 1*A* near wing margin,
    fig. 294*K* .... Bombyliidae
    Wing having Cu_{1b} fused with 1*A* considerably before wing margin, as in fig. 294*H*
    Empididae

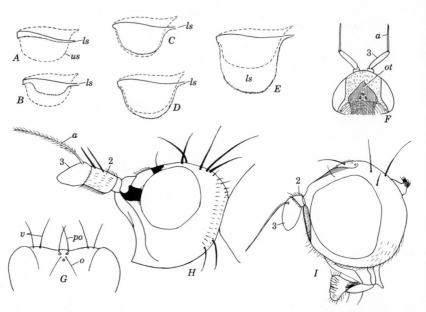

Fig. 296. Parts of Diptera. *A*, squamae of *Scopeuma*, Scopeumatidae; *B*, of *Pegomyia*, Anthomyiidae; *C, D*, of *Fannia*, Fanniidae; *E*, of *Musca*, Muscidae; *F*, head of *Agromyza*, Agromyzidae; *G*, diagram of upper part of head showing ocelli and associated bristles; *H*, head
of *Euthycera*, Sciomyzidae; *I*, of *Nemopoda*, Sepsidae. *a*, arista; *ls*, lower squama; *o*, ocellar bristle;
ot, ocellar triangle; *po*, postocellar bristles; *us*, upper squama; *v*, vertical bristle. In *A* to *E*
the upper squama is shown in broken outline. (*H, I*, after Curran)

28. Wing with Cu_{1b} straight and long, angled only slightly from stem of Cu, fig. 294L
.... 29

Wing with Cu_{1b} markedly angled from its parent vein, appearing like a cross-vein, fig. 295B–D; spurious vein never developed 30

29. Wing with M_1 sinuate; a linear, veinlike thickening (the spurious vein) usually present between R_s and M, fig. 294L .... **Syrphidae,** p. 382

Wing with M_1 short and straight, appearing to be a crossvein; no spurious vein present .... **Conopidae,** p. 384

30. Lower squama large and platelike, nearly twice as long as upper squama and projecting considerably beyond it, fig. 295H, 296E 31

Lower squama at most slightly longer than upper squama, fig. 296C, D, frequently appearing as only a cordlike band, fig. 296A, B 35

31. Oral opening either a minute circle or a small triangular cleft, as in fig. 297P; body fuzzy with long dense hair .... **Oestridae,** p. 388

Oral opening large, fig. 297 Q; body usually not fuzzy but often spiny in appearance 32

32. Hypopleura with only weak, scattered hairs or none .... **Muscidae,** p. 386

Hypopleura with a row of bristles, fig. 295I 33

33. Mesopostnotum having a convex bump below the scutellum, fig. 295F
Tachinidae, p. 389

Mesopostnotum having no well-developed extra bump, at most a gentle convexity below scutellum; the flesh flies 34

34. Arista of antenna with feathering not extending much beyond middle; body dull black or striped with gray and black .... **Sarcophagidae**

Arista feathered to tip, fig. 297J, or body entirely metallic blue or green
Calliphoridae

35. Oral opening round and minute, mouthparts vestigial, cheeks inflated, fig. 297P
Gasterophilidae, p. 387

Oral opening large, mouthparts well developed 36

36. Front of head lacking sutures or ridges running ventrolaterad from base of antennae, fig. 297O 37

Front of head having a pair of frontal sutures or ridges, fr, each running from near the antenna toward the oral excavation, fig. 297Q 38

37. Crossvein r–m situated at the base of the wing, and inconspicuous, sometimes difficult to find; free end of M_{3+4} lacking; predaceous species in moist habitats
Dolichopodidae

Either crossvein r–m situated at least one-third distance from the base to apex of wing, or M_{3+4} present, fig. 295C .... **Empididae**

38. Sc distinctly sclerotized for its entire length, curving gently to join C before R_1 does, fig. 295B, D (this character best seen from anterodorsal view of wing)
.... 39

Sc either partially atrophied, or fused at tip with R_1, or abruptly angled, fig. 295E
.... 47

39. Head nearly spherical, eyes occupying most of lateral aspect, and second antennal segment small, fig. 296I .... **Sepsidae,** p. 384

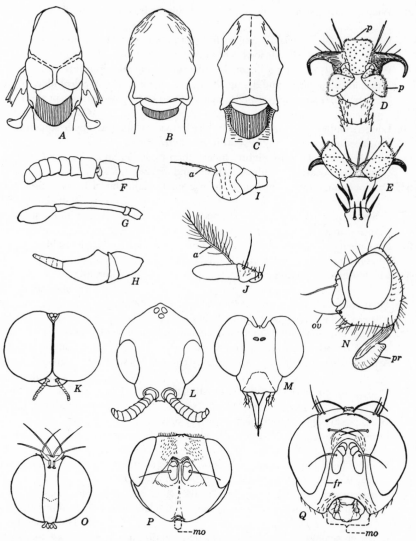

Fig. 297.   Diagnostic parts of Diptera.   *A*, mesonotum of *Helobia*, Tipulidae; *B*, mesonotum of *Palpomyia*, Ceratopogonidae; *C*, mesonotum of *Chironomus*, Chironomidae; *D*, tarsal claws and pads of *Tabanus*, Tabanidae; *E*, tarsal claws and pads of *Zenillia*, Tachinidae; *F*, antenna of *Bibio*, Bibionidae; *G*, antenna of *Mydas*, Mydaidae; *H*, antenna of *Tabanus*, Tabanidae; *I*, antenna of *Geosargus*, Stratiomyidae; *J*, antenna of *Pollenia*, Calliphoridae; *K* and *L*, face of *Bibio*, Bibionidae, male and female; *M*, face of *Asilus*, Asilidae; *N*, head of *Hylemya*, Antho-myiidae; *O*, face of *Dolichopus*, Dolichopodidae; *P*, face of *Gasterophilus*, Gasterophilidae; *Q*, *Tephrita*, Tephritidae.   *a*, arista; *fr*, frontal ridge or suture; *mo*, mouth opening; *ov*, oral vibrissa; *p*, pulvillar pads; *pr*, proboscis.

Either head not spherical, or eyes situated a considerable distance from ventral margin, or second antennal segment massive, fig. 296*H*. 40

40. Dorsum of head and mesonotum flat, all but the lateral areas clothed only with abundant, short, stiff setae; seashore species . . . .**Coelopidae**

Dorsum of mesonotum convex and arched, often with long bristles scattered over much of its area. 41

41. Oral vibrissae present (compare fig. 297*N*). 42

Oral vibrissae absent 45

42. Lower squama differentiated as a flap of the axillary cord, fig. 296*B–D*. 43

Lower squama not differentiated, being simply a thin, fuzzy edge of the axillary cord, fig. 296*A* 44

43. Anal vein reaching wing margin, although the end of the vein may be faint, fig. 295*B* . . . . **Anthomyiidae**, p. 386

Anal vein reaching only about halfway to wing margin. . . . . **Fanniidae**, p. 386

44. Postocellar bristles long and convergent, fig. 296*G*; body never hoary

Helomyzidae, p. 384

Postocellar bristles either short, absent, or divergent; body often hoary with long, thick hair . . . . **Scopeumatidae**

45. Some or all of the tibiae with a preapical dorsal bristle. 46

Tibiae without preapical bristles. . . . . **Otitidae**, p. 384

46. Postocellar bristles convergent; second antennal segment always small; small flies, rarely over 6 mm. long. . . . . **Lauxaniidae**, p. 384

Postocellar bristles parallel, divergent, or absent; second antennal segment often massive, as large as third, fig. 296*H*; moderate-sized flies

Sciomyzidae, p. 384

47. Wing having costal cell wide and having *Sc* ending abruptly or angled abruptly much before apex of cell, either *Sc* beyond this point weak or atrophied, or *C* with a distinct break, at which point there are several stout bristles, fig. 295*E*

Tephritidae, p. 384

Wing either having costal cell narrow, or *Sc* gradually fading out towards its apex, or fusing with R_1, or absent. 48

48. Posterior basitarsus short and enlarged. . . . . **Borboridae**, p. 384

Posterior basitarsus little if at all thicker than succeeding segments and usually longer than second tarsal segment. 49

49. Wing with 1*A* and Cu_{1b} almost entirely atrophied, cell Cu_{1b} therefore open or absent 50

Wing with 1*A* and Cu_{1b} present and enclosing cell Cu_{1b}. 51

50. Dorsum of head with a large, triangular, sclerotized area (the ocellar triangle, fig. 296*F*, *ot*) flanked by wide membranous areas; wing having vein Cu_{1a} slightly sinuate . . . . **Chloropidae**, p. 384

Dorsum of head with sclerotized areas either very small, or quadrangular, or occupying all of dorsal aspect; wing with vein Cu_{1a} not sinuate

Ephydridae, p. 384

51. Postocellar bristles convergent, as in fig. 296*G*. . . . . **Drosophilidae**, p. 385

Postocellar bristles parallel or divergent. . . . . **Agromyzidae**, p. 384

SUBORDER NEMATOCERA

The North American fauna contains representatives of about twenty families of Nematocera. Midges, crane flies, mosquitoes, and black flies are examples of the more abundant and conspicuous families. The larvae of all families have a well-defined sclerotized head, fig. 303, which is retracted into the thorax only in the Tipulidae.

CECIDOMYIIDAE (ITONIDIDAE), THE GALL GNATS. The adults, fig. 298, are inconspicuous fragile flies, having greatly reduced wing venation, and elongate beadlike antennae. The adults feed only on aqueous material such as sap; the larval period of the life cycle is the part in which most of the food is consumed. Many genera of the family are gall makers, and it is by these galls, especially on willows, deciduous trees, and many herbs, that the family is known to most observers. The family exhibits a wide range of other habits. Some larvae are predaceous, feeding on mites and small insects; others feed on developing plant seeds; some feed on decomposing organic matter; and still others feed in or on the tissues of leaves or stems of plants. To the latter category belongs the most notable economic species of the family, the hessian fly *Phytophaga destructor,* fig. 298; the larvae feed in the lower stems of grasses and are especially injurious to wheat and barley.

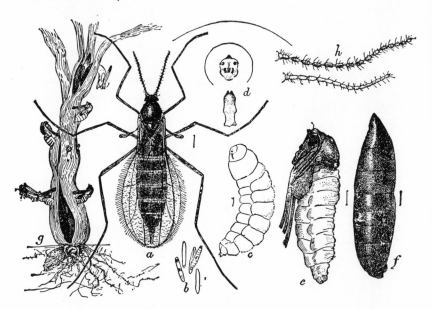

Fig. 298. The hessian fly *Phytophaga destructor. a,* female fly; *b,* eggs, one at hatching; *c,* larva; *d,* head and breastbone of same; *e,* pupa; *f,* puparium; *g,* infested wheat stem showing emergence of pupae and adults; *h,* antennae, male and female. (From U.S.D.A., E.R.B.)

Fig. 299. Adult black fly *Simulium vittatum.*
(After Knowlton and Rowe)

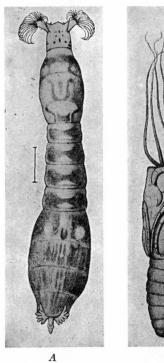 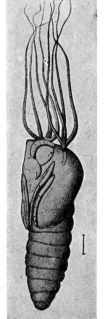

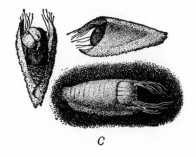

C

Fig. 300. *A*, larva, and *B*, pupa
of a black fly *Simulium johannseni*
and *C*, pupal cases of *S. venustum.*
(From Illinois Nat. Hist. Survey)

A   *B*

SIMULIIDAE, BLACK FLIES.   This family, often referred to as buffalo gnats, has aquatic larvae and pupae and bloodsucking adults, like the Culicidae. The black flies are short, stubby, and humpbacked, with short legs and compact, usually 11-segmented antennae, fig. 299.   Like the mosquitoes, only the females bite.   Horses, cattle, ducks, and many wild animals are attacked.   Certain species attack man also, gathering around the ears, eyes, and exposed areas of the face, hands, and ankles.   They draw a great amount of blood and produce burning welts on the victim's skin. These welts may itch and burn for a week or more.

The eggs are laid in clusters near the water edge or in the water.   The larvae are sedentary, elongate, and slightly vasiform, fig. 300, and occur only in running water; the posterior end is anchored by a hooked sucker-disc to some support such as a rock, log, or trailing leaves; the head has a pair of feathery branched rakes which are supposedly used as strainers to

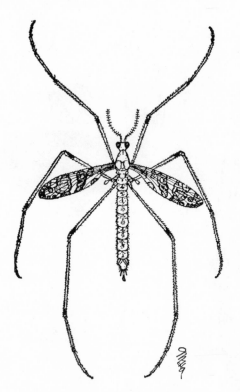

Fig. 301.   A cranefly *Epiphragma fascipennis,* adult female.

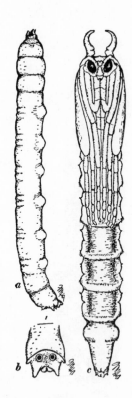

Fig. 302.   *Epiphragma. a,* larva; *b,* end of larva from above; *c,* pupa. (After Needham)

obtain food from the running water, primarily microorganisms and organic material. The pupae are also aquatic, formed in a boot-shaped cocoon, fig. 300. On the prothorax of the pupa are a pair of long respiratory processes, usually branching into many slender filaments.

TIPULIDAE, CRANE FLIES. The adults are long-legged slender-winged flies, fig. 301; the antennae are threadlike, with many distinct segments, and the mesonotum has a V-shaped transverse furrow or suture. The adults are extremely abundant in moist woods and sheltered ravines, and along wooded stream banks. The larvae are elongate and wormlike, fig. 302; many are aquatic, living especially in submerged clusters of rotting leaves; many feed in leaf mold; and a few feed on living plant roots or mine in leaves. They vary greatly in size and appearance but have in common a stout head which is partly retracted within the thorax (in certain species it is further retractile and may be completely hidden in the thorax when the insect is disturbed) and strong toothed mandibles which work from side to side.

CULICIDAE, MOSQUITOES. Long-legged slender insects, fig. 303, having beadlike antennae (in the males they are plumose as in the Chironomidae) and many-veined wings. The mouthparts form a beak, composed of a highly modified assemblage of piercing-sucking parts (see fig. 53, p. 69).

In certain genera eggs are laid on water, either singly or glued together to form rafts; they hatch in a few days. In other genera eggs are laid in humus or just above the water line on the sides of containers; these eggs hatch at some later time when the water rises and inundates them. The larvae are aquatic, most of them living in still water, but a few species live in slowly moving water. In other parts of the world there are species which breed in rapid streams, quite in contrast with the habits of nearctic species. The larvae are called wrigglers; they have large heads with fairly long antennae, a large swollen thorax, and a cylindrical abdomen. In all the mosquitoes and some of their relatives the abdomen bears a dorsal breathing tube or plate on the eighth segment. In the main, mosquito larvae feed on microorganisms and organic matter in or on the water. A few groups are predaceous and feed solely on other mosquito larvae. The pupae, which are called tumblers, are also aquatic, free living, and active. Their breathing tubes are situated on the thorax. Adult females of a few species and males of all mosquito species feed only on nectar or water. But unfortunately in most species the females seek a blood meal, which under natural conditions is necessary for reproduction.

Economically, mosquitoes are of tremendous importance to man. Some species, principally those of the genera *Anopheles* and *Aedes,* transmit an imposing list of human diseases, including malaria, dengue, yellow fever, and filariasis. For this reason mosquitoes are of utmost importance

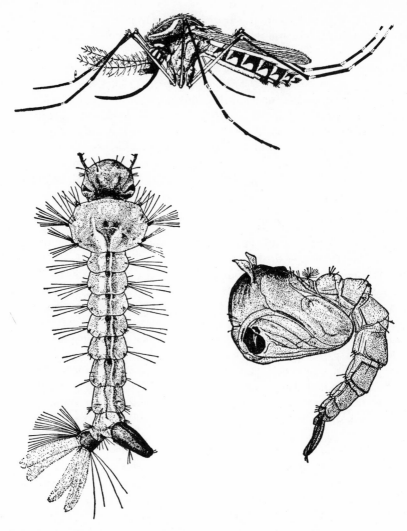

Fig. 303. The yellow fever mosquito *Aedes aegypti*. Adult above; larva at left; pupa at right. (From U.S.D.A., E.R.B.)

in a consideration of medical entomology. As a direct nuisance mosquitoes have an economic influence. The severity of their attacks decreases property values, especially in resort areas, and has undoubtedly had an influence in the settlement of extreme northern areas where mosquitoes are unusually abundant.

CHIRONOMIDAE (TENDIPEDIDAE), MIDGES. These are frail insects, fig.

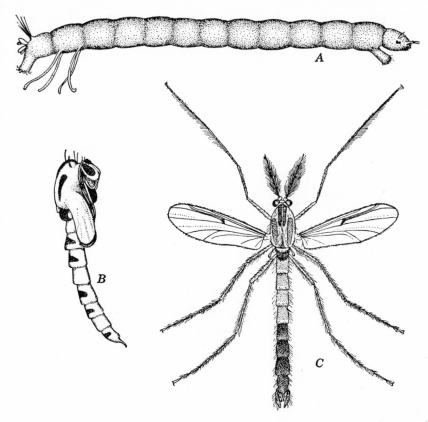

Fig. 304. Life history stages of Chironomidae. *A*, larva of *Chironomus tentans;* *B*, pupa of *Cricotopus trifasciatus;* *C*, adult male of *Chironomus ferrugineovittatus.* (From Illinois Nat. Hist. Survey)

304, frequently mistaken for mosquitoes, but they do not bite and have several structural differences which set them off. The male antennae are plumose with whorls of long silky hair. The larvae are all aquatic, some free living, others spinning a loose web of bottom particles and silk and in certain genera making a definite case. They are slender and worm-like, and a small but distinct sclerotized head, and a 12-segmented body. In some groups the prothorax, or the last body segment, or both, may have a pair of nonjointed leglike protuberances or pseudopods. The larvae feed on organic matter on the bottom of bodies of water and are found in rivers, lakes, ditches, and stagnant ponds. The pupae are also aquatic, some of them free living, but most of them staying in the web or case made by the larva. In many bodies of water these midge larvae are tremendously abundant and form one of the principal items of fish food. When the adults emerge, they appear at night in clouds and blanket near-by lights in a humming mass. The eggs are laid in water and hatch in a few days.

SUBORDER BRACHYCERA

The adults of this suborder are mostly larger and stouter bodied than the Nematocera and are stronger on the wing. The larvae have hooked or blade-shaped mandibles which work up and down, rather than sideways; the head is frequently much retracted within the thorax, fig. 305, and may have stout internal supports extending far into the body. About fifteen families are represented in North America; some of them

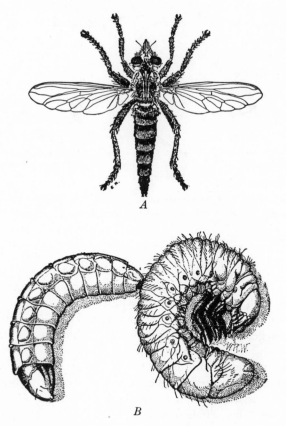

Fig. 305. A robber fly *Promachus vertebratus. A*, adult; *B*, legless larva attacking white grub. (From Illinois Nat. Hist. Survey)

are abundant, and in one, the Tabanidae or horse flies, the adults are voracious bloodsuckers.

Other families of the Brachycera contain species of great diversity as regards structural peculiarities and habits. Most of them, for example, the family Dolichopodidae, are predaceous on other insects in both adult and larval stages and have free pupae; larvae of most species live in rotten logs or in soil and pupate there.

ASILIDAE, ROBBER FLIES. These are also large flies, fig. 305, usually with a humped thorax and elongate abdomen, but in some genera the body is stout, hairy, and brightly colored, resembling a bumble bee. In many genera also the lower part of the face has a beardlike brush of hair called a mystax. These flies are predators on other insects, capturing and eating bumble bees and dragonflies in addition to smaller forms. The adults are not easily frightened and are both conspicuous and easily taken in a sweep net. The larvae are found chiefly in soil and rotten wood and are predaceous on insects found there. Some of them are important natural enemies of white grubs and other soil-inhabiting species attacking cultivated crops, fig. 305.

MYDAIDAE, MYDAS FLIES. This is a small family, but it contains some of our largest and showiest species. The antennae are clubbed, and the wings have a modified venation, the ends of many veins bending forward towards the anterior margin of the wing. The common eastern species *Mydas clavatus,* is black with an orange band and has a wing spread of 55 mm. The larvae, resembling large asilid larvae, are predaceous and found in rotten logs.

TABANIDAE, HORSE FLIES AND DEER FLIES. The adults, fig. 306, are large-headed stout-bodied insects, often attaining a length of 25 to 30 mm., having strong wings with fairly complete venation and frequently a striking color pattern. The venation features a wide V-shaped cell R_4, which embraces the apex of the wing. The females are bloodsuckers; their mouthparts are developed for cutting skin and sucking the blood which oozes from the wound. The males feed on nectar. Eggs are laid in masses on stems or other objects growing over water; the newly hatched young crawl or drop into their breeding place. The larvae live in swamps or sluggish streams, staying in the bottom mire, and are often abundant in damp pasture soils. They are predaceous, feeding on snails, insects, and other aquatic organisms. The larval body is tough and leathery, usually white or banded; the head is completely retractile within the thorax. The pupae are cylindrical and brown, normally formed in a mass of peat or dead vegetation which is damp but above the free-water level.

Farm livestock, especially cattle and horses, are often bothered by adult horse flies. Locally these pests may assume major importance in reducing

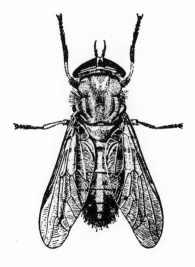

Fig. 306. A horse fly *Tabanus lineolus*.
(From U.S.D.A., E.R.B.)

condition of stock. Many species of horse flies and deer flies abound in marshy areas of the northern and humid montane areas of the continent and discourage vacationers and settlers.

Closest relatives of the Tabanidae are the Rhagionidae (snipe flies) that have predaceous larvae and adults. Some species of the genus *Symphoromyia*, fig. 50, have blood-sucking adults that attack man and other vertebrates. More distant relatives of the Tabanidae include the Stratiomyidae (soldier flies) whose larvae are predators or scavengers and whose adults are predaceous, and the Empididae and Dolichopodidae, both of whose larvae and adults are predaceous on other insects.

SYRPHIDAE, FLOWER FLIES, SYRPHID FLIES. Small to large flies, fig. 307, characterized by the upturned ends of some of the wing veins, and the presence of a veinlike thickening or *spurious vein* in front of media. These flies feed almost exclusively on flowers and have a remarkable ability for hovering apparently motionless in the air. Many are brilliantly striped or marked with yellow, red, white, and black; within the family are mimics of various wasps, bumble bees, and other bees. The larvae are extremely diverse. Many feed on aphids, crawling from colony to colony and devouring huge numbers of individuals. Other species feed in decaying liquid organic matter, for example the rat-tailed maggots of the genus *Eristalis*, fig. 308.

In these maggots the posterior spiracles are situated at the end of extensile "tails" that can be extended to the surface of the food and so provide a contact with air. Other larval foods include debris in burrows of wood-boring insects, leaves and sheaths of grasses, and debris in nests of

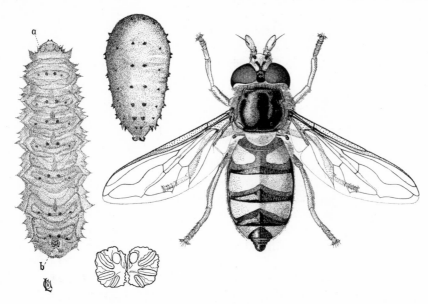

Fig. 307. A flower fly *Didea fasciata*. Right, adult; left, larva; center, puparium; *a*, anterior spiracle; *b*, caudal spiracles. (After Metcalf)

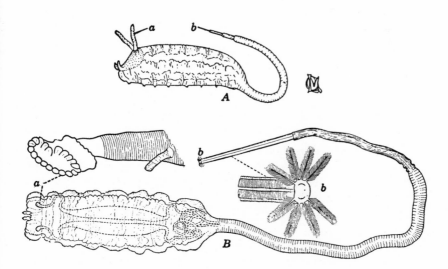

Fig. 308. Rat-tailed maggots. *A*, puparium of *Eristalis aeneus; B*, larva of *Eristalis tenax*. *a*, anterior spiracle; *b*, posterior spiracles on long "rat-tail." (After Metcalf)

bumble bees and wasps. Larvae of two species are serious pests of bulbs, eating out the centers of planted tulips and related species.

Allied families include the Conopidae, whose larvae are internal parasites of other insects, and the Phoridae, small species of varied habits, but principally with saprophytic larvae. These two families and the Syrphidae belong to a group of families transitional between the Brachycera and the Cyclorrhapha, having larval mouthparts more like those of the Cyclorrhapha but otherwise being typical Brachycera.

SUBORDER CYCLORRHAPHA

This is by far the largest of the three suborders of Diptera. It contains over forty families, many of them composed of a large number of species. For the most part the adults are relatively short and stout bodied, having broad wings in which vein R_{4+5} is undivided, M_2 has atrophied, and Cu_{1b} is only a short recurved stub. The antennae are usually three-segmented, the third segment bearing a style or arista. The larvae have practically no external sclerites of the head capsule remaining, so that the head appears to be only a conical membranous anterior segment of the body. The two mouth hooks (mandibles) are connected with a complex internal cephalic skeleton, which provides the attachment basis for muscles controlling feeding. The fourth larval instar and the pupa are formed within the larval skin of the third instar, which hardens to become a puparium. The adult emerges by pushing off the anterior end of the puparium.

The Cyclorrhapha may be divided into three series of families, the Acalyptrata, Calyptrata, and Pupipara.

Series Acalyptrata. In this group are included a large number of families of varied habits. Primitive families include the Otitidae (picture-winged flies), Lauxaniidae, Coelopidae, and Heleomyzidae, whose larvae feed on various decaying organic material; Sepsidae, whose larvae feed in dung; and Sciomyzidae, whose larvae are aquatic and feed chiefly on aquatic or amphibious snails. More specialized families include the Borboridae, the little dung flies; Agromyzidae, the leaf miners; Chloropidae, chiefly scavengers; Ephydridae, the shore flies and brine maggots; and the two following.

TEPHRITIDAE (TRYPETIDAE, TRUPANEIDAE), FRUIT FLIES. Here belong many brightly colored species frequently with mottled or banded wings. The larvae of several species live in the seed heads or fruits of plants. Important commercially are the cherry fruit fly *Rhagoletis cingulata*, fig. 309, whose maggots tunnel in cherry fruits, and the apple maggot *Rhagoletis pomonella*, which tunnels through apples. The Mediterranean

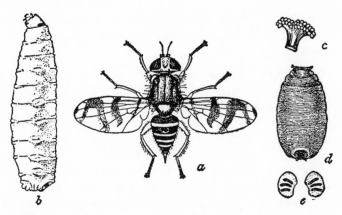

Fig. 309. The cherry fruit fly *Rhagoletis cingulata*. *a*, fly; *b*, maggot; *c*, anterior spiracles of same; *d*, puparium; *e*, posterior spiracular plates of pupa. (From U.S.D.A., E.R.B.)

fruit fly *Ceratitis capitata* is a tropicopolitan species that feeds in citrus fruits; it was discovered in Florida in 1929 but the infestation was exterminated in 1930 through a vigorous eradication campaign costing about six million dollars. The Medfly, as the insect is now generally called, was re-introduced into Florida and detected in April, 1956. A concerted Federal-State eradication program has again been initiated.

DROSOPHILIDAE, VINEGAR GNATS. These include small flies, fig. 310, which are common at times in most houses and stores. The larvae breed in decaying fruits and other organic materials. *Drosophila melanogaster*

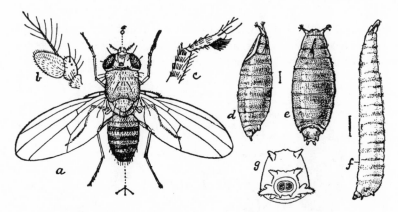

Fig. 310. Vinegar gnats *Drosophila melanogaster*. *a*, adult; *b*, antenna of same; *c*, base of tibia and first tarsal joint of same; *d*, puparium, side view; *e*, puparium from above; *f*, full-grown larva; *g*, anal spiracles of same. (From U.S.D.A., E.R.B.)

has been used as an experimental subject by geneticists and from the standpoint of chromosome and gene mapping is undoubtedly the best-known animal in the world.

Series Calyptrata. This series is characterized most conspicuously by the development of the upper and lower squamae in the adults and the elongation of the posterior spiracles in the larvae. The series contains about ten families occurring through a wide ecological range. More primitive members include the Fanniidae (scavenger flies), Anthomyiidae, and Muscidae; more specialized families include the Sarcophagidae and Calliphoridae (flesh flies) and other families mentioned below.

ANTHOMYIIDAE. Members of this family are similar in general appearance to the house fly. Economic species in the family include the onion maggot *Hylemya antiqua,* fig. 311, and the cabbage maggot *Hylemya brassicae,* which feed on the roots of their respective hosts.

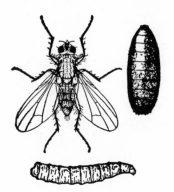

Fig. 311. Life history stages of the onion maggot *Hylemya antiqua,* adult, larva, and puparium. (From Illinois Nat. Hist. Survey)

MUSCIDAE, THE HOUSE FLY AND ITS ALLIES. This family contains probably the world's commonest and most ubiquitous insect, the house fly *Musca domestica,* fig. 312. House fly larvae are white maggots which breed

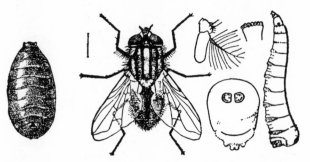

Fig. 312. Stages of the house fly *Musca domestica.* Puparium at left; adult next; larva and enlarged parts at right. (From U.S.D.A., E.R.B.)

in many types of decaying matter. The adult flies transmit several dangerous and widespread diseases, including typhoid fever, several kinds of dysentery, cholera, and trachoma. Another widespread member of this family is the stable fly *Stomoxys calcitrans,* the adult of which inflicts a painful bite and attacks man and domestic animals. Its larvae breed in decaying organic matter; rotting piles of new grass and lawn clippings and manure are high on the list of favorites.

GASTEROPHILIDAE, BOT FLIES. These are moderate-sized flies about the

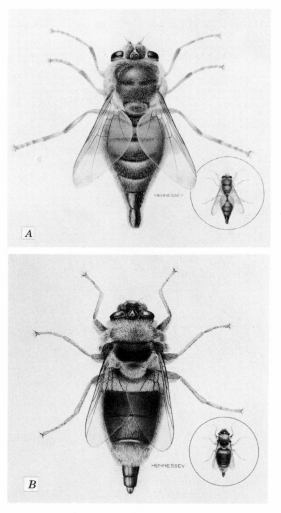

Fig. 313. Bot flies. *A, Gasterophilus intestinalis; B, G. haemorrhoidalis.* (From Canadian Department of Agriculture)

size of a honeybee, somewhat hairy in appearance, and banded with black, yellow, or red, fig. 313.   The larvae are internal parasites of horses, mules, and some of the larger wild mammals.   In North America we have only the single genus *Gasterophilus,* represented by four species which infest horses.   The horse bot fly *G. intestinalis* lays its eggs on the hairs of the horse's legs and forequarters.   When the horse licks these eggs, they hatch, and the young larvae work their way through the mouth and throat tissues into the horse's stomach.   Here the young bots attach to the lining and feed, growing into stout spiny grubs 15 mm. or more in length.   They pass the summer and winter in the horse; in spring they loosen their hold, are passed by the horse, and pupate in the ground.   The adults emerge in a few weeks.

OESTRIDAE, OR WARBLE FLIES.   These, fig. 314, comprise a small group of

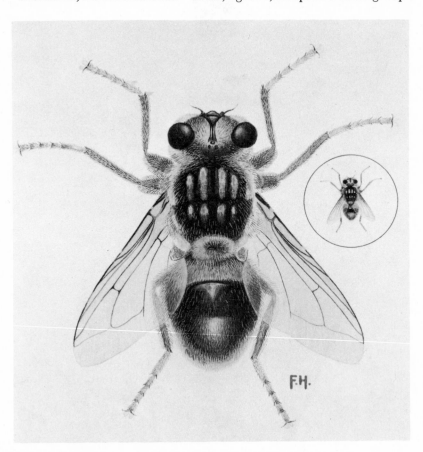

Fig. 314.   A warble fly, the adult of the common cattle grub *Hypoderma lineatum.*   (From Canadian Department of Agriculture)

fast-flying fuzzy bumble bee-like flies, parasites of mammals; some of them have a life cycle as specialized as that of the Gasterophilidae. Many species have a simple life cycle. For example, the sheep bot fly *Oestrus ovis* deposits young larvae in the nostrils of sheep; the larvae migrate into the sinuses and horns, where they mature. The mature larvae escape to the ground through the sheep's nostrils and pupate. Much more complex is the life cycle of the northern cattle grub *Hypoderma bovis*, which attacks cattle. The fly lays eggs on the hairs of the hind legs or flank of the animal. The larvae soon hatch, crawl down the hair, and burrow beneath the skin, making their way slowly through connective tissue to the oesophagus; this journey takes about 4 months. After 3 months' development in the oesophagus the maggots (still fairly small) journey through connective tissue again and come to rest beneath the skin in the lumbar region. Here the larvae attain most of their growth, each causing a swelling called a warble, provided with a small hole through the skin. Through this the larva first obtains air while maturing and then escapes to the ground for pupation.

TACHINIDAE (LARVAEVORIDAE), TACHINA FLIES. This, fig. 315, is one of the largest families of Diptera. All its members are parasitic on insects and attack larvae or adults of many orders, especially lepidopterous larvae. Many members of this family have been propagated and introduced into various parts of the world in an effort to hold in check some injurious insect. The adult females of many species lay their eggs directly

Fig. 315. Tachina flies. At right, *Winthemia quadripustulata,* a parasite of Lepidoptera larvae; at left, *Trichopoda pennipes,* a parasite of Hemiptera. (From U.S.D.A., E.R.B.)

on the host; when they hatch, the maggots bore into the host body and feed on its tissues.   In other species the eggs are laid on foliage; if this foliage is eaten by a larva of the host species, the parasite eggs hatch in the alimentary canal of the host, and from there the young larvae bore into the body cavity where development takes place.   When mature, the parasite larvae either pupate in the host body or leave it and pupate in the ground.

Series Pupipara.   Here belong three small families living as ectoparasites on the bodies of birds, mammals, or bees.   So far as known, most members of the Pupipara have the feature, very unusual among insects, of uterine development of the young.   The larvae are retained in the body of the female in a special uterine pouch and nourished on glandular secretions.   When mature, the larvae are "born" and glued to the hair of the host, in which position they pupate immediately.   The best-known species in North America is the sheep ked *Melophagus ovinus*, fig. 316.   This is a wingless ticklike insect common locally on sheep.   It belongs to the family Hippoboscidae.   Also famous is the minute parasite of the honey bee, *Braula caeca*, the sole member of the Braulidae.   This species lays eggs, and may not belong to the Pupipara.

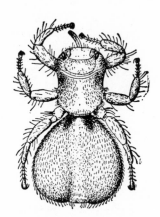

Fig. 316.   The sheep ked *Melophagus ovinus*.   (From Kentucky Agricultural Experiment Station)

References

Aldrich, J. M., 1905.   A catalogue of North American Diptera.   *Smithsonian Inst. Publs., Misc. Collections,* **46:**1–680.

Crampton, G. C., ET AL. 1942.   The Diptera or true flies of Connecticut.   *Connecticut State Geol. Nat. Hist. Survey,* **64:**1–509.

Curran, C. H., 1934.   *The families and genera of North American Diptera.*   New York: The Ballou Press.   512 pp.

Frost, S. W., 1924. A study of the leaf-mining Diptera of North America. *Cornell Univ. Agr. Expt. Sta. Mem.,* **78**:1–228.

Herring, E. M., 1951. *Biology of the leaf miners.* The Hague: Junk. 420 pp.

Roback, S. S., 1951. A classification of the muscoid calyptrate Diptera. *Ann. Entomol. Soc. Amer.,* **44**:327–361.

Order TRICHOPTERA: Caddisflies

Mothlike insects, having aquatic larvae and pupae. The adults, fig. 317, vary in length from 1.5 to 40 mm. They have chewing-type mouthparts, with all parts greatly reduced or subatrophied except for the two pairs of palpi; long multisegmented antennae; large compound eyes; and long legs. Except for the wingless or brachypterous females of a few spe-

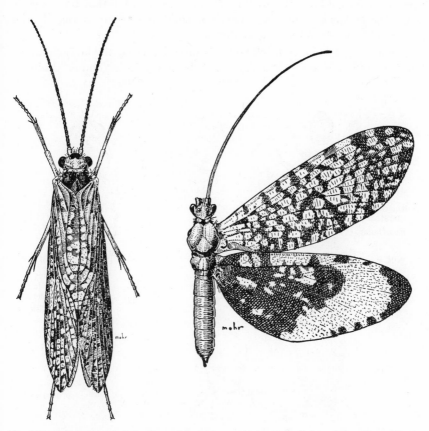

Fig. 317. Caddisflies. At left, *Rhyacophila fenestra;* at right, *Eubasilissa pardalis.* (From Illinois Nat. Hist. Survey)

cies, the adults have two pairs of large membranous wings, with a fairly complete set of longitudinal veins, but a reduced number of crossveins. In most species the body and wings have hair but no scales; in a few species the antennae, palpi, legs, and wings may have patches of scales or a scattering of scales among the longer hair. The larvae, figs. 318 and 319, are diverse in general appearance and habits. The eyes are each represented by a single facet, the antennae are small and one-segmented, the mouthparts are of the chewing type and well developed. The thorax has three pairs of strong legs, the abdomen a pair of strong legs bearing hooks and frequently a set of finger-like gills.

Most of the adults are somber in color or tawny, but a few have wings which are marked with yellow or orange, or have silvery streaks.

Caddisfly larvae live in both lakes and streams, showing a definite preference for colder and unpolluted water. As a group they have a wide ecological tolerance but a more restricted one than midge larvae (Chironomidae) in relation to pollution.

Casemaking has been developed to a high degree by the larvae of most families of these insects. The cases, which are portable "houses" built by

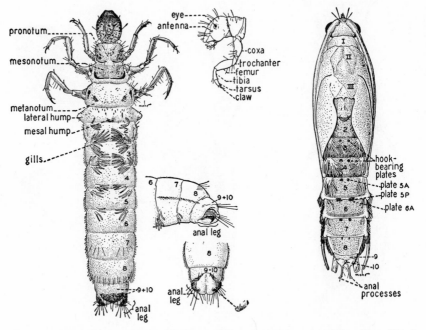

Fig. 318. Larva and pupa of a casemaking caddisfly *Limnephilus submonilifer*. (From Illinois Nat. Hist. Survey)

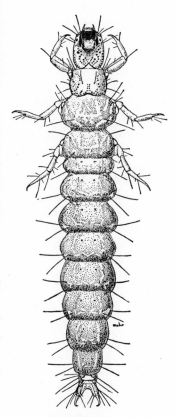

Fig. 320.   Purselike caddisfly case and larva, *Ochrotrichia unio.* (From Illinois Nat. Hist. Survey)

Fig. 319.   Free-living caddisfly larva *Rhyacophila fenestra.* (From Illinois Nat. Hist. Survey)

Fig. 321.   Caddisfly larva in cylindrical case, *Limnephilus rhombicus.* (From Illinois Nat. Hist. Survey)

the larvae, have fascinated observers of fresh-water insect life. Cases, figs. 320, 321, are of varied shapes, ranging from a straight tube to the coiled case of *Helicopsyche,* which resembles a snail shell. Many types of materials are used in the construction of these cases. Small stones, sand grains, bits of leaves, sticks, conifer needles, and frequently small snail shells may be utilized. In most instances a given genus or species constructs a case of characteristic shape, but genera in different families often make very similar cases.

Caddisfly females lay from three hundred to one thousand eggs each. In some species the eggs are discharged in strings, fig. 322*B*; the female enters the water and lays the eggs on stones or other objects, grouping the strings into irregular masses containing up to eight hundred eggs.

Females of other groups extrude the eggs and form them into a large mass at the tip of the abdomen before depositing them, fig. 322*A*. The eggs are incased in a gelatinous matrix which swells on absorbing moisture. These masses are attached to sticks or stones which are submerged in, adjacent to, or overhanging water.

Females of the family Limnephilidae frequently deposit egg masses on branches overhanging water. On numerous occasions observers have reported that these egg masses swell and liquefy, and the eggs hatch during rain. The gelatinous drops formed by this process run down the twigs and drop into the water, carrying the larvae along.

The larvae are all active, most of them feeding chiefly on small aquatic animals or microorganisms which encrust decayed organic matter in the water. A few genera, notably *Rhyacophila* and *Oecetis*, are predominantly predaceous, feeding on small insect larvae. Larvae of the genus *Rhyacophila*, fig. 319, are free living, constructing no larval case. Three families, the Philopotamidae, Psychomyiidae, and Hydropsychidae, have larvae that weave a fixed net and shelter, the net presumably being used to trap small aquatic organisms which constitute their food. In all other families the larvae construct portable cases of various types. The larva uses these to protect the greater part of the body, which has thin integument. Only the heavily sclerotized head, legs, and anterior portion of the thorax are extruded from the case when the larva is actively moving about, figs. 320, 321.

Prior to pupation, the larvae of all caddisflies spin a cocoon. The case-makers form this very simply by spinning a silken lining inside the case and closing the ends of the case with a barred or slit membrane. The free-living and net-making species spin an ovoid cocoon of silk and sand, stones, or bits of debris, which is firmly attached to a stone, log, or other rigid support. The pupae develop in the cocoon until the adult structures

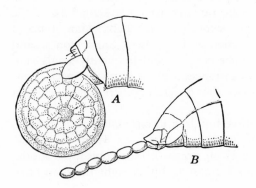

Fig. 322. Caddisfly eggs. *A, Triaenodes tarda; B, Cyrnellus marginalis.* (From Illinois Nat. Hist. Survey)

(except wings) are completely formed and fully sclerotized. The pupae, fig. 318, are unusual in possessing strong mandibles. With these, the mature pupa cuts its way out of the cocoon. It then swims to the surface, crawls on a log or stone, and transforms into an adult.

Usually the complete life cycle requires a year, most of it spent in the larval stage. The egg stage lasts only a short time; the pupal stage requires 2 to 3 weeks; and the adults live about a month.

Of interest among caddisflies are the larval habits of "micro" caddisflies, or Hydroptilidae. This family contains only small individuals, ranging from 1.5 to 6 mm. in length. The first instars are minute free-living forms, with small abdomens; the last instar builds a portable case and develops a swollen abdomen. Information is available only on two North American genera, *Mayatrichia* and *Ochrotrichia*. In the latter there are differences in the structure of the claws between the free-living and case-making instars. This dimorphism is similar in many respects to typical examples of hypermetamorphosis.

The caddisfly fauna of the nearctic region contains over eight hundred species grouped in seventeen families.

References

Betten, Cornelius, 1934. The caddis flies or Trichoptera of New York State. *N. Y. State Museum, Bull.,* **292**:576 pp., 61 text figs., 67 pls., bibl.

Denning, D. G., 1956. Trichoptera. *Aquatic insects of California.* Berkeley: Univ. Calif. Press. Pp. 237–270.

Nielsen, Anker, 1948. Postembryonic development and biology of the Hydroptilidae. *Kgl. Danske Videnskab., Biol. Skrifter,* **5**(1):1–200.

Ross, Herbert H., 1944. The caddis flies or Trichoptera of Illinois. *Illinois Nat. Hist. Survey, Bull.,* **23**(1):1–326, 957 figs., bibl.

———— 1956. *Evolution and classification of the mountain caddisflies.* Urbana: Univ. Ill. Press. 213 pp.

Order LEPIDOPTERA: Moths and Butterflies

Insects with two pairs of wings, fig. 323, except for a few species which have apterous females. The body, wings, and other appendages are covered with scales which are often brilliant in color and arranged in showy patterns. Adult mouthparts are greatly reduced; in most forms only the maxillae are well developed. These are fused and elongated to form a coiled tube for sucking up liquid food. The large compound eyes, the long antennae, and the legs are all well developed. The species have complete metamorphosis. The larvae, fig. 335, are cylindrical (the familiar caterpillars); most of them have a definite head, thoracic legs, and five pairs of larvapods (including the end pair). The pupae are usually hard and brown, and the appendages appear cemented onto the body.

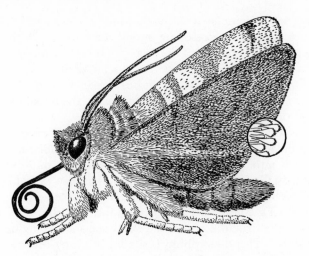

Fig. 323. A typical moth, showing scales on wings and body, and sucking tube which is coiled under head when not in use. (From Illinois Nat. Hist. Survey)

The order is a large one, embracing about ten thousand North American species. The adults of these vary greatly in size from minute forms a millimeter or two long to large species with a wing span of 6 inches or more. The wings are distinctively patterned, and the order as a whole presents an attractive array of color and beauty.

The larvae of the Lepidoptera are plant feeders except for a scattering of species which are predaceous, scavengers, or feed on stored products. The great bulk of the species feed externally on foliage; a large number of the minute species mine inside leaves or leaf petioles; and another large group, including both large and small species, bore inside stems, trunks, or roots. A great number of these species attack cultivated plants and cause a high annual loss of crops and stored products. The order is therefore one of great economic importance.

In addition to its destructive species, the order Lepidoptera contains one of great commercial value—the silkworm *Bombyx mori.* This insect is the sole source of natural silk. The propagation of the species and the harvesting of the silk, known as sericulture, is an important industry in many parts of the world, with an annual harvest in oriental countries of many million dollars' worth of silk. At the turn of the century a sericulture industry was established in California. This failed because of high labor costs. In 1945 a new venture was started in Texas in which a large part of the work was done mechanically.

Phylogeny. In three families of Lepidoptera the front and hind wings are similar in size and venation and the front wing has a lobelike coupling

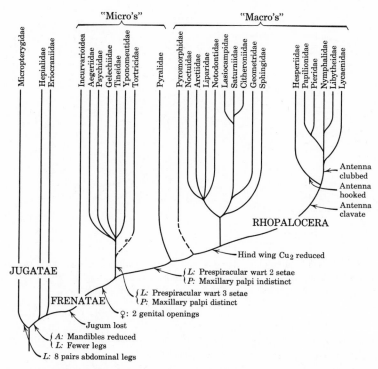

Fig. 324. Simplified family tree of the order Lepidoptera (based chiefly on Forbes, 1923). *A*, adult; *L*, larva; *P*, pupa.

device called the *jugum*, fig. 325*A, J*, at the base of the hind margin. These characters resemble the condition in Trichoptera, and indicate that these three families, called the suborder Jugatae, are the most primitive group of the Lepidoptera. In the family Micropterygidae the larvae have legs on eight abdominal segments (reminiscent of some Mecoptera larvae) and the adults have functional mandibles. These conditions indicate that the Micropterygidae are descendants of the most archaic known branch of the suborder, indeed, of the entire order. In the other line of the Jugatae the adult mandibles were reduced, then lost, and the larval abdomen lost at least one pair of legs. This branch of the Jugatae gave rise to two typical jugate families, the Hepialidae (swifts) and the small moths of the family Eriocraniidae.

From the base of the Eriocraniidae line arose a lineage in which the jugum was lost and a long hair or *frenulum*, fig. 325*E, F*, at the base of the hind wing was the chief wing-coupling structure. In this lineage the front and hind wings became dissimilar in both shape and venation. This was the beginning of the suborder Frenatae. Our only living representa-

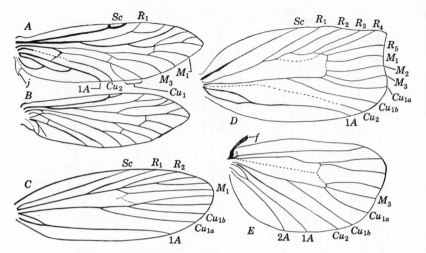

Fig. 325. Wings of Lepidoptera. *A* and *B*, front and hind wings of *Mnemonica*, Eriocraniidae; *C*, front wing of *Achroia*, Pyralidae; *D* and *E*, front and hind wings of *Archips*, Tortricidae. *f*, frenulum; *j*, jugum. (*C* loaned by Dr. Kathryn M. Sommerman)

tives of the early Frenatae are a group of seldom-collected minute moths comprising the superfamily Incurvarioidea.

From an early ancestral Frenatae evolved a form having two genital openings in the female, one for insemination and one for egg-laying. This form gave rise to two major branches. One branch that changed little gave rise to many families of small or minute moths, such as the Tineidae and Tortricidae, and to certain highly distinctive families such as the Psychidae (bagworms) and Aegeriidae (clear wings).

In the other branch certain body setae of the larvae atrophied and the maxillary palpi of the pupae became indistinct. The more primitive members of this branch may be the Pyralidae and the rarer Pyromorphidae, in which Cu_2 of the hind wing remained distinct. More specialized members appear to be the larger moths and the butterflies, in which Cu_2 of the hind wing became indistinct. More specialized members appear to be the larger moths and the butterflies, in which Cu_2 of the hind wing became reduced. The large-moth lineage evolved into the abundant families Noctuidae, Arctiidae, Saturniidae, and many others.

From the base of the large-moth branch evolved a lepidopteran whose adult was diurnal and had antennae enlarged at the tip. This lineage gave rise to the suborder *Rhopalocera*, which includes the skippers (Hesperiidae), having the adult antennae hooked at the tip, and the butterflies (Papilionidae and allies), having the antennae clavate.

KEY TO COMMON FAMILIES

1. Wings reduced to small pads or entirely lacking, fig. 338*b*................ 2
   Wings well developed, at least nearly as long as abdomen, fig. 323.......... 4
2. Legs lacking or reduced to short stubs; usually associated with a baglike case,
   fig. 330 .... **Psychidae,** p. 403
   Legs elongate and normal in appearance, fig. 338*b*..................... 3
3. Abdomen having closely set scales or spines, or bristling dark-gray hair; usually
   not found near cocoon.......................... **Geometridae,** p. 411
   Abdomen smoothly clothed with fine light woolly hair; moth usually found
   clinging to cocoon............................... **Liparidae,** p. 413
4. Front wing having a *jugum,* a lobe at the base of the posterior margin for use in
   wing coupling, fig. 325*A*; front and hind wings similar in venation and shape
   (**Jugatae**) .... **Hepialidae,** p. 403
   Front wing without a jugum; either anterior margin of hind wing having an
   enlarged lobe at base, fig. 327*D, H, I,* or with a long basal spine, or *frenulum,*
   fig. 325*E,* both used for wing coupling; or front and hind wings markedly
   different in shape and venation................................... 5

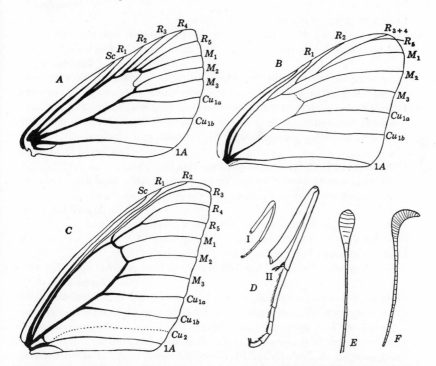

Fig. 326. Diagnostic parts of Lepidoptera. *A,* front wing of *Pamphila,* Hesperiidae; *B,* front wing of *Pieris,* Pieridae; *C,* front wing of *Papilio,* Papilionidae; *D,* front (small) and middle legs of *Brenthis,* Nymphalidae; *E,* tip of antenna of same; *F,* tip of antenna of *Thanaos,* Hesperiidae.

5. Hind wing without a frenulum, and antenna clubbed or hooked at apex, fig. 326*E*, *F* (**Rhopalocera**)................................... 6
Either hind wing having a frenulum, or antenna not clubbed or hooked; instead either threadlike, or serrate, or pectinate (**Frenatae**)................. 11
6. Front wing having each of the 5 branches of radius and 3 of media arising from the discal cell, fig. 326*A*; antennae usually hooked at apex, fig. 326*F*
Hesperiidae, p. 415
Front wing having some of these fused at base, branching beyond discal cell, fig. 326*B*, *C*... 7
7. Front wing having Cu_1 appearing 4-branched, fig. 326*C*, because both M_2 and M_3 are more closely associated with it than with R_5..... **Papilionidae**, p. 417
Front wing having Cu_1 appearing 3-branched, fig. 326*B*, because M_2 is more closely associated with R_5.. 8
8. Labial palps longer than thorax, thickly hairy, and extending forward
Libytheidae
Labial palps shorter than thorax.................................... 9
9. Front legs reduced, and much shorter than the other legs, fig. 326*D*
Nymphalidae
Front legs larger in proportion to the others......................... 10
10. Front wing having M_1 fused for a considerable distance with posterior branch of radius, fig. 326*B*; colors white, yellow, or orange, plus black marks
Pieridae, p. 417
Front wing having M_1 either not fused with R, or only slightly so, thus arising from discal cell or very near it; colors coppery, blue, or brown.... **Lycaenidae**
11. Hind wing having a posterior fringe as long as wing is wide, figs. 328, 332; wing usually lanceolate. A large number of families of small moths difficult to identify, including Yponomeutidae, Gelechiidae, and Tineidae
Not keyed further here
Hind wing markedly wider than its fringe, figs. 333, 334................ 12
12. Front wing narrow, more than four times as long as wide; hind wing and sometimes front wing having transparent areas devoid of scales, fig. 333
Aegeriidae, p. 407
Front wing wider, hind wings usually entirely covered with scales........ 13
13. Hind wing having 3 veins posterior to Cu_{1b}, fig. 325*E*.................. 14
Hind wing having 1 or 2 veins posterior to Cu_{1b}, fig. 327*E*............... 15
14. Front wing with Cu_2 fairly well developed, at least towards apex, fig. 325*D*
Tortricidae, p. 407
Front wing with Cu_2 atrophied, fig. 325*C*................. **Pyralidae**, p. 408
15. Front wing with 1*A* evenly bowed anteriorly, the vein coming close to central portion of Cu_1 or apex of Cu_{1b}, fig. 327*F*.............. **Sphingidae**, p. 409
Front wing with 1*A* straight or only slightly sinuate, fig. 327*C*............ 16
16. Front wing having both M_2 and M_3 associated closely with Cu_1, which therefore appears 4-branched, fig. 327*C*.................................. 17
Front wing having M_2 either midway between M_3 and M_1, or closer to M_1, so that Cu_1 appears 3-branched, fig. 327*A*........................... 20

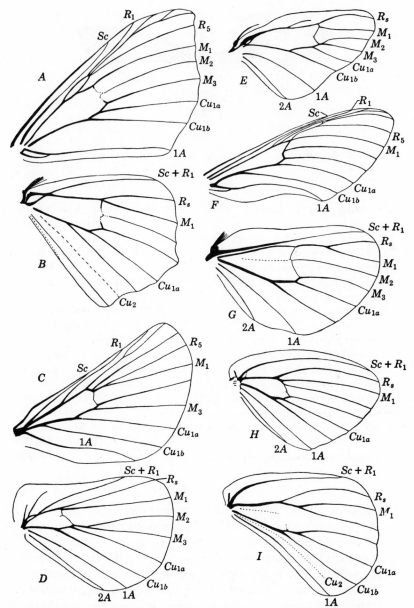

Fig. 327. Wings of Lepidoptera. *A, B,* front and hind, *Acidalia,* Geometridae; *C, D,* front and hind, *Malacosoma,* Lasiocampidae; *E,* hind, *Halisidota,* Arctiidae; *F,* front, *Protoparce,* Sphingidae; *G,* hind, *Hyperaeschra,* Notodontidae; *H,* hind, *Citheronia,* Citheroniidae; *I,* hind, *Samia,* Saturniidae.

Fig. 328.   The tapestry moth *Trichophaga tapetzella,* one of the Microlepidoptera.   (From U.S.D.A., E.R.B.)

17. Hind wing without a frenulum, the base of the front margin greatly expanded, fig. 327*D* .... **Lasiocampidae,** p. 408
    Hind wing either with a frenulum or base of wing not expanded 18
18. Hind wing having $Sc + R_1$ and R_s fused for about half length of discal cell, then the two veins separating, fig. 327*E*..................... **Arctiidae,** p. 414
    Hind wing having $Sc + R_1$ fused with R_s for only a short distance, or the veins not at all fused 19
19. Head having two ocelli............................. **Noctuidae,** p. 413
    Head without ocelli................................. **Liparidae,** p. 413
20. Hind wing having $Sc + R_1$ arcuate, curving forward from its base and well separated from R_s for its entire distance, fig. 327*H, I;* frenulum obsolete... 21
    Hind wing having $Sc + R_1$ either fused for a distance with R_s, fig. 327*B,* or running close to it, fig. 327*G;* frenulum present, often tuftlike.......... 22
21. Hind wing having two anal veins, fig. 327*H*........... **Citheroniidae,** p. 411
    Hind wing having only one anal vein, fig. 327*I*......... **Saturniidae,** p. 411
22. Hind wing with Sc making a short sharp angulation at the base of the wing, fig. 327*B* .... **Geometridae,** p. 411
    Hind wing with Sc not angulate at base, fig. 327*G*.............. **Notodontidae**

SUBORDER JUGATAE

The Jugatae are the most primitive present-day Lepidoptera.   In addition to possessing a jugum, adults of this suborder have front and hind wings of very similar shape and venation.   In one group the pupae have long stout mandibles used for cutting an exit from the cocoon; the adult mouthparts include well-developed mandibles and lobelike maxillae and labium, which are not elongated and appressed to form the usual sucking tube.   These characters are of great interest, because they are also found in the Trichoptera, and they demonstrate the close relationship between primitive members of the Trichoptera and of the Lepidoptera.

The Jugatae are represented in North America by only a few species. The family Micropterygidae has mandibulate adults of small size; its larvae feed on moss. The Eriocraniidae also includes small species but these have vestigial mandibles and a sucking tube; the larvae are leaf miners. The Hepialidae includes a few larger species (with a wing expanse up to 2 inches) called swifts because of their rapid flight; their larvae are wood borers or root feeders.

SUBORDER FRENATAE

Here belong the greater number of the Lepidoptera, representing a great variety of shapes, sizes, and habits. The adult, in addition to lacking a jugum, has hind wings which differ in shape from the front ones, and which also have the radial sector usually reduced to a single vein; in the front wings the radial sector usually has three or four branches. The pupa is obtect; that is, the appendages appear embedded in the body of the pupa and are incapable of movement.

In North America we have fifty families represented, many of them containing species of great economic importance. The diagnosis and identification of these families are difficult and require making special preparations of the wings after their scales have been removed, a process known as denuding.

The "Micros." Many of the more primitive families of the Frenatae comprise small slender moths having the hind wings with vein 1A present, or bearing a long posterior fringe, or extremely narrow and pointed in outline. These families are frequently termed the Microlepidoptera or the Microfrenatae. A few members of the Microfrenatae are quite large, so that size is only an average characteristic for the group. Although it is impractical to give defining characters for them here, a few families are mentioned which contain common species of importance.

TINEIDAE. The larvae feed chiefly on fungi or fabrics, or as scavengers. Included in the Tineidae are our common clothes moths, fig. 329. The webbing clothes moth *Tineola bisselliella* has a larva which makes an indefinite web; in addition to fabrics, it has a marked liking for old feathers. The larva of the casemaking clothes moth *Tinea pellionella* makes a portable case of silk and fabric fragments. Both species are common pests of fabrics of animal origin, such as wool, furs, and feathers.

PSYCHIDAE, THE BAGWORMS. This family has relatively large species, some attaining a wing spread of 30 mm. The larvae construct a case or bag of silken fabric and bits of leaf or bark, fig. 330. Only the heavily sclerotized head and thorax project from the case. These bags are a familiar sight hanging from the twigs or leaves of many species of coniferous and deciduous trees. The common bagworm in the eastern states is *Thyridopteryx ephemeraeformis*, fig. 331. The males have fuzzy dark bodies,

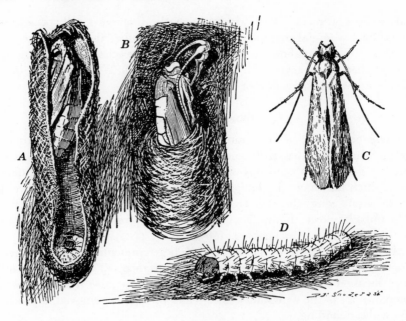

Fig. 329.   The casemaking clothes moth *Tinea pellionella*.   *A*, cocoon cut open showing fully formed pupa within; *B*, empty pupal skin projecting from door of cocoon after the moth has emerged; *C*, adult moth; *D*, the larva which does the damage to clothes.   (After Snodgrass)

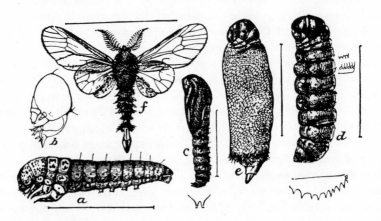

Fig. 330.   Stages in the life cycle of the bagworm *Thyridopteryx ephemeraeformis*.   *a*, full-grown larva; *b*, head of same; *c*, male pupa; *d*, female pupa; *e*, adult female; *f*, adult male.   (From U.S.D.A., E.R.B.)

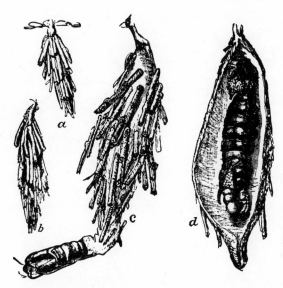

Fig. 331. Bagworms at successive stages (*a*, *b*, *c*). *c*, male bag; *d*, female bag. (From U.S.D.A., E.R.B.)

pectinate antennae, and usually clear wings. The females are larviform and have almost lost the power of locomotion. The life history has some peculiarities. By late summer the larva is full grown, whereupon it fastens its bag to a twig and pupates inside the bag. When mature, the male pupa emerges partially out of the bag, and the adult male emerges from it in this position. The female pupa stays within the bag; the female adult works itself partway out of the pupal skin and awaits fertilization. The males fly about looking for bags containing mature females and mate with them by means of an elongate extensile copulatory apparatus that can be extended deep into the bag containing the female. Soon after fertilization the female lays eggs, simply allowing them to fall into her old pupal skin, which becomes half filled with them. This completed, the spent female crawls out of the bag, falls to the ground, and dies. The eggs lie dormant over winter and hatch the next spring. The newly hatched larvae crawl out of the old bag, disperse, and begin feeding.

GELECHIIDAE. The larvae of this family exhibit a wide diversity of food and hosts. Several eastern species of *Gnorimoschema* make large stem galls on goldenrod and related Compositae; the larva feeds within the galls and when full grown pupates there. The pink bollworm *Pectinophora gossypiella* tunnels through the developing cotton boll and feeds on the seeds. It is a

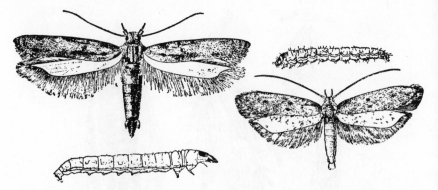

Fig. 332.   Left, the potato tuberworm *Gnorimoschema operculella;* right, the eggplant leaf miner *G. glochinella.*   (From U.S.D.A., E.R.B.)

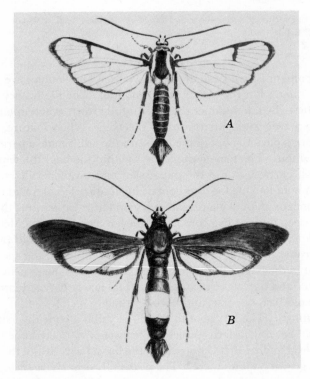

Fig. 333.   The peach tree borer *Sanninoidea exitiosa.*   *A*, male; *B*, female.   (From U.S.D.A., E.R.B.)

native of Asia that has spread to all the cotton-growing regions of the world and is one of the worst pests of this crop. Also included in the family are species which mine the needles of pine and other conifers and attack potatoes, fig. 332, and tomatoes, and some which bore in twigs and fruits of certain trees. One species is a world-wide pest of stored grain, the Angoumois grain moth *Sitotroga cerealella.*

AEGERIIDAE, THE CLEAR-WINGS. These are moderate-sized narrow-winged forms in which the front and hind wings are coupled by a series of interlocking spines situated near the middle of the wing margins. In most species the wings have definite window-like areas free from scales; hence the name "clear-wing" moths. The adults are diurnal and extremely rapid in flight, and in many the body and wings are banded with purple, red, and yellow, apparently mimicking some of the common wasps. The larvae are stem borers, attacking herbs, shrubs, and trees. Two of notable economic importance are the peach tree borer *Sanninoidea exitiosa,* fig. 333, and the squash vine borer *Melittia cucurbitae.*

TORTRICIDAE. A very large family of small to medium-sized moths, having wide wings, the front pair with the apical margin truncate or even concave or excised. The larvae feed on nuts, fruits, and leaves, and in stems. In many species the larvae make nests by rolling and tying leaves, from which they get their common name, leaf rollers. The group includes the red-banded leaf roller *Argyrotaenia velutinana,* fig. 334, which feeds on many wild and cultivated trees and shrubs. Some other economic species in the family are the oriental fruit moth *Grapholitha molesta,* which feeds in the twigs and fruits of peaches, plums, and related fruit trees; the pine shoot moths, several species of *Rhyacionia,* which destroy the terminal buds of pine; and the grape berry moth *Polychrosis viteana,* which feeds on the leaves and in the fruits of grapes.

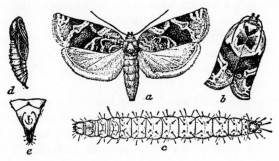

Fig. 334. The red-banded leaf roller *Argyrotaenia velutinana. a, b,* moth; *c,* larva; *d,* pupa; *e,* tip of pupal abdomen. (From U.S.D.A., E.R.B.)

Fig. 335.   The garden webworm *Loxostege similalis,* adult and larva.   (From U.S.D.A., E.R.B.)

PYRALIDAE.   This is a large family, economically one of the most important in the order.   In taxonomic position the pyralids are intermediate in many characters between the "micros" and the "macros."   The moths vary greatly in size and shape, but are usually delicate, trim, and have a rather detailed and soft coloration pattern.   The larvae exhibit a wide range of food habits: feeding on leaves, fruits, and flowers; boring in stems or stalks; some being saprophagous and a few predaceous.   Many spin an extensive web over their food and surroundings and are called webworms, such as the garden webworm *Loxostege similalis,* fig. 335.   The family includes some of the most troublesome pests of agricultural crops. The European corn borer *Pyrausta nubilalis* and the southwestern corn borer *Diatraea grandiosella* are serious pests of corn; the greenhouse leaf tier *Udea rubigalis* is a pest of chrysanthemums and other greenhouse crops; the greater wax moth *Galleria mellonella* is often a serious pest of beehives. Several species, including the Indian-meal moth *Plodia interpunctella,* attack stored grain and prepared foods, and *Ephestia elutella* feeds on stored tobacco and other dried vegetable products.   Many other species attack a wide assortment of crops.

The "Macros."   In the Frenatae, the more specialized moth families containing the larger species, such as the miller moths, the various native "silkworm" moths, and the hawk moths, are usually referred to as the Macrofrenatae, or "macros."   In these the hind wing usually is broad, has only a short fringe, and lacks vein 1*A*.   As is the case with the "micros," the "macros" contain many families which are difficult for the nonspecialist to identify.

LASIOCAMPIDAE.   To this small family belong a few species of moderate size that have hairy larvae and velvety large-bodied adults.   Tent caterpillars of the genus *Malacosoma,* fig. 336, are the best known of these.   The larvae feed on a variety of deciduous trees, including fruit trees, and periodically occur in outbreak numbers, the hordes of caterpillars defoliating thousands of acres of trees.

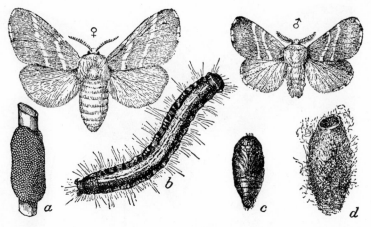

Fig. 336.   Stages of the eastern tent caterpillar *Malacosoma americanum.*   *a*, egg mass; *b*, larva; *c*, pupa; *d*, cocoon; ♀, female moth; ♂, male moth.   (From U.S.D.A., E.R.B.)

In the genus *Malacosoma,* winter is passed in the egg stage.   The eggs hatch in late spring, and the young caterpillars of an egg mass spin a colonial webbed nest, usually around the fork of a branch.   The caterpillars leave this to feed on foliage, each individual returning via a silken thread left in its wake.   The larvae leave the nest and pupate singly in a protected spot under bark or debris where the cocoons are constructed. Adults emerge in late summer, and, after mating, females deposit eggs in bands around small twigs.   Each egg mass contains several hundred eggs, the whole incased in a secretion that hardens and becomes impervious to the elements.

During an outbreak year the adults may be attracted in huge numbers to the lights of towns.   On a summer night in 1925 I witnessed a tremendous flight in Edmonton, Alberta.   Throughout the entire business section the moths were about 6 inches deep, their greasy bodies completely stopping streetcar and automobile traffic, and making it difficult to walk. Under each street light and show window the moths formed piles reaching an apex from 2 to 3 feet high.   The great majority died from suffocation by their fellows.   Their rotting bodies gave a distinctive odor to the city streets for some time.

SPHINGIDAE, HAWK OR SPHINX MOTHS.   These moths are all large, most of them having a wing spread of over 65 mm. (2½ inches).   The body is stout and spindle-shaped, frequently tapering to a sharp point at the posterior end.   When spread, the posterior margins of the hind wings are seldom back as far as the middle of the abdomen; the front wings are

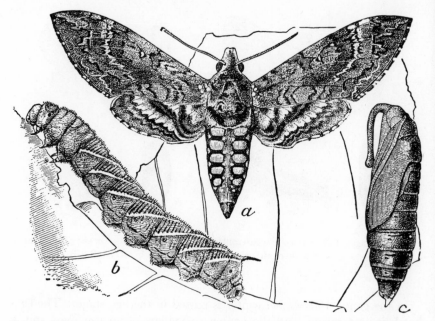

Fig. 337. A hornworm infesting tomato and tobacco, *Protoparce sexta.* *a*, adult; *b*, larva; *c*, pupa. (From U.S.D.A., E.R.B.)

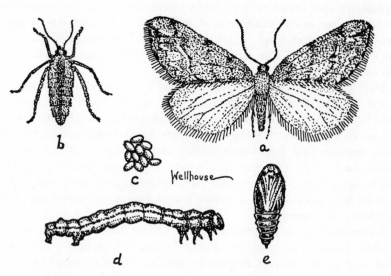

Fig. 338. The spring cankerworm *Paleacrita vernata.* *a*, male; *b*, female; *c*, eggs; *d*, larva; *e*, pupa. (From Wellhouse, *How insects live,* by permission of The Macmillan Co.)

long and proportionately narrow. The antennae are long and simple, frequently slightly thickened towards the tip. The moths are extremely rapid fliers and feed on nectar. The larvae are leaf feeders; most forms have a sharp horn on the eighth abdominal segment and are commonly called hornworms. In the main our species feed on a wide variety of herbs, vines, and trees. A few are of economic importance, particularly the tomato hornworm *Protoparce quinquemaculata* and the tobacco hornworm *Protoparce sexta*, fig. 337.

GEOMETRIDAE, MEASURING WORMS, GEOMETERS. These are fragile moths, with slender or pectinate antennae, large delicate wings, and slender bodies, fig. 338. The larvae are well known for their peculiar walking habit, consisting of a series of looping movements. They have long slender bodies but only two or three pairs of large abdominal legs, located near the end of the body. When they are walking, the abdomen is raised in a high loop, and the hind legs are brought forward to grasp a supporting position close to the thoracic legs; these latter are then released, the body stretched forward, and a new grip taken by the thoracic legs at the end of the reach. The hind legs then let go, the body arches, and the operation is repeated. Geometers make up a large family; their host list includes many plant families and genera. Certain species are common defoliators of deciduous and coniferous trees. Some species, known as cankerworms, fig. 338, are locally very destructive to shade and fruit trees; these species have normal winged males but completely apterous females.

SATURNIIDAE, GIANT SILKWORM MOTHS. Here belong the largest North American moths and caterpillars. The adults are velvety or woolly, with broad wings having a showy pattern. The hind wings have no trace of a frenulum; instead the basal portion of the anterior margin is enlarged and projects under the front wing in the flight position, thus synchronizing the two pairs. The antennae are feathery in the males and frequently in the females also. The caterpillars are leaf feeders; the larger ones have a voracious appetite and will consume an entire large leaf with astonishing speed. The larvae are stout bodied, bear spiny tubercles or tufts of stiff hairs, and may attain a length of 100 mm. (nearly 4 inches). The full-grown larvae spin large brown cocoons on branches or twigs near the ground, pupate, and pass the winter in this stage. A common eastern species is the promethea moth *Callosamia promethea*, fig. 339, which has a reddish or brown adult; the larva is green with yellow, red, and blue spiny tubercles.

The related family Citheroniidae contains the green-striped mapleworm *Anisota rubicunda*, fig. 340, a widespread species east of the Rocky Mountains.

Fig. 339.   The promethea moth *Callosamia promethea*.   (From Comstock, *Introduction to entomology*, by permission of the Comstock Publishing Co.)

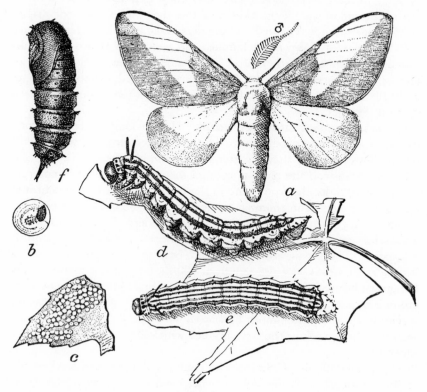

Fig. 340.   The green-striped mapleworm *Anisota rubicunda*.   *a*, female moth and antenna of male; *b*, egg showing embryo within; *c*, egg mass; *d*, *e*, larva; *f*, pupa.   (From U.S.D.A., E.R.B.)

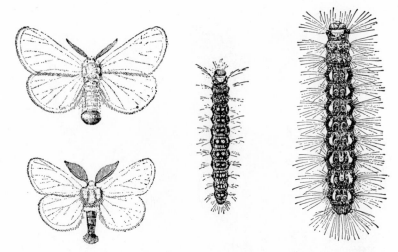

Fig. 341.   The brown-tail moth *Nygmia phaeorrhoea;* female above, male below, larva in center, enlarged larva to right.   (From U.S.D.A., E.R.B.)

LIPARIDAE.   This also is a small family containing species of only moderate size.   The caterpillars are hairy, the adults frequently fuzzy.   In several genera the females cannot fly, having only small padlike wings. To this family belong a few species of extreme economic importance, including the gypsy moth *Porthetria dispar* and the brown-tail moth *Nygmia phaeorrhoea,* fig. 341.   Both these gained entrance to the United States from Europe and are destructive enemies of shade trees in northeastern United States.   Another species of Liparidae attacking a wide variety of deciduous trees is the white-marked tussock moth *Hemerocampa leucostigma,* fig. 342.   The tussock moth larvae has tufts of long nettling hairs at each end of the body, and "pencils" of white hairs on some of the central segments.   It makes a cocoon on the bark of trees.   The female is grublike; she lays a large group of eggs encased in a foamy white secretion that forms a protective covering for them.

NOCTUIDAE (PHALAENIDAE), OWLET OR MILLER MOTHS.   This is by far the largest family of the Lepidoptera.   The adults vary greatly in size, shape, and color; the structural characters are also diverse, so that the family can be distinguished from its relatives only on the basis of a combination of several critical differences.   The larvae are usually leaf feeders or stem or root borers and for the most part are unadorned with horns or conspicuous processes.   From the standpoint of agriculture the family is an important one.   It includes many species whose larvae, called cutworms, attack a wide variety of grain, truck, and field crops.   Other economic species are the armyworm *Pseudaletia unipuncta* and the fall army-

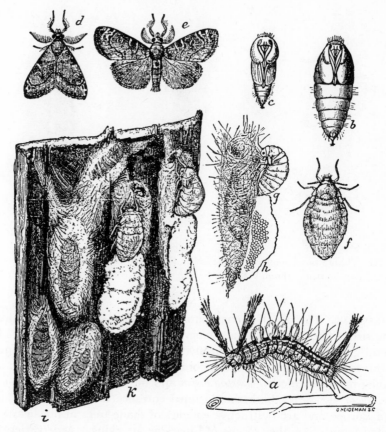

Fig. 342. The white-marked tussock moth *Hemerocampa leucostigma*. *a*, larva; *b*, female pupa; *c*, male pupa; *d*, *e*, male moth; *f*, female moth; *g*, same ovipositing; *h*, egg mass; *i*, male cocoons; *k*, female cocoons, with moths laying eggs. (From U.S.D.A., E.R.B.)

worm *Laphygma frugiperda*, fig. 343, which attack pasture grasses, corn, and small grains; the corn earworm or cotton bollworm *Heliothis zea*, which attacks cotton, corn, tomatoes, and other crops; and the cabbage looper *Trichoplusia ni* which feeds on cruciferous crops and "loops" like a geometrid larva, but has only two middle pairs of abdominal legs. In all these the larvae do the damage.

ARCTIIDAE, THE ARCTIID MOTHS. This family is a close relative of the Noctuidae, differing from it mainly in that the adults are usually white or yellow or have intricate bright or yellow patterns. The larvae have thick tufts of hairs, and hence many of them are called woolly bears. These larvae are a common sight in late summer, hurrying along the ground looking for a sheltered place to make a cocoon and pupate.

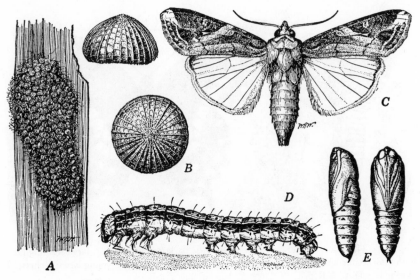

Fig. 343.   The fall armyworm *Laphygma frugiperda.   A*, egg mass; *B*, eggs; *C*, adult; *D*, larva; *E*, pupa.   (From U.S.D.A., E.R.B.)

OTHER MOTHS.   There are many other moths, some of striking appearance, others similar in general characteristics to the few just mentioned. To identify these the student is referred to the works listed at the end of the section on Lepidoptera.

SUBORDER RHOPALOCERA

BUTTERFLIES AND SKIPPERS.   In this suborder the hind wings have no frenulum, and in both front and hind wings the stem of media has been lost, resulting in a large central discal cell.   Most of the species are brightly patterned.   The adults are diurnal in habit and lovers of sunshine, in marked contrast to the crepuscular or nocturnal habit of most of the moths.   When at rest, the Rhopalocera hold their wings upright over the body instead of folding them flat on the body as do the moths. The suborder is divided into two well-marked groups: the skippers and the butterflies.

Skippers.   These are very rapid on the wing, able to fly in a straight line like a wasp or hawk moth.   Most of our species belong to the Hesperiidae; they are dull colored, with yellow and brown predominating, and are less than 30 mm. from wing tip to wing tip.   Except for a few species, the larva has a large head accentuated by a small necklike prothorax.   Most of the species live in a nest made by sewing together a few leaves of the host plant.   A common eastern species is the silver-spotted

Fig. 344.   The celery swallowtail *Papilio polyxenes.* *a*, larva from side; *b*, larva showing head with odoriferous appendages; *c*, male butterfly; *d*, outline of egg; *e*, young larva; *f*, chrysalis. (From U.S.D.A., E.R.B.)

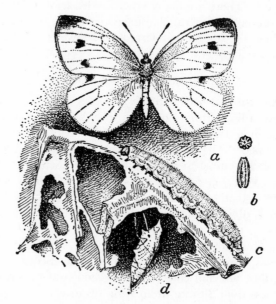

Fig. 345.   The imported cabbageworm *Pieris rapae.* *a*, female butterfly; *b*, egg (above as seen from above, below as seen from side); *c*, larva, or worm, in natural position on cabbage leaf; *d*, suspended chrysalis.   (From U.S.D.A., E.R.B.)

skipper *Epargyreus tityrus,* which feeds on several legumes; *Wisteria* is a favorite.

Butterflies. The inversion "flutterbys" is descriptive of the flight of most members of this group. They have a very slow rate of wing stroke and hence fly in a series of up-and-down movements producing an erratic course. The butterfly group is represented in North America by seven families, most of which have members well known to the naturalist.

PAPILIONIDAE, THE SWALLOWTAILED BUTTERFLIES. In this small family, fig. 344, the margin of the hind wing is usually notched, and the vein M_3 ends in a finger-like projection or tail. These are large butterflies, many of them gaudily spotted or striped with many colors, yellow predominating in several species. The larvae are leaf feeders and may be as conspicuously marked as the adults. They are unique in having a forked eversible stink gland, or *osmeterium,* fig. 344*b,* on the dorsum of the pronotum. This is shot out when the larva is alarmed; it is usually bright orange and emits a pronounced odor.

PIERIDAE, THE WHITES AND SULPHURS. These are predominantly white, yellow, or orange butterflies, some with extensive black markings. The larvae have an abundant supply of stiff hairs and look bristly. The imported cabbageworm *Pieris rapae,* fig. 345, whose adult is white marked with black dots, occurs commonly in the central and northern states. The larvae are green and are pests of cabbage and related plants. This is an introduced species from Europe. A native species of the same genus, *P. protodice* the southern cabbageworm, is a pest of cruciferous crops in the southern states. Another pierid that is often a pest locally is *Colias philodice eurytheme* the alfalfa caterpillar or orange sulphur. It is a highly variable species in color and occurs over most of the continent. The larva feeds on alfalfa, clover, and certain other legumes.

OTHER BUTTERFLIES. In North America there are many species of butterflies in addition to the few just listed. Every locality on the continent has a selection from strikingly colored to somber forms, many of them abundant locally. For more information regarding their identification characters, hosts, and range, the student is referred to Dr. Holland's *Butterfly Book* and Dr. Klots' *Field Guide.*

References

Forbes, W. T. M., 1923–1954. The Lepidoptera of New York and neighboring states. Parts 1–3. *Cornell Univ. Agr. Expt. Sta. Mem.,* **68:**729 pp.; **274:**263 pp.; **329:**433 pp.

Holland, W. J., 1913. *The moth book.* New York: Doubleday & Co. xxiv and 479 pp., 48 col. pls.

1931. *The butterfly book.* New York: Doubleday & Co. xii and 424 pp., 77 col. pls.

Klots, A. B., 1951. *A field guide to the butterflies.* Boston: Houghton Mifflin Co. 349 pp.

McDunnough, J., 1938–1939. Check list of the Lepidoptera of Canada and the United States of America. Pts. I and II. *Mem. So. Calif. Acad. Sci.,* **1:**1–272; **2:**1–171.

Chapter Eight

Geological History
of Insects

T HE existing insect fauna that we see about us is the result of eons of change and evolution and a struggle for existence among the insects. When they first arose as a group, insects were undoubtedly few in numbers of both individuals and species; furthermore, they had only a limited range of ecological conditions under which to live. As the entire organic world developed, there evolved in the insect group members having biological and physiological characteristics enabling them to become adapted to new ecological niches. If we look over the existing insect fauna as a whole, it is apparent that certain orders, such as the Lepidoptera and Hymenoptera, are large and flourishing, exhibiting a great diversity of structure and habits, but they are relatively young orders in geologic time. We have other orders, for instance, the Dermaptera and Ephemeroptera, that are represented now by relatively few remnants; yet their geological records indicate a beginning of great antiquity. These remnants, like the horsetails and ginkgoes of the plant world, are lonely survivors of what may have been once great orders. Some ancient orders, such as the Paleodictyoptera, were dominant insect groups in the geologic past, but apparently were unable to change with climate or compete with new forms and are long since extinct.

The principal avenue of discovery for knowledge of this past history is through fossil remains from strata, or layers of rock, representing the various periods of geologic time. Insect fossils are not found in so many places as are fossils of some other groups because of special conditions needed to insure adequate preservation. Because of their small size, the delicacy of their parts, and the minute nature of identification criteria, insect remains must be preserved in a medium of extremely fine texture providing a grainless matrix. Satisfactory materials are mud and vol-

418

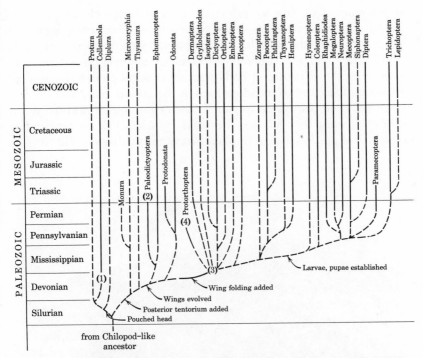

Fig. 346. Family tree of the insect orders plotted against geological time scale. Solid lines, branch of order after earliest known fossil; broken lines, portion of dating suggested by various methods of deduction. (1) This point refers to *Rhyniella praecursor;* (2) this branch also contains the Protohemiptera, Megasecoptera, and possibly other extinct orders; (3) the Russian Devonian fossil *Eopterum devonicum* probably belongs here; (4) this branch also contains the Protoelytroptera (Protocoleoptera), Caloneurodea, Protoperlaria, and possibly other extinct orders. (See also fig. 159, p. 204.)

canic ash, resulting in shales; concretions; fine humus, such as coal; and resin of coniferous trees, giving amber. Fossil-bearing deposits have been found in scattered localities all over the world. Some in North America are productive of valuable additions to the record. Along Mazon Creek, near Morris, Ill., are found iron nodules, or concretions, containing Pennsylvanian (Paleozoic) insects. In Kansas and Oklahoma deposits have been explored containing large numbers of Permian insects. In several localities in the western states occur deposits containing abundant fossils of Cenozoic insects, and other deposits containing a few Mesozoic forms. In Alaska and northern Manitoba there have been located deposits of amber containing many insect remains thought to be Cretaceous in origin. New records from these periods have recently been clarified from Russia.

In the main, wings are the principal insect structures clearly preserved in fossils. In amber insects other structures are frequently seen with great clarity, resembling a balsam preparation on a microscope slide, fig. 348. In shale and concretion specimens, however, it is only occasionally that the structure of body and appendages can be determined.

Although there are tremendous gaps in the known insect fossil record, enough has been discovered to gain a fair idea of the general trend. In the brief account that follows, the main emphasis has been placed on the biological and ecological implications of the evidence. This is given

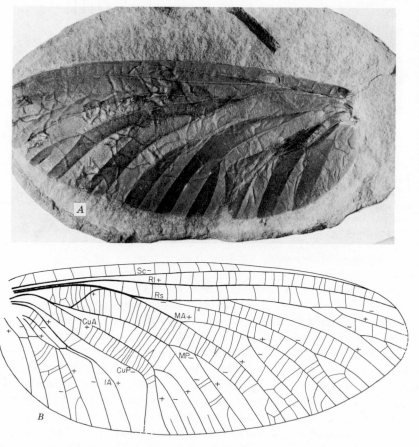

Fig. 347. A fossil from an iron nodule, Mazon Creek, Ill.; hind wing of an ancestral mayfly *Lithoneura mirifica*. *A*, photograph of fossil impression; *B*, reconstruction of venation. (After Carpenter)

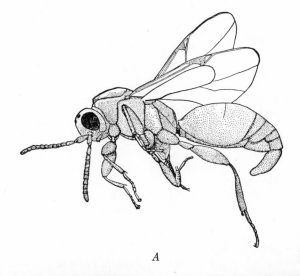

A

B

Fig. 348. *Cryptoserphus succinalis,* a hymenopteran described from Baltic amber. *A*, drawing made from specimen in the amber block shown in *B*. (After Brues)

in relation to the geological timetable in general use, Table 5, with which the student should become familiar.

Table 5 represents only the period of evolution for the last 600 million years during which a relatively continuous and decipherable fossil record

Table 5. Condensed geologic timetable represented by the fossil record

Cenozoic Era—Modern life		
Quaternary Period		
Pleistocene Epoch	1	Includes the Ice Age
Tertiary Period		
Pliocene Epoch	13	
Miocene Epoch	25	
Oligocene Epoch	36	Dominance of flowering plants; diver-
Eocene Epoch	58	sification of mammals
Paleocene Epoch	63	
Mesozoic Era—Medieval life		
Cretaceous Period	135	Dominance of cycads, tree ferns, and
Jurassic Period	181	conifers; era of dinosaurs; origin of
Triassic Period	230	mammals
Paleozoic Era—Ancient life		
Permian Period	280	Evolution of modern insect orders
Pennsylvanian Period	310	The Carboniferous; first great tropical
Mississippian Period	345	forests
Devonian Period	405	Beginnings of land animals
Silurian Period	425	
Ordovician Period	500	Predominance of marine inverte-
Cambrian Period	600	brates; rise of land plants

Geologic timetable. Names in the left column designate the various time periods for which there are relatively satisfactory fossil records. Each number in the middle column refers to the number of millions of years ago that a particular time period began. Comments in the right column indicate some of the notable events that occurred during the last 600 million years.

is available. Life, however, originated long before that, certainly over 2½ billion years ago. If we take the thickest known deposits laid down since then and add them together, it would result in deposits 100 miles deep. Such a mass of material means the wearing away to sea level, one after another, of more than 20 ranges of mountains like the present European Alps or American Rockies. Visualize the amount of time it would take erosion to do this, and then add the much longer quiescent periods when the low-lying lands furnished little sediment, and we have some idea of this length of time.

EARLY GEOLOGIC ERAS

Several long eras passed in the geological record before insects appeared. The first forms of life were unicellular organisms which probably began over 2½ billion years ago. From these evolved the many kinds of multicellular plants and animals. Through pre-Cambrian times for which no good fossils are known, life was presumably marine. The same was true of the animals through the Cambrian and Ordovician periods of the early Paleozoic, when the seas were shallow, extensive, and warm. At this time the Arthropods were well developed, represented by Trilobita, complex Crustacea, and Eurypterida. All the animal phyla as we know them today are represented in the fossils of these periods, including the first vertebrate fossils, primitive fish. It is interesting to speculate on the appearance of the world at this time. The land was barren, at most with encrustations or local pockets of lower plants, and away from the seas probably there were not even these signs of living matter.

In the Silurian came the first feeble beginnings of a land biota, with the appearance in the fossil record of a fair variety of land plants and the first known air-breathing animals, scorpions, and myriapods. It seems almost certain that primeval wingless insects also must have evolved in this era. All these animals were probably tropical and shade-loving, occurring along the beaches and in the swamps in the warm moist Silurian climate.

The Devonian and Mississippian periods are termed the "Age of Sharks" by geologists. Although these animals attained great evolutionary development, the event of most interest from an entomological view-

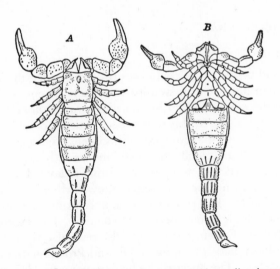

Fig. 349. Restoration of two Silurian scorpions, among the earliest known air-breathing animals. The similarity between these and recent forms is remarkable. *A*, dorsal view of *Palaeophonus nuncius; B*, ventral view of *P. hunteri*. (From Pirsson after Pocock)

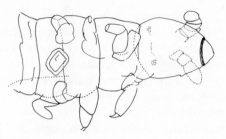

Fig. 350. *Rhyniella praecursor,* a Devonian fossil belonging to the order Collembola. (After Aubert)

point was the rise of land biotas. Forests flourished along stream edges and in low-lying swamps. This development, providing shelter, humid conditions, and food, was essential for the extension of animal life beyond the limited areas of the beaches. The fossil record of the land fauna is fragmentary but demonstrates the presence of amphibians, diplopods, and a few insects (the collembolan *Rhyniella praecursor,* fig. 350, from the Devonian of Scotland and the winged insect *Eopterum devonicum* from the Devonian of Russia). We may infer from the scarcity of terrestrial animal fossils in this period that, even after the forests had become stabilized, their successful invasion by land animals was a slow process.

LATE PALEOZOIC ERA

Pennsylvanian

In this second period of the coal measures, often called the Upper Carboniferous, luxuriant, tropical swamps were more extensive than at any other time in the earth's history, supporting a varied flora and fauna of ancient types, fig. 352. In this setting occurred the oldest extensive insect fauna of which we have record. Specimens from even the earliest Pennsylvanian beds are chiefly of winged insects fully as well developed regarding general structure as are some existing today. The 1500 species of fossil insects known from this period show further that already two large groups of orders had developed, the ancient flying insects or Paleoptera and the folding-wing insects or Neoptera. The Paleoptera were represented by the Ephemeroptera and by several orders now extinct, including the Paleodictyoptera, fig. 351*B,* and Protodonata (ancestral to the true Odonata). The ancestral Ephemeroptera, fig. 347, were primitive forms having the front and hind wings the same size, whereas modern forms have greatly reduced hind wings. The Neoptera were represented

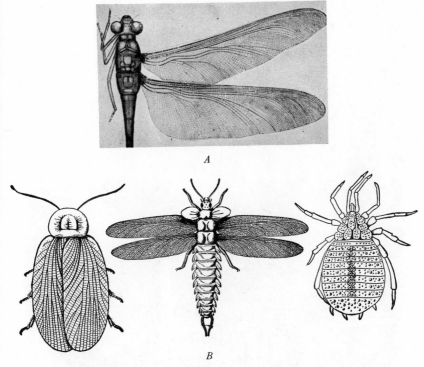

Fig. 351. Pennsylvanian insects and a spider. *A*, the largest known insect of all time, *Meganeuron monyi*, from the coal measures of Belgium. A reconstruction by R. J. Tillyard. About one-seventh natural size. *B*, left, a cockroach *Aphthoroblattina johnsoni;* center, a paleodictyopteran *Stenodictya lobata;* right, a spider *Eophtynus prestwichii.* (After Schuchert and Dunbar, *A textbook of geology.* Pt. II: *Historical geology.*)

by the Dictyoptera (cockroaches) and by a group of orders now extinct but considered close relatives of the orthopteroid orders, including the Protorthoptera, Caloneurodea, and Protelytroptera (includes the Protocoleoptera which are not related to the beetles). So abundant are cockroach fossils in these strata that the Pennsylvanian is called the "Age of Cockroaches."

It is thought that the nymphs of several of these insect orders were aquatic or semiaquatic and lived in the swamp pools which were extensive in many areas of the earth's surface. Here also insects attained their greatest size, including Protodonata (primitive dragonflies) with a wing span of 29 inches.

Insects and amphibians dominated the swamps. The competition between these two groups must have been ferocious for, whereas the adult

Fig. 352.   Pennsylvanian flora and amphibia, as restored by J. Smit.   In the background are *Sigillaria,* with tree ferns and conifers in the middle distance.   In the foreground are *Cala-mites* and seed-bearing fernlike plants.   Amphibia are represented by a small four-limbed microsaurian (*Keraterpeton*), a large-headed form (*Loxomma*), and a snakelike gill-bearing stegocephalian (*Dolichosoma*) from Linton, Ohio.   (From Schuchert after Knipe, *A textbook of geology.   Pt. II: Historical geology.*)

amphibians undoubtedly fed on insects, many of the large predaceous insect nymphs surely fed on the vulnerable larval amphibians. This competition, together with competition among the insects themselves, may have exerted a strong evolutionary pressure toward the development of large size in certain insect groups. The climate would not discourage this tendency but rather encourage it by providing uniform growing conditions throughout the year. In spite of the glamor surrounding these giant insects, they were nevertheless a minority in their class. The great bulk of the insects from the Pennsylvanian were smaller than their closest living relatives.

Permian

During this period many new types of both animals and plants first made their appearance, at least as far as the fossil record is concerned. They include forms thought to have been better adapted than any of the Pennsylvanian insects to cooler and drier climates. This increased diversity of biota was associated with climatic changes in many areas, following extensive mountain making in late Mississippian and culminating in the Permian.

Thousands of Permian insect fossils have been collected in Kansas and Oklahoma. Evidence from these and other sources indicates that all the

Fig. 353. Photograph of a fossil of a primitive insect *Dunbaria fasciipennis,* from the Early Permian period (Late Paleozoic). (From Mavor, *General biology,* by permission of The Macmillan Co.)

common Pennsylvanian insect orders continued into the Permian, and many new ones appeared. These include the Plecoptera and three hemipteroid orders, the Psocoptera, Thysanoptera, and Hemiptera. Here also appear five neuropteroid orders, the Coleoptera, Rhaphidiodea, Megaloptera, Neuroptera, and Mecoptera. Some members of the Megaloptera and Neuroptera were remarkably similar in basic features to families existing today. It is plain, therefore, that complete metamorphosis had evolved before the end of the Pennsylvanian, and that the basic ancestors of our modern fauna evolved before or during the Permian.

Few of the older orders prospered beyond the Permian. The Ephemeroptera, Protodonata, and Dictyoptera continued, but, judging from the fossil record, all other archaic coal measure orders became extinct at the end of the Permian. Some investigators believe that widespread arid or cold conditions caused this extinction of primitive forms, whereas others believe that competition with new forms caused the change in biota. More likely a combination of many factors brought about these results.

MESOZOIC ERA

Relatively few early Mesozoic fossil insect deposits have been discovered; hence we have a very imperfect picture of insect evolution in this era. In addition, apparently a long period elapsed between known deposits of late Paleozoic and early Mesozoic, with the result that there is a marked advance in many Mesozoic insects in comparison with their relatives of the Paleozoic.

Triassic

In this period great areas were desert or semidesert, caused in North America by the extensive emergence of the continent. The medieval floras of the Triassic were strange in their varied types of rushes, ferns, cycads, and conifers. This plant world was populated in the main by a great variety of reptiles, including dinosaurs, a group of reptiles which became dominant and for which the Mesozoic is called the "Age of Reptiles." These were the most extraordinary animals the world had ever seen, as diversified in form and size as are living mammals. Among them were huge swift carnivorous forms; also sluggish duck-billed types feeding on swamp vegetation; other vegetarians had great snakelike necks for browsing on tall trees. In addition to the dinosaurs there were many small reptiles of varied habit and a few very primitive mammals. The

A

B

Fig. 354. Feeding tunnels of *A*, a buprestid beetle, and *B*, a scolytid beetle in petrified Triassic trees, Petrified Forest National Monument, Ariz. (After Walker)

varied dinosaur types were the ecological equivalents in the Mesozoic of the herbivorous, carnivorous, and insectivorous mammals of today.

Fossils of the true bugs, the suborder Heteroptera, are first found from this period, but these forms are so well developed that the group undoubtedly occurred in the Permian. By this time most of the ancient orders such as the Paleodictyoptera had apparently died out completely, and the general character of the fauna was more like that of today. Many groups had developed specialized habits of feeding familiar to us in present-day insects. A striking example of this is the preservation in petrified Triassic trees of feeding tunnels typical of modern Buprestidae and Scolytidae, fig. 354. The Orthoptera and Trichoptera apparently evolved at this time, and existing primitive families of Hymenoptera had evolved.

Jurassic

In the early part of this period conditions were somewhat as in the Triassic. At this time arose the curious assemblage of flying reptiles called Pterodactyls. Some of these were huge, soaring like buzzards. Many frequented ocean shores and probably fed on fish, but small types flitted about over the ground and probably were insect feeders.

The insects were smaller than their present-day relatives. The order Dermaptera appears for the first time, most of its representatives belonging to modern families. In middle Jurassic the climate moderated over the entire world, and insects increased in size and numbers. It is probable that the plant-feeding habit became better established on the varied floras which developed at this time. Some plant flowering bodies were conspicuous, and it is likely that many insects visited them.

Toward the close of Jurassic time, extensive mountain uplifts increased the areas of the world having varied climatic zones and definite winters. These conditions probably helped to establish the plant and animal land groups that possessed provision or adaptations for overwintering.

Cretaceous

Much of the modern plant and insect life was established in this period. The Angiosperm flowering plants had begun their ascendancy over the spore-bearing floras and the archaic gymnosperms (the cycads and their allies), so that by the end of the period 90 per cent of the plant genera were of the woody kinds known today, including elms, oaks, figs, magnolia, beech, birch, and maple. Sedges and grasses appeared. Although most of these were wind-pollinated groups, some of those mentioned and many other genera known from this horizon were undoubtedly insect-pollinated as they are today. The evolution of insect-pollinated plants and pollen- and nectar-feeding insects were complementary developments. This association, which probably started in the Jurassic, was the beginning of a partnership which has proved highly successful for both. Today over 65 per cent of the flowering plants are insect-pollinated, and 20 per cent of the insects, at least in some stage of their development, depend on flowers for food.

The relatively few Cretaceous insect fossils are varied in taxonomic composition and include stoneflies, dragonflies, cockroaches, springtails, midges, aphids, caddisflies, and a large number of parasitic Hymenoptera. Many of these are close relatives, or actually members, of existing genera; some others represent genera of a primitive nature. A few are of exceptional interest because they represent families intermediate between existing families; an example is the collembolan family Protentomobryidae described from the Cretaceous amber of Manitoba, a family having distinctive characteristics yet intermediate in antennal structure and body segmentation between the Recent families Entomobryidae and Poduridae.

The rise in the Cretaceous of many favorable host plants must have encouraged great evolutionary development among phytophagous insects. The appearance of numerous parasitic Hymenoptera indicates that populations of their insect host species were at least fairly large and that the present-day interrelations of the groups had already been established. It is interesting to note that reptiles, which had dominated the land life and much of the aquatic up to this time in the Mesozoic, were rapidly being supplanted by the primitive land mammals. A circumstance in the Cretaceous of great importance in determining later composition of the world faunas was the widespread dispersal of many genera.

Throughout geologic history the shape of the continents as well as their topography has changed. For instance, North America has sometimes been reduced to half its present size and also has been increased far beyond the present margins and has had various connections with Northern Eurasia. North and South America have been connected broadly at some times, widely separated at others. One has only to allow a little freedom of imagination to visualize what this has done to the migration and isolation of floras and faunas.

On the basis of both geologic and biogeographic data it seems that during the latter part of the Cretaceous mountain chains and many land bridges, fig. 355, provided avenues for both temperate and tropical forms to disperse between Australia, Eurasia, North America, and South America. Changes in land bridge conditions caused populations of vari-

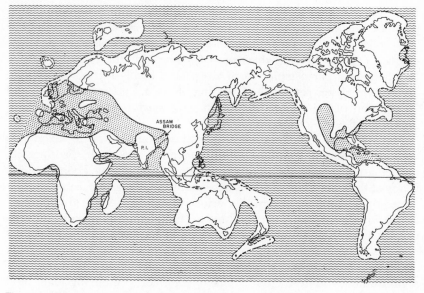

Fig. 355.   Suggested geography of Upper Cretaceous.   Stippled areas are principal epeiric seas.   P. I. indicates Peninsular India, the possible area of isolation of many tropical genera. Not shown are probable land dispersal routes between Australasia and South America through Antarctica.   (Compiled from various sources)

ous widespread forms to become isolated in different areas.   Each isolate gave rise to a new evolving line.   Some of these isolated lines have remained fundamentally little changed to the present; others have evolved at a more rapid rate and are now quite different from their Cretaceous progenitors.

CENOZOIC ERA

The Cenozoic prior to the glacial age consists of five distinct periods, the Paleocene, Eocene, Oligocene, Miocene, and Pliocene, collectively called the Tertiary.   The last period of the Cenozoic is the Pleistocene or glacial age.   Of great significance in the present distribution of insects were the continental and climatic conditions of the Cenozoic.

Through the Tertiary as a whole there occurred a vast development of plant species and diverse kinds of modern birds and mammals.   Certain insects had unquestionably become associated with them back in the Mesozoic and continued to evolve along with their hosts.   For example, bugs of the family Polyctenidae (ectoparasites of bats) have been associated with bats so long that the parasite bug genera essentially parallel the host bat genera as regards phylogeny and distribution.

Paleocene

This was a time of uplifted mountains, after the Laramide orogeny reached its peak. Temperate areas were cooler than in the Cretaceous. It is probable that warm-adapted groups were restricted to small equatorial areas, but that cool-adapted forms spread freely between Eurasia and North America.

Eocene

In this period conditions were apparently reversed. General temperatures reached a higher point than at any other time in the Cenozoic. It is estimated, for example, that isotherms in North America were nearly 30° north of their present location. As a result, cool-adapted forms were probably restricted to the higher altitudes in mountainous areas, and warm-adapted groups probably dispersed between the continents more freely than at any other time. It is likely that the greatest number of our tropicopolitan genera became world-wide in distribution at this time. For the world as a whole, practically every large genus or tribe of insects had evolved by the end of this period.

Oligocene

More moderate temperatures appear to have followed the Eocene, together with land connections between Asia and North America. Of especial importance at this time was the formation of a broad belt of temperate deciduous forest, composed of oaks, linden, sycamore, and other trees, which extended from the Atlantic coast of North America, across the continent and completely across Eurasia. This temperate belt was apparently north of its present range in America, and possibly covered much of the northern half of the continent. South of the deciduous forest belt the continent was subtropical to tropical. What is now the Great Plains area was arid or semiarid scrub, possibly as far north as Nebraska, much like that now covering much of northeastern Mexico and western Texas. The stream banks and marshes of this area supported luxuriant forests, which were probably more extensive towards the southeastern part of the continent. This plains area was much lower in elevation than at present, much of it only a thousand feet in elevation where it is now five or six thousand. The older ranges of the Rocky Mountains flanked the semiarid area westward, but most or all of the more northern ranges were low or not yet formed. Such a situation would account for the continuous transcontinental belt of temperate deciduous forest.

At Florissant, Colorado, large numbers of insect fossils have been found, preserved in fine sediments of volcanic ash that settled in the ponds and backwaters at the foot of the old Rocky Mountains along the edge of the generally semiarid area. The genera represented by this Oligocene fauna indicate that the fossilized insects came from an extremely varied assortment of habitats, ranging from tropical to cool temperate, and show that the Oligocene insect fauna contained fully as many genera as does that of the present day.

Some of the insect genera found in the Oligocene beds at Florissant no longer occur in North America but are now known only from other regions. Species of the genus *Glossina* (the tsetse flies, carriers of trypanosomiasis) were fossilized in these beds, but at present the genus occurs only in equatorial Africa. Fossils of the curious lacewing family Osmylidae occur in the Florissant Oligocene, fig. 356; living species are restricted to tropical areas of the world.

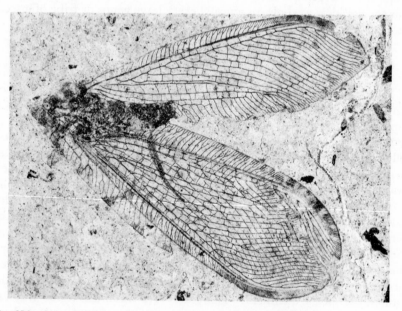

Fig. 356. An osmylid lacewing (Neuroptera) *Lithosmylus columbianus*, from the Oligocene shales of Florissant, Colo. (Photograph loaned by F. M. Carpenter)

Miocene

It is thought that little change from Oligocene conditions occurred in the earliest part of this period. Soon, however, majestic orogenic movements took place in many parts of the world. In Asia the highest ranges of the Himalayas arose, and in western North America many ranges were re-elevated and others pushed up. The subsequent arid conditions to leeward of the new mountains disrupted the northern trans-Holarctic forests and brought about many features of the landscape that exist today.

The Great Plains area became increasingly elevated and winter temperatures dropped, probably over the whole continent. The temperate deciduous forest became restricted to the eastern part of the continent and has never been reconnected with any of the Old World remnants. In this isolated condition the insect fauna of our eastern deciduous forest has evolved many distinctive genera or species groups, many of them still moderately or completely restricted to this forest. The Great Plains area gradually became grassland, which replaced the temperate deciduous forest to the north, and much of the semiarid scrub flora to the south. It is believed that the grasslands of the Great Plains had reached their present extent by the end of the Miocene.

Pliocene

The mountain making of the Miocene continued through this period, as did the southward movement of tropical and temperate boundaries. There is evidence of some faunal interchange of cool-adapted insect groups between northwestern North America and northeastern Asia, but apparently very little if any other faunal interchange involving insects of other ecological characteristics. Very likely the tropical elements in North America had shrunk to about their present areas by the end of the Pliocene.

Pleistocene, the Great Ice Age

After the extensive mountain elevation of the Pliocene, there was a temperature reduction over the entire world. Glaciers were formed in many continental regions of the world, especially northern North America and northern Europe, and spread southward. The ice over large areas of these huge glaciers or ice caps was at least several thousand feet thick. In their extension southward they blotted out all possibility of life over vast areas. Four main glacial periods are recognized, and between each there was a longer interval with a climate warmer than our present one.

Fig. 357.   Many insect lineages, such as the scorpionfly genus *Panorpa*, evolved primarily in North America.   In other insect genera, the North American fauna reached this continent at different times and from various directions, as shown in the above map for many of the nearctic species of the mosquito genus *Culex*.   *Culex, Melanoconion,* and *Neoculex* are subgenera of *Culex*.   (Courtesy of *Mosquito News* and The Illinois Nat. Hist. Survey.)

We are living at present in the receding portion of the last glacial period, and remnants of the northern ice cap are now found only as isolated glaciers or as island masses such as that on Greenland.   This alternation of warm and cold climates caused waves of first one type of flora and fauna and then another to spread first southward, then northward, over the glaciated areas.   For instance, during the coldest periods such Arctic animals as the musk ox ranged as far south as Georgia, and during the warm interglacial periods the elephant group inhabited the circumpolar area as far north as Alaska.

The effect of these glacial movements on the biota is imperfectly under-stood. Some investigators believe that the advancing glaciers reduced the tundra and northern coniferous forest bands into narrow strips but had little effect on climates much south of the glacier's edge. Others believe that all the North American ecological bands moved southward ladder-like in front of the glacier. There is also disagreement on how much species multiplication was caused by the glaciers. It is certain that many cool-adapted species did spread between North America and Eurasia, presumably through a dispersal route in the Bering Sea region. Such east-west and west-east dispersals, coupled with the south-north-south biotic movements caused by successive glacial advances and retreats, undoubtedly resulted in the fracturing and subsequent isolation of many populations in different areas of the Northern Hemisphere. Certain isolated populations dating from some of the earlier Pleistocene events have undoubtedly evolved into distinctive species, and others may have evolved into races. A goodly number of our insect species which today occur across the entire Holarctic region owe their widespread dis-tribution to more recent Pleistocene events. Although we are gradually acquiring some sound general ideas concerning these insect events, the details are proving difficult to determine.

As more data become available, however, it seems fairly certain that only cool temperate genera of insects have increased in number of species through intercontinental Pleistocene dispersal. There is some evidence that a comparable dispersal effect within individual mountain systems, as for example in the Rocky Mountains, may have led to some species multiplication in other insect groups. There is little doubt that the glacial changes have been responsible for a small amount of dispersal between distant mountain systems on the same continent, as for example the movement of some species between the Rocky Mountain and Allegheny Mountain systems. A slight interchange of Arctic or cool temperate forms between Europe and northern North America has probably also oc-curred. On the other hand, it is obvious that there has been no extensive east-west movement of insects associated with warm temperate forests, either deciduous or coniferous. It is also being found that many phe-nomena formerly attributed to effects of glaciation undoubtedly had their origin in earlier periods of geologic time.

The preceding brief outline is admittedly sketchy. For many periods of time, such as the Pliocene and pre-Cretaceous, we have little informa-tion on insect movements from one part of the world to another. The evolution of desert and grassland insects is poorly understood. The future integration of fossil and biogeographic evidence is a field of study which should eventually answer many of these fascinating problems.

REFERENCES

Carpenter, F. M., 1953. The geological history and evolution of insects. *Am. Scientist,* **41:**256–270.

Dillon, L. S., 1956. Wisconsin climate and life zones in North America. *Science,* **123:**167–176.

MacGinitie, H. D., 1953. Fossil plants of the Florissant beds, Colorado. *Carnegie Institution Washington Publ.* **599.** 198 pp.

Matthew, W. D., 1915. Climate and evolution. *Ann. N. Y. Acad. Sci.,* **24:**171–318.

Ross, H. H., 1953. On the origin and composition of the nearctic insect fauna. *Evolution,* **7:**145–158.

——— 1956. *The evolution and classification of the mountain caddisflies.* Urbana: Univ. Ill. Press. 213 pp.

Schuchert, C., and C. O. Dunbar, 1941. *A textbook of geology.* Pt. II: *Historical geology.* New York, John Wiley & Sons. 544 pp.

Shrock, R. R., and W. H. Twenhofel, 1953. *Principles of invertebrate paleontology,* 2nd ed. New York: McGraw-Hill Book Co. 816 pp.

Chapter Nine

Ecological
Considerations

WHEN we look at life in natural surroundings, we see that the landscape is broken up into different types of interlocking areas, such as prairies, forests, deserts, lakes, and streams. The type of vegetation, that is, whether it is desert, prairie, or forest, is determined by climatic factors of temperature, rainfall, and evaporation. In the main, forests occur in regions with a high rainfall, prairies in regions having lower rainfall, and deserts where rain is scant and evaporation high. Types of aquatic habitat depend on slope, rainfall, and a large variety of local factors including acidity and leaching qualities of the soil, drainage, seepage, and temperature. The vegetation type of the landscape is therefore a reflection of the climate, and widely separated areas having similar climate have the same kind of landscape aspects. These principal landscape aspect areas are the most inclusive ecological units and are called biomes, fig. 358. Each of these is divided into smaller units. A forest, for instance, has an edge area and may have small open areas or glades scattered through it; in one place the forest may be well drained and high, with a preponderance of oaks and hickories, in another place it may be low and swampy, having elms, gums, and other trees different from those in the better-drained areas. Each of these fairly uniform areas is considered by the ecologists as the biological unit of natural areas and is called a community. Each community has a definite set of animal species living in it, a set that persists year after year with only minor change. The animal species living in similar communities are practically the same. Thus oak-hickory communities in Wisconsin, Indiana, Missouri, and Oklahoma are each populated by very nearly the same species of animals.

Although an elm or gum forest community contains a fair proportion

439

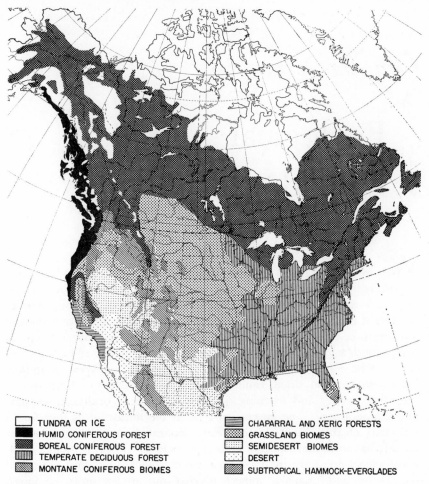

□	TUNDRA OR ICE	▤	CHAPARRAL AND XERIC FORESTS
■	HUMID CONIFEROUS FOREST	▦	GRASSLAND BIOMES
▓	BOREAL CONIFEROUS FOREST	▥	SEMIDESERT BIOMES
▤	TEMPERATE DECIDUOUS FOREST	⋅	DESERT
▨	MONTANE CONIFEROUS BIOMES	▨	SUBTROPICAL HAMMOCK-EVERGLADES

Fig. 358. Principal areas of terrestrial biological zones or biomes of North America. (From Illinois Nat. Hist. Survey.)

of the species found in an oak-hickory community, it lacks many species found there but possesses in addition species distinctive to itself. If we go further afield, a prairie community has a species make-up differing greatly from that of a forest community, and both have almost nothing in common with aquatic communities.

Examining communities more closely, we see that the animal species in each are stratified in various ways. In terrestrial communities some of the animals live in the soil, some on the herbs, and some in the trees, if the habitat is a forest. There is a vital relationship between various organisms in the community, as between herbivorous animals and the plants they eat, or between predatory animals and their prey. Altogether these coordinated relations make a network of dependency that binds all the diverse individuals of a community into a biological whole.

There are two ways of looking at these phenomena. One is from the standpoint of the community as a whole, studying its development, population, the interrelationships of its component species, and the distribution of the kinds of communities over the face of the earth. Study from this viewpoint is *synecology*. The other way is from the standpoint of individual species involved, to find out in what communities they are distributed, what niches they occupy, and why. This study is *autecology*. Synecology and autecology together are the broader field of ecology, which may be defined as the study of the relations between living organisms and their environment. A treatment of ecology as a whole is beyond the aim of this book. There is, however, a great deal of important information about insects in relation to their environment that can best be organized according to ecological factors. This material having a direct bearing on insects is dealt with in this chapter. Ecological considerations are becoming of increasing importance in insect studies and are giving valuable aid to taxonomy, biogeography, and insect control.

AUTECOLOGY

Each insect species is specially adapted to live in a particular "niche" in the community. In a sense, the species is a prisoner in its abode, because there are various environmental factors such as weather and food that restrict the species to its particular type of habitat. The limits of these factors within which the species can exist are spoken of as its ecological tolerance, which varies for different species in regard to the several factors involved. Correlated with its ecological tolerance, the individuals of a species have instinctive reactions which tend to insure that the individuals will always move to the place in the community that affords optimum conditions for their success. Both ecological tolerance and behavior are distinctive for each species; hence it is highly important to obtain the accurate identification of species which are subjects of ecological study.

Most ecological studies concerning insect species fall into one or more of three important categories: (1) environmental factors that determine where the species may live; (2) instinctive reactions or tropisms that enable the insects to find suitable living conditions; and (3) the effect of the sum of all these on the distribution and abundance of the species. This third category is termed *population dynamics*.

Environmental Factors

The most important environmental factors concerning the distribution and abundance of insects are weather, physical and chemical conditions of the medium, food, enemies, and competition.

WEATHER

The weather forms a blanket over the entire community and directly or indirectly affects conditions and organisms in practically all parts of the community. Weather is a composite condition of which light, temperature, relative humidity, precipitation, and wind are the most important ecological components. It is not the annual averages of these components (climate) which affect the species populations, but conditions from day to day. A single night's frost, for instance, may decimate a population of a subtropical insect, although the average temperatures for that year may be high. Similarly local conditions differ notoriously, and may result in great population differences in a short distance. In hilly country the single night's frost would be most severe in the valleys and might not affect the portion of our insect population on the hill crests.

Light. Little definite information is available concerning the ecological effect of wavelengths making up the greater part of sunlight (ultraviolet to red). In most experimental work on the subject there is considerable doubt if light is the only variant factor involved. A great number of insects normally diurnal in habit have been reared successfully for many generations either in artificial light deficient in many wavelengths or in total darkness. It would appear, therefore, that the effect of light on most insects is indirect and expressed through quality of food caused by plant reactions to light. Light, however, is an extremely important factor in insect behavior and is considered in this relation in the section on tropisms.

Temperature. In the lives of insects temperature is one of the most critical factors. Insects are cold-blooded, so that within narrow limits their body temperatures are the same as that of the surrounding medium. Except for a few unusual instances, insects are unable to control the temperature of their medium; instead they have physiological adjustments that enable each species to survive temperature extremes normally occurring in its ecological niche (see p. 140).

The honeybee is the best-studied example of an insect that regulates the temperature of its surrounding medium, in this case, the air within the hive. In summer the hive is maintained at about 95°F. If the temperature rises above this point, bees at the hive entrance set up ventilating currents by fanning their wings, and other bees may bring water and put it on the comb to obtain the cooling effect of its evaporation. In winter the bees keep the hive up to a safe temperature by heat obtained through oxidation of foods in the insects' bodies. Other social bees and ants exercise a certain amount of control over nest temperatures.

Effects of temperature are shown in two ways, the effect on rate of development, and the effect on mortality.

EFFECT ON DEVELOPMENT. Because they are cold-blooded and their body temperature reflects that of their medium, the temperature of insects is not constant. The chemical reactions of metabolism therefore automatically speed up with an increase in temperature. As a result we find that temperature has a marked effect on insect development and activities. Now all chemical reactions do not respond at the same rate to temperature increase, and certain physical factors, such as the solubility of gases in liquids, tend to produce unfavorable metabolic conditions as temperature increases. As a result, insect development is not equally responsive to changes over the entire temperature scale. There is a definite low point at which development stops, called the threshold temperature; this point may be 10° to 50° above the actual point of death from low temperature. There is also a definite high point for each species at which development stops; this point is usually very close to that of lethal high temperature.

Between these two points, rate of development responds to temperature changes. But the response is not uniform throughout the insect world. Each species has its own individual rate of development. Figure 359

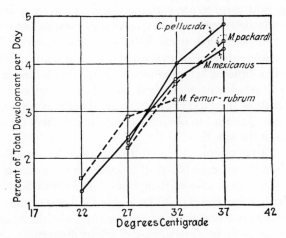

Fig. 359. Rates of development of four species of grasshoppers at constant temperatures from 22° to 37°C. (= 71° to 98°F.). (After Chapman, *Animal ecology*, by permission of McGraw-Hill Book Co.)

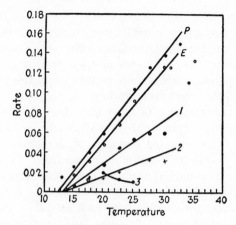

Fig. 360. Comparison of the rates of development of each stage of the Japanese beetle. Temperature scale in centigrade, 10° to 40°C. (= 12° to 104°F.). Numbers 1, 2, and 3 refer to the three larval instars. (After Ludwig)

illustrates differences in rate of development for four species of grass-hoppers. Within a species each developmental stage may have a different rate of development. This is well illustrated by the various stages of the Japanese beetle as graphed in fig. 360. The eggs and pupae have a much higher rate than do the larval stages, at identical temperatures.

An interesting example of dissimilar developmental rate is shown by eggs and nymphs of the red-legged grasshopper. The rates of development for eggs and nymphs are extremely different from each other. For the nymphs the developmental rate increases steadily with increase in temperature to a point close to the lethal high temperature. The rate for the eggs increases with the lower range of increased temperature and then decreases with additional temperature increase. With the eggs this point of decreased development is reached far below the lethal temperature.

These cases show the necessity of studying separately the various stages of the life history in order to obtain accurate information on the developmental phase of the species.

SEASONAL COORDINATION. The different rates of growth of species feeding on plants or cold-blooded animals are correlated extremely closely with the growth rate of their hosts. The result achieved is that, when the host has reached a point favorable for a certain insect to attack it, that insect has reached the proper stage to make the attack. Let us examine this relation in two species of Hymenoptera, a sawfly (Tenthredinidae) and its ichneumonid wasp parasite (Ichneumonidae). The sawfly adults emerge first early in spring at a time when the host plants have young leaves suitable for oviposition. The eggs hatch 1 or 2 weeks later when the plant is in the midst of vigorous growth and is providing a bountiful supply of food for the larvae. The ichneumon wasp has either a slower development or one that starts at a higher temperature, so that the adult ichneumonid emerges about 3 or 4 weeks after the adult sawfly. At this time the sawfly larva is nearly full grown and at the right stage for the ichneumon adult to lay eggs on it.

Another example is a group of aphids or plant lice (Aphididae) feeding in the spring on apple. The developmental rate of the overwintering eggs is such that the young aphids hatch at almost the exact time the apple buds first begin to open in spring. The aphids feed immediately on the minute leaves of the opening buds.

So constant is this coincidence of certain insect events with definite plant events that the plant phenomena (which are easy to see) are used as guides in many control programs. There are "bud sprays" for early aphid control, "petal-fall sprays," "calyx sprays," and so on, in which plant development is taken as a criterion for insect development.

EFFECT ON MORTALITY. The temperature range that insects can withstand varies tremendously with the species. The most heat-resistant insects known die at temperatures of 118° to 125°F. Probably the great majority have a high lethal point from 100° to 110°F. Species that live in cool places have correspondingly lower heat tolerances, such as the mountain genus *Grylloblatta*. The optimum for this group is about 38°F., and normal activity occurs between the approximate range 30 to 60°F.; heat prostration occurs at about 82°F.

Temperatures low enough to cause death vary as much as lethal high temperatures. Insects of tropical origin usually succumb as the temperature drops near freezing, fig. 361. The confused flour beetle, for example, will die in a few weeks at 44°F. Many insects die at temperatures only a few degrees below 32°F. (0°C.). Hibernating stages of most northern insects are remarkably resistant to cold. The hibernating pupa of the promethea moth, for example, can survive continued exposure to −31°F., and some other insects are known to survive −58°F.

Insects occurring in regions having freezing winters almost invariably exhibit a different temperature tolerance in each stage of their life cycles. At least one stage is resistant to low temperatures, and in this stage the species is able to withstand the winter temperatures (see p. 140). Parasites of warm-blooded animals are exceptions. The resistant form may be the egg, nymph, larva, pupa, or adult. In most cases only a single stage is cold-resistant; when winter arrives, the resistant form lives, and

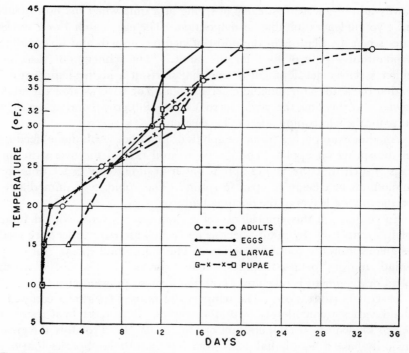

Fig. 361. Days of exposure required for assuring complete mortality of eggs, larvae, pupae, and adults of the cigarette beetle at various temperatures ranging from 15° to 40°F. (After Swingle)

individuals in any other condition die. Thus in chinch bugs only the adults are cold-resistant; when extremely low winter temperatures occur, the adults live, and any nymphs still remaining in the field die.

In their natural environment insects are well adjusted to prevailing usual temperatures. Temperature operates as a restricting factor in unusual or unseasonable periods of hot or cold weather. Generally unusual temperatures modify or control the range of a species along some frontier. The southern house mosquito *Culex quinquefasciatus* may migrate northward and extend its range during years with mild winters, but is cut back southward during severe winters.

Unseasonable temperatures, such as early or late frosts, may be as effective as temperature extremes in this action, because unfavorable conditions may occur before a species has entered the stage at which it is immune to them, or after it has passed to a susceptible stage. For instance, in the north-central states hibernating chinch bugs cannot withstand many alternate periods of freezing and thawing. A winter that has a number of unusual warm thawing periods, each followed by a zero or subzero period, produces this alternation of freezing and thawing and is extremely destructive to chinch-bug populations. Unseasonable temperatures would affect any one species more frequently at some periphery of its range, and thus contribute to restricting its distribution. Such temperatures would also occur hit-and-miss anywhere over the range of the species and affect abundance in local areas throughout the main body of the range.

Precipitation. Insects are ordinarily not affected directly by normal precipitation, but indirectly through the effect of precipitation on humidity, soil moisture, and plant food supply. Snow has an unusually important effect on soil temperatures. Bare soil is responsive to temperature changes to a depth of 2 feet; when covered with snow, even surface soil is remarkably insulated from changes in air temperature, fig. 362. Thus a snow cover has a marked effect on both the extremes of temperature and average temperature to which insects in the soil are subjected.

Certain expressions of precipitation, however, have a direct effect on

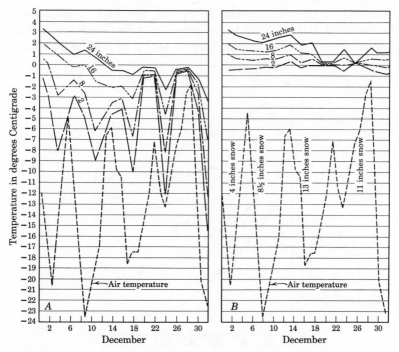

Fig. 362. Air temperatures and soil temperatures at different depths in bare (*A*) and snow-covered (*B*) ground for December in Montana. (After Mail)

insects. Excessive precipitation may inflict severe physical damage to insects. An inch of rain coming as a gentle sustained rain in one area may cause no harm, but coming as a sudden pelting downpour in another area may beat into the ground and kill most of the aphids or early stage chinch bug nymphs. Hail inflicts the same type of physical damage.

Humidity and Evaporation. It is difficult to separate the factors of humidity and evaporation in their effect on insects, either experimentally or geographically. Humidity pertains to the amount of moisture in the air, and evaporation to the actual water loss of a surface. In experimental work, if insects are subjected to low humidities, the evaporation from their bodies increases. Because of their small size, increased evaporation quickly depletes the water content of insects' bodies. In prolonged experiments, to test the effect of humidity it is therefore necessary to allow the insects to replenish their water supply by feeding. Unless this precaution is taken, effects due to desiccation may be attributed to humidity conditions of the medium. The graph in fig. 363 delineates the relation between evaporation and humidity for a common grasshopper under conditions of starvation.

Available data indicate that, in general, humidity is important but not so critical a factor as is temperature, and that each species has an optimum, which may be different for various stages of the life cycle. In

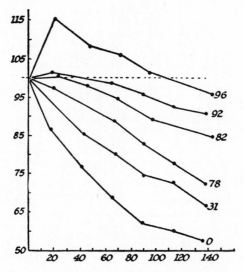

Fig. 363. Rate of loss of weight in *Chortophaga viridifasciata* at different relative humidities indicated at the end of each curve. Vertical figures represent weight as percentage of original weight, horizontal figures indicate time in hours. (From Wigglesworth, after Ludwig)

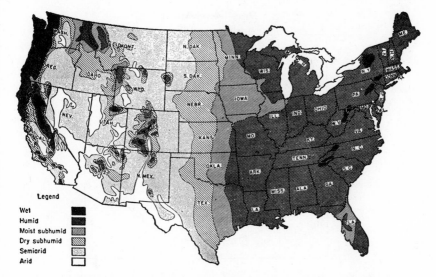

Fig. 364. Climatic moisture bands in the United States. (Adapted from U.S.D.A.)

the bean weevil the larvae develop faster at high humidities, but the eggs and pupae develop more rapidly at low humidities. In many cases, however, the rate of growth has been found to be practically constant over a wide range of humidity conditions.

Humidity also affects mortality rate. Low humidity has been found to increase mortality of *Drosophila*, and high humidities are recorded as interfering with hatching and molting in some species of aphids. In certain cases it has been found that high humidities apparently reduce the resistance of a species to fungus attack and act unfavorably to the insect in this manner.

There seems little doubt that humidity and evaporation constitute the barrier that restricts the geographic range of many species of insects along some periphery. There are many species occurring in eastern North America whose ranges extend westward to about the Mississippi River. The less humid conditions to the west appear to be the factor that prevents further extensive spread of these species in that direction. Conversely, there are other species occurring in the Great Plains area that do not extend much further eastward, probably because their optimum humidity requirements are lower than those of eastern species. The assumption that humidity is the limiting factor in these cases is based on the fact that, in general, lines of equal rainfall go from north to south in the area east of the Rocky Mountains, the heavy rainfall bands occurring to the east, and the scant rainfall bands to the west, fig. 364.

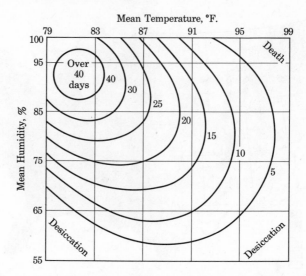

Fig. 365. Zones of maximum life of adult cerambycid beetles *Hoplocerambyx spinicornis* at different combinations of temperature and relative humidity. (From Linsley, after Beeson and Bhatia).

Temperature and Humidity. Together these two have a marked effect on both general development and distribution of insect species. Their action is frequently critical on different phases of a species and at different times of the year. Critical cold temperatures, for instance, might operate in winter against the hibernating mature larvae, whereas adverse humidities might operate during the summer against eggs or actively feeding larvae. The results of the interaction of a species to various combinations of temperature and humidity seem to comprise orderly patterns of survival, fig. 365.

Investigations on flour beetles of the genus *Tribolium* indicate that different combinations of temperature and humidity may have a profound effect on the competitive survival of different species. Thus differences in abundance or ranges of competing species may reflect, not the empirical responses of these species occurring singly to the two factors, but the result of a three-fold interaction between each species and the two weather factors plus the additional competitive species in the environment.

Daily Rhythm. During the 24-hour cycle of day and night there is a daily rhythm of temperature and humidity characteristic of each area. Except during diapause activities of most insects are correlated very definitely with this rhythm. The most conspicuous example of this is found in areas having hot days. During the heat of a summer day, when the humidity is depressed, many insects will be relatively inactive, frequenting cooler and moister niches. Toward dusk there is a drop in temperature and a sudden increase in humidity. During this period a great number of insects emerge from daytime hiding and swarm over the ground and foliage, and in the air.

Air Movement. As it concerns physiological effects, air movement has little direct action on insects. It acts indirectly by influencing evaporation and humidity; by causing evaporation it is an aid in reducing body temperature. As drafts or wind it plays a remarkable role in insect dissemination. The upward drafts caused by dawn and dusk air-convection currents carry an astonishing diversity of insects hundreds of feet in the air. Insects caught up by these currents include not only a large array of winged insects but also small wingless forms such as springtails (Collembola). It is principally on this group of air-borne insects that the swifts and nighthawks feed.

Many cases are on record of strong-winged insects such as the *Erebus* moth being blown by storms a thousand miles or more north from their tropical homes. Occasional specimens of the *Erebus* moth have been found, still alive, in Canada, the end of a journey started in Mexico, the West Indies, or southward. Shorter wind dispersals of large numbers of butterflies and moths are fairly common.

Periodic Dispersal. Several widespread species of insects which become abundant each year in northern states either die out during the winter or become greatly reduced in the northern part of their range. Overwintering populations persist in the extreme southern states, and the species move back into the northern states each spring. The army worm *Pseudaletia unipuncta*, the six-spot leafhopper *Macrosteles fascifrons*, the potato leafhopper *Empoasca fabae*, and the potato psyllid *Paratrioza cockerelli*, fig. 366, are examples of species which behave in this manner. Evidence to date indicates that the spring northern dispersal is windborne. The manner in which depleted overwintering populations are built up in the South is still a mystery. It is possible that this wind-borne type of dispersal occurs in many more species than the few for which it is known.

PHYSICAL AND CHEMICAL CONDITIONS OF THE MEDIUM

The medium in which insects live may either temper or accentuate weather conditions and, in addition, impose definite conditions peculiar to itself on the organisms living within it. From a practical standpoint, three media are of paramount importance: terrestrial, subterranean, and aquatic.

Terrestrial Medium. For the purposes of this discussion, the terrestrial medium is considered as the land surface of the earth and everything above it. This includes the aerial and arboreal regions, but it is difficult to draw a satisfactory line between these, because so many insects move frequently and rapidly from one to another.

In the case of free-living insects, conditions of the terrestrial medium are essentially those of the atmosphere. Differences from it depend on cover. In exposed areas such as treetops, the upper foliage of desert and grassland plants, and bare unshaded soil or rock, sun temperatures

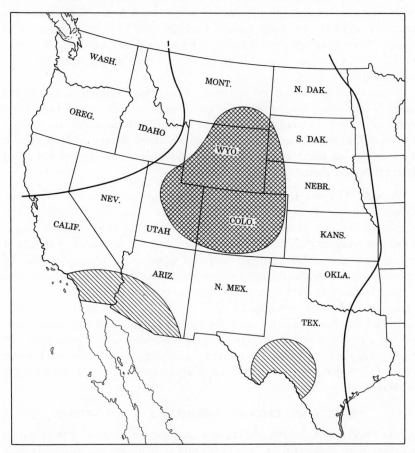

Fig. 366. Distribution of the potato psyllid *Paratrioza cockerelli* in the United States. The crosshatching indicates the area of greatest injury, the diagonal shading the approximate overwintering areas, and the heavy lines the eastern and western limits of summer occurrence. (After Wallis)

rather than shade temperatures (which are official weather temperatures) tend to prevail. Insects under leaves or other cover enjoy a moderation of temperature extremes, as do insects living within and under the tree canopy of a forest. The diurnal and seasonal rhythm of temperature and humidity is greatest in the more exposed areas and progressively less in the more shaded or protected areas.

Microhabitats of various types have peculiar conditions. Rotting organic matter produces heat of fermentation that adds to the temperature. Insects in fungi, plant galls, leaf mines, and tunnels in living trees enjoy a high humidity that approaches an aquatic environment. Insects in rotten logs find a moderation of extreme temperatures due to insulation, and their medium approaches the subterranean in character.

Subterranean Medium. No insects live in rock, so that we may consider as the subterranean medium for insects only that part of the substratum that classifies as soil and sand. This medium reflects the general climate but tempers its extremes and at the same time possesses several important characteristics of its own.

TEMPERING OF CLIMATE. Depending on circumstances, soil acts as a sponge, an insulator, and a radiator. It stores rain, giving it up slowly, so that its humidity, or moisture content, fluctuates over a much narrower range than does that of the air. Its surface layers soak up heat and insulate the part beneath; the absorbed heat is also given up slowly, so that diurnal rhythm and temperature extremes are greatly moderated in comparison to those of the terrestrial medium.

SOIL PROPERTIES. Many properties of soil are characteristics of the soil itself and are not superimposed by the immediate climate. Important among these are texture, moisture, drainage, chemical composition, and physiography.

These characteristics are almost entirely a direct result of the geological history of an area and reflect the type of strata exposed, glacial action, wind-carried material, or volcanic activity. To these factors are added the accumulated effects of the vegetation over a long period of recent time. Prairie plants, for instance, have built up thick black soils in many areas; forests tend to build thinner and lighter soils.

Man's activities have disturbed natural soil conditions more than any other element of the ecological pattern. Not only does cultivation change the original condition of the soil, but also ploughing and tilling keep changing it at various intervals, drainage or irrigation decreases or increases moisture content, and methods of farming can increase or decrease chemical constituents and organic content, the latter sometimes influencing texture profoundly. These changes have been detrimental to many insect species but have allowed others to increase and become major crop pests.

TEXTURE.   Soil texture varies from hard-packed clays to loose sands. Few insects occur in the harder-packed types, because they are unable to push or dig their way through them. The loams are probably the favorite soils for insect use. These allow digging and burrowing operations and are usually favorable in other characteristics, such as moisture content, drainage, and organic content.

The critical effect of soil texture on species abundance was demonstrated by the pale western cutworm *Porosagrotis orthogonia*. A species of the northern prairies, it was a collectors' rarity in the early days of collecting in central North America. After extensive breaking of the prairie sod and cultivation of the land in northern United States and Canada, this cutworm increased sharply in numbers, and the larvae became extremely destructive to grain crops. Investigation of the cause of increase revealed this situation. The larvae live only in soil of fairly light texture, in which they move around freely in response to daily or seasonal temperature and moisture changes. In the primeval northern prairie they occurred only in local sandy areas having loose soil, but not in the unbroken prairie sod. Cultivation transformed the sod into soil of optimum texture for the cutworm larvae, and opened to them thousands of square miles of new habitat.

DRAINAGE AND MOISTURE.   The moisture content of the soil is affected greatly by drainage. Impervious layers of substrata, such as clay or rock, may retard natural drainage, resulting in permanent or temporary semimarsh conditions or wet soils. In such situations occur only those insects that are at least partially modified for aquatic existence, such as many larvae of Diptera. In other cases impervious substrata may cause the water to percolate a considerable distance underground and, as seepage water, affect moisture conditions in other areas.

More open types of subsoil, such as sand, gravel, or shale, allow free drainage, contributing to the maintenance of better-aerated soils and more rapid restoration of normal moisture content after rains. Well-aerated soil is a prerequisite of all soil insects that have no modifications for aquatic or semiaquatic existence.

An interesting demonstration of the effect of soil moisture on an insect species is the case of the Colorado corn rootworm *Diabrotica virgifera*. Collecting records indicate that until recent years this species was a fairly rare one occurring in the arid regions of New Mexico, Colorado, and Nebraska. The larvae feed on corn roots and, if present in large numbers, may destroy the root system of the plant and cause great reduction in yield. Since about 1890 the species has been a constant pest in south-central Nebraska but practically disappeared during the drought years. Since that time, however, the species has become of major importance in the irrigated portion of the Platte River valley, owing to the increase in soil moisture resulting from more widespread irrigation.

Wireworms in the Pacific Northwest afford another striking example of changes in species composition and abundance due to changes in soil moisture. In that region four species of wireworms, *Limonius californicus, infuscatus, canus,* and *subauratus,* are wet-land pests, normally restricted to swamp and river-bottom areas. When arid land was irrigated and used for farming, these wet-land wireworms became important pests of potatoes, corn, lettuce, onions, and many other crops grown in irrigated fields. The high soil moisture maintained by irrigation apparently allows the wireworms to increase in great numbers, fig. 367.

Drainage and texture together exert considerable influence on the distribution of insects that live part of their life in the soil. The range of

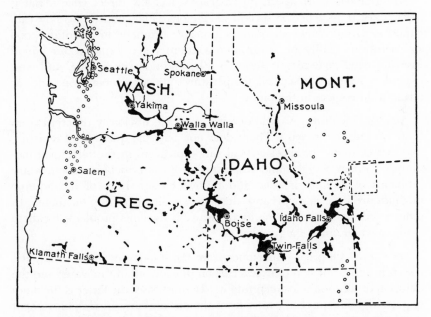

Fig. 367. Distribution of wet-land wireworms in the Pacific Northwest. Black areas represent irrigation projects on which one or more species of wireworms cause serious injury to crops annually. Circles represent localities where wireworms are known to occur naturally, without benefit of irrigation, and to cause occasional injury. (From U.S.D.A., E.R.B.)

the destructive Texas leaf-cutting ant *Atta texana* extends into Louisiana, and there the species nests only in fine sandy loams having light subsoils and excellent drainage.

CHEMICAL COMPOSITION. Chemicals naturally present in the soil affect both the abundance and distribution of phytophagous insects. Deficiencies of mineral elements, resulting in similar plant deficiencies, inhibit the growth of some insects but seemingly not others. Nitrogen deficiency lowers the productivity of some species of insects, but seems to contribute to outbreak numbers of others. In some cases the results are caused by changes in plant morphology, such as leaf toughness, which might make it difficult for the insect to eat or pierce the plant tissues. Many studies on this subject have shown little correlation between soil chemistry and insect growth, but few studies have been made in sufficient detail to be considered conclusive.

Soil chemistry also determines the species of plants that grow naturally in an area. This determines the hosts available for phytophagous insects, and in this fashion the distribution of many host-specific plant feeders.

PHYSIOGRAPHY. In itself, physiography has few direct effects, but it has a marked influence on several soil factors. Flat country has slow rain runoff and must have adequate subsurface drainage to maintain good soil aeration. Hilly or mountainous country has rapid water runoff insuring good general soil aeration. In addition south, east, and west exposures have a higher soil temperature, greater evaporation, and, as a result, differences in the biota.

Aquatic Medium. Conditions in water are obviously different from those on land or in soil. There does not exist the question of moisture and evaporation, the most critical single problem in the terrestrial medium. Instead, oxygen and respiration are the critical complementary problems of aquatic insects, and many characteristics of the aquatic medium are important because they have a direct bearing on these. In other words, for insects living in air, water is the chief problem; for those living in water, air is the problem.

AERATION. Of great importance from the standpoint of many aquatic insects are the diffusion of excess carbon dioxide out of the water and the diffusion or solution of oxygen into it. In most cases the latter is the more important. The oxygen comes from the air, and any stirring movement that brings more water into direct contact with the air increases the oxygen supply. In lakes and ponds wind action is the chief agent. Waterfalls, rapids, or movement of current are stirring agents in streams, in order of greatest efficiency.

Temperature has a direct bearing on aeration, because the colder the water, the greater is the amount of gases (including oxygen) that will dissolve in a given volume of water. High temperatures greatly decrease the solubility of gases in water.

An important distinction must be made regarding aeration among aquatic insects. Many groups, such as mosquito larvae, horsefly larvae, and certain aquatic bugs have extensile respiratory tubes that reach the surface, or the individuals periodically come to the surface to breathe; others, such as the water boatmen or adult diving beetles, take a bubble or film of air into the water with them, coming to the surface to replenish it from time to time (see p. 133). These groups are almost independent of the aeration factor in water, and many live in water almost devoid of oxygen.

Aquatic insects without modifications for obtaining direct contact with air are dependent for respiration on oxygen in the water. As with other ecological factors, various insects have different aeration requirements and are limited in distribution by it. Certain dragonfly nymphs and midge larvae are examples of forms able to tolerate very poor aeration and are often found in stagnant ponds. Larvae of the midge family Blepharoceratidae have extremely high oxygen requirements and occur only in rapid mountain cascades.

TEMPERATURE. Aquatic temperatures do not have the same range as air temperatures, but in most bodies of water there is a definite temperature response to air conditions. Insect species usually show a definite restriction to water of a certain temperature range. In many cases this is undoubtedly correlated directly with aeration, but in some cases temperature is probably the factor. Certain mosquito larvae, for instance, live and transform normally in water at 65°F. but die during molting in water at 80°F. Since mosquitoes are not dependent on water for oxygen, it appears that in this case temperature is a factor.

A few aquatic insects have been found in hot springs with temperatures ranging from 110°F. to 120°F. These are chiefly aquatic beetles and fly larvae that obtain oxygen directly from the air.

Temperature plays the same part in relation to growth and activity in aquatic insects as in other insects, as discussed on page 140. Through adjustment to temperature, various aquatic species appear at various times throughout the year. For the most part development and activity increase as water temperatures rise, resulting in the production of great swarms of adults through late spring and summer.

In the stonefly genus *Allocapnia* the reverse is true. The nymphs live in streams throughout the north-central and eastern states. During sum-

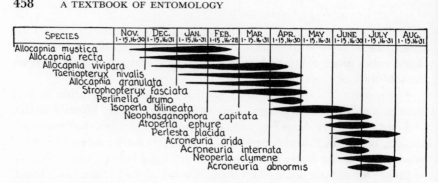

Fig. 368.   Seasonal succession of adult emergence for 15 stonefly species at Oakwood, Ill.   The widest part of the spindle indicates the time of maximum abundance of adults for each species.   (From Illinois Natural History Survey)

mer, nymphal growth seems to be retarded but increases with the advent of cool autumn weather.   As a result the adults emerge during the winter months, beginning during late November and continuing until February or March, fig. 368.

DEPTH.   In lakes and ponds depth has a marked influence on oxygen, temperature, and light; in running water the effect of depth is less marked owing to the stirring action of the current.

Water absorbs heat and light passing through it, converting the latter to heat.   Heat rays and the red end of the light spectrum are absorbed first; at greater depths the other wavelengths are gradually absorbed until, even in very clear lake water, almost all light is absorbed at about a hundred feet.   This has a profound effect on plant life.   Practically none exists below the 60-foot level, and most of it is concentrated in the first 6 or 10 feet of water where there is a good supply of light for photosynthesis.   This produces a zonation of food supply that in turn limits the distribution of many insects.

Depth in deep lakes (100 feet or more deep) is accompanied by another phenomenon of great biological interest, the thermocline.   In summer the surface waters of a large lake are appreciably warmer than the water at the bottom, which remains near its point of greatest density, 38–40°F. (= 4–5°C.), fig. 369.   This bottom layer being the heaviest, the upper warmer waters "float" on it, rather than mixing with it.   Between the two is a relatively narrow dividing area, the *thermocline,* intermediate in general conditions between the fairly uniform upper and lower strata. The upper stratum, the *epilimnion,* is churned and agitated by wind action so that it is almost uniform in temperature and well aerated.   The bottom stratum, the *hypolimnion,* is stagnant, and its oxygen is gradually used up by organic oxidation. Almost all the life in a deep lake occurs in the epilimnion; the hypolimnion is practically a biological desert.

TURBIDITY. Minute particles of earth or "blooms" of algae and other organisms usually cloud water to some extent. The clouding or turbidity has an indirect effect on insects, because it reduces light penetration and therefore plant production on the bottom. Under conditions of continuous high turbidity, there is persistent settling of suspended material on the bottom, thus modifying its character and its fauna.

BOTTOM. The great proportion of aquatic insects live on or in the bottom, and most species will live only where the bottom is of a particular type. The most useful categories for purposes of classifying the bottom types are based on size of particle: namely, muddy, sandy, and rocky. Mud bottoms are highest in organic material that serves as food; sandy bottoms and rocky bottoms have the least. Mud bottoms in streams, however, are usually associated with slow current, lower oxygen content of the water, and higher temperatures. Sandy bottoms are relatively unstable and usually have a small fauna. Rocky bottoms afford the most stable footing and are the favorite habitat for a large number of groups.

VEGETATION. To some insects aquatic vegetation is primarily food; to others it is a haven. Vegetation beneath the water provides shelter and footing, especially valuable to species that are the prey of other animals or that have no special adaptation for swimming. Aquatic vegetation is especially abundant in lakes, and it is there that it is most useful. Relatively few lake-inhabiting insects frequent open water; they stay on the

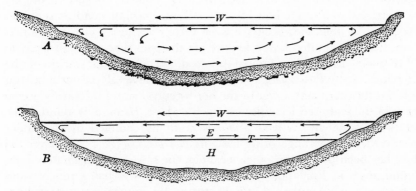

Fig. 369. Diagrams illustrating a thermocline. The circulation of the water (*A*) in a lake of equal temperature, (*B*) in a lake of unequal temperature. *W*, represents the direction of the wind; *T*, thermocline; *H*, hypolimnion. (From Ward and Whipple, after Birge)

bottom and in the weed beds most of the time. Those that move about freely in the open water usually do so only at dusk or night, hiding on the bottom during the daytime.

FOOD

Food is one of the most important factors influencing the distribution and abundance of insects. For many insect species it is a factor that has been changed radically by man's agriculture, travel, and transportation.

As a Factor in Distribution. Often the range of the host plant for a given species of insects extends much beyond that of the insect, demonstrating that some other factor such as temperature or soil condition is the actual factor limiting the insect's success and distribution. In other cases it is obvious that the insect is usually present wherever a suitable food is present and is prevented from extending its range because of the food factor. This is graphically shown in the case of many plant lice that have become successfully established on agricultural crops far beyond the range of the native hosts of the insects.

GENERAL FEEDERS. Many insects have a wide assortment of acceptable hosts or prey or feed on material such as decayed or dead organic matter that is widely distributed through most biotic communities. In these, food is only infrequently a limiting factor of distribution. Areas in which such species are absent usually have food material, but other factors such as temperature or moisture may be intolerable for the species.

SPECIFIC FEEDERS. A great number of insect species, including chiefly plant feeders and parasites, feed on only a small number of diverse host species, or are restricted to a group of closely related host species, or may be restricted to a single host species. All intermediate conditions of host specificity or tolerance occur between these extremes. The species that have the most limited host tolerance are the ones that are most likely to have their distribution limited by food. The over-all range of the Hepatica sawfly *Pseudodineura parvula* appears to cover all the north-central and northeastern states; yet in the north-central states the sawfly occurs only in the scattered localities in which its host, *Hepatica*, is found. The black-locust sawfly *Nematus tibialis* is normally confined by the rather restricted eastern range of its host, *Robinia pseudoacacia*. Whenever the host has been planted in other localities, the sawfly has ultimately been found, even in England, indicating that the sawfly has a much wider ecological tolerance than that indicated by the natural range of its host.

HOST CROSSOVER. In the case of insects feeding on a definite species of plant host, this latter may become defoliated, and it is necessary for the species to adopt a new host or to have its numbers reduced to the carrying capacity of the original host. Experimental investigations of this crossover of an insect from one host to another have brought out some exceedingly interesting information:

1. Some species, such as the forest tent caterpillar, make the change to closely related hosts with ease and without evident ill effects. Chinch bugs, for instance, will feed on oats or wheat until nearly full grown and can transfer to corn readily and without noticeable mortality due to food reactions.

2. Other species will make a change from one host to a close relative with the greatest difficulty, either after compulsion of a period of starvation (in the case of immature stages) or under circumstances of extreme necessity (in the case of ovipositing females). In most cases of this kind the transfer cannot be made by advanced larvae. They will eat the new food but develop symptoms of intestinal disturbance, such as diarrhea, and die. If first-instar larvae are put on the new host, they will eat but suffer an extremely high mortality. During succeeding instars the mortality rate decreases until the pupal stage is reached, and here the mortality is again high. But out of thousands of first-instar larvae a few adults will finally be obtained. These will prefer to oviposit on the new host, and the resulting larvae will feed on it readily and without ill effects.

An interesting case of this type was supplied by the satin moth *Stilpnotia salicis*. A European pest of Lombardy poplar, it was introduced on the Pacific Coast about 1922 and in a few years became very abundant, completely defoliating Lombardy and other introduced poplars in each locality to which it spread. In British Columbia by the late 1920's it had spread up the valley of the Lower Fraser River. There it built up a large population and practically exterminated the Lombardy poplars in that region. When Lombardy poplars were no longer available, the satin moths began laying eggs on the native cottonwood *Populus trichocarpa hastata*, an abundant tree in this region. At first only sporadic colonies became established on this new host which is also a member of the poplar genus, but in a few years the satin moth was the most serious insect enemy of cottonwood in this area. Laboratory rearings demonstrated that in making this host crossover the satin moth went through the initial high mortality process just described.

An example of a species that does not follow this behavior is the gypsy moth *Porthetria dispar*. The gypsy moth larvae are plant feeders and will feed on over four hundred and fifty species of plants. Of these forty-two are favored hosts, including willows, birches, and oaks; later instars of the larvae may switch to one of the other four hundred and fifty host species, one of the favorites being white pine. They suffer no ill effects and develop to maturity. For normal development, however, it is necessary for the first two larval instars to feed on one of the forty-two favored species.

The complete host relationships of only a relatively small number of insect species have been studied comprehensively. Undoubtedly much interesting material will continue to be discovered on this subject.

3. A third group of species appear to be tied irrevocably to a single species of host. If put on even a closely related species, they will die before they will feed, or all die if they do feed.

These phenomena of host crossover have a great effect on the distribution and abundance of the insects involved and represent an important type of ecological tolerance so far as food is concerned. They may be of economic importance if crossover cases involve agricultural species. There is always the possibility of this happening when new plants are brought into the national agricultural economy, as, for instance, the soy bean. When the soy bean was introduced into the central states it was virtually unattacked by any of the native insect species. Gradually some of these began using it as a host, and now the soy bean in this country has several serious insect enemies, including grasshoppers, root maggots, and white grubs.

As a Factor in Abundance. Amount of available food is an important factor affecting the population of a species in a given community. It is not uncommon for a species to utilize its entire available food supply with a resulting sharp reduction in population due to starvation.

Although there may be considerable variation in size between different individuals of the same species, there seems to be a definite minimum amount of food required for the normal development of an individual. Housefly maggots, for instance, die during pupation if the larvae are removed prematurely from their food supply. Most sawfly larvae will die without further development if removed from their food only two or three feedings prior to completing their full food intake. If, therefore, an excessive number of individuals are feeding on an insufficient amount of food, those with a head start complete their normal food intake and mature, and many of the remainder run out of food and fail to develop.

The most notable exceptions to this are found in certain parasitic species having a wide range of hosts. In these, the size of the individual parasite is determined by the size of the respective host species. An excellent example of this is the mutillid wasp *Dasymutilla bioculata;* larvae feeding on small prey species develop into small individuals, those feeding on large prey species develop into large individuals, fig. 370.

Agriculture has changed the insect food factor in several ways: (1) by providing suitable food when or where it would not be present under

Fig. 370. Correlation in size between *Dasymutilla bioculata* and its hosts *Microbembix monodonta* (left) and *Bembix pruinosa* (right). Each vertical row has *Dasymutilla* male and female above, host wasp at bottom. (Material, courtesy of C. E. Mickel, photo by W. E. Clark)

natural conditions, (2) by providing better food, and (3) by simply providing a greater food supply.

EXTENSION OF FOOD SUPPLY. A striking example of this factor came to light in the extensive sampling of grain in granaries in the late 1930's and early '40's. It was discovered that in many states, especially in the corn belt, a tenebrionid beetle *Cynaeus angustus* had become extremely abundant and widespread and had developed into a major pest of stored grain. This beetle was first described in 1852 from California and prior to 1938 remained a collector's rarity. It was known to occur at the base of yucca plants. In 1938 the beetle was encountered as a stored-grain pest in Washington, Kansas, and Iowa. By 1941 it was known to be widely distributed through the corn belt. The species reached population peaks in the man-made conditions of grain storage which are in astonishing contrast to the scarcity of the species in its natural habitat.

BETTER FOOD. It has been shown in some cases that certain introduced agricultural crops increased the fecundity and thereby the abundance of native insect species. The two grasshoppers, *Melanoplus differentialis* and *M. bivittatus,* showed marked increases in fecundity on a diet of soy beans and alfalfa, respectively. This is correlated with field observations of the increase in population of the two species following the planting of large acreages of the two crops mentioned.

ADDITIONAL HOST MATERIAL. Practically every crop favors at least a few insects in supplying more food. The Colorado potato beetle, the corn aphids, and other pests on potatoes and corn certainly have flourished on the thousands of acres of host crops that man has planted and have built up huge populations that dwarf the scattered colonies that existed before agriculture, when their hosts were relatively sparse and the individual plants not so luxuriant as those of improved agricultural varieties.

ENEMIES

A wide array of organisms prey on or parasitize insects. Some of the parasites, such as the malarial organisms *Plasmodium* sp., seem to do the insect no harm, but the majority have a harmful effect on the insect host. These enemies constitute an environmental factor having a definite effect on the abundance, and sometimes the distribution, of the host species. Each stage of the host species may be subject to attack by a different set of enemies, or several stages may be attacked by the same one. As a rule, predaceous enemies and plant enemies such as fungi are more general in their attack on various stages, and internal parasites are restricted regarding the stage they attack.

Internal Parasites. Insects are attacked by several groups of internal parasites, of which the most important are certain groups of insects, parasitic worms, bacteria, and fungi. Other groups also parasitize insects.

INSECTS. The larvae of many families of Hymenoptera (Ichneumonidae, Chalcididae, Scelionidae, and many others) and a few families of Diptera (Pyrgotidae, Tachinidae) are entirely endoparasitic on insects or closely allied arthropods. A few Lepidoptera have endoparasitic larvae, and several Coleoptera, including the entire small series Stylopoidea. On the basis of rough estimates there are about eleven thousand species of parasitic insects known at present in North America. Most of these are fairly specific at least as to what group they attack. Some, for instance, will attack a wide variety of lepidopterous caterpillars; others will attack only certain primary parasites in these caterpillars (see p. 306).

OTHER ANIMALS.   Some species of Protozoa and invertebrate metazoan parasites pass one stage of their life cycles in insects.   Examples of such protozoan parasites are the malarial organisms *Plasmodium* sp. and sleeping-sickness organisms *Trypanosoma* sp.   Among the parasitic worms that spend part of their life cycle in insects are trematodes, nematodes (for example, *Filaria*), and Acanthocephala (for example, *Macracanthorhynchus hirudinaceus* the thorny-headed worm of swine).   In each case only one of the early stages of development is passed in the insect, which is an intermediate host for the parasite.   This group of parasites does not appear to have a deleterious effect on the insect, at least not the fatal effect of the insectan parasites.   It is probable, therefore, that this class of non-insectan parasites is a negligible factor in relation to insect populations. Certain Protozoa, especially the microsporidians, have insects as their only host; these parasites frequently kill the host.

FUNGI, BACTERIA, AND VIRUSES.   Species of these groups attack insects in various stages and at times are destructive to their hosts.   Anyone who has carried on rearing experiments with insects can well attest this, for cultures are very susceptible to attack by fungous and bacterial organisms. The reason for this is that the best development of both types of these

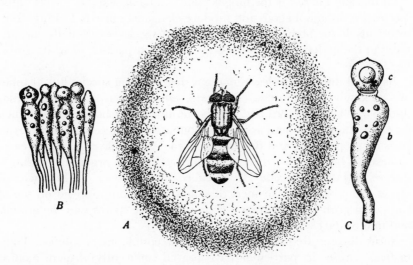

Fig. 371. *Empusa muscae* the common fly fungus. *A*, house fly (*Musca domestica*), surrounded by fungus spores (conidia); *B*, group of conidia in several stages of development; *C*, basidium (*b*) bearing conidium (*c*) before discharge. (From Folsom and Wardle, Entomology, by permission of P. Blakiston's Son & Co.)

Fig. 372. Fruiting structures of a fungus *Cordyceps ravenelii*, arising from the body of a white grub *Phyllophaga*. (After Riley)

parasitic organisms is attained under conditions of relatively high humidity and temperature which are frequently increased to an unnatural degree in caged experiments.

Among common fungus diseases of insects is *Empusa muscae* the house fly fungus, fig. 371. Other members of the same genus attack a large variety of insects, including grasshoppers, aphids, and chinch bugs. A famous fungus disease is *Beauveria globulifera*, often referred to as *Sporotrichum globulifera*, the white fungus of the chinch bug. During warm and humid seasons this fungus kills large numbers of chinch bugs and other insects in late spring and early summer and at times has controlled the bugs to the point of local extermination. Entomophagous fungi of the genus *Isaria* are the chief species attacking insects under artificial conditions.

Of unusual interest is the fungus family Laboulbenaceae. Most of the species are entomophagous and produce elongate or ornate fruiting structures outside the body of the host insect. A species occasionally encountered in the eastern states is *Cordyceps ravenelii*, a parasite of white grubs, fig. 372.

Bacterial diseases are less numerous than fungi in species but at times are strikingly devastating. Flacherie, an infectious and highly fatal disease of silkworms, is caused by a bacterium. Grasshoppers and chinch bugs are attacked by similar bacteria, but attempts to control these insect pests by propagating and disseminating the disease have failed. Greater success has been achieved with *Bacillus popilliae*, the organism causing a disease of Japanese-beetle larvae called milky disease. The bacterial spores are mixed with an inert dust and the mixture applied on top of the soil in grub-infested areas. Rain washes the spores into the ground and into contact with the grubs.

Virus diseases are extremely toxic to susceptible insect species. Polyhedrosis viruses in particular have proved sufficiently virulent against certain sawfly and lepidopterous larvae to be employed successfully as control agents.

External Parasites. Insects have few ectoparasites of the type of the lice or fleas, in which the adult stage or both immature and adult stage use the body of the host as a home. A few mites infest various insects, but only scant information is known regarding the ecological significance of the groups. An unusual ectoparasite is the bee louse *Braula caeca,* a curious minute fly that is ectoparasitic in the adult stage on honeybees.

In the Hymenoptera, some families whose larvae are mostly endoparasites, such as the Braconidae, contain genera whose larvae are attached externally to their host larvae. These parasites have the same host relation as their endoparasitic allies, in that normally only one parasite individual lives on one host individual, the latter almost always dying when or before the parasite is mature.

Predators. As in the case of internal parasites, so in this category too insects are their own worst enemies. Carabidae and Staphylinidae are two very large beetle families that feed in both adult and larval stages almost exclusively as predators on other insects. Many families of wasps are predaceous, as are larvae of Tabanidae, Dolichopodidae, and some other large families of Diptera. Odonata (damselflies and dragonflies) are predaceous as both nymphs and adults. The same is true of certain families of Hemiptera such as Pentatomidae (stink bugs), Reduviidae (assassin bugs), and Phymatidae (ambush bugs); in some other families of Hemiptera, such as the Miridae (plant bugs), most genera are phytophagous, but some are predaceous. There are many other small groups of predaceous forms.

Noninsectan predators of insects include members of several large groups. Spiders are primarily insectivorous; there are about three thousand species of spiders in North America, and each spider population takes its toll of insects. Centipedes feed on insects to a large extent also.

Vertebrates contain many groups that are insectivorous. Among the fish, perch, sunfish, crappies, bass, and sheepshead use insects for a large share of their diet. Reptiles and amphibians are largely insectivorous, as are bats and moles; other mammals such as mice, skunks, shrews, and raccoons eat large numbers of insects.

Birds are the outstanding vertebrate insect eaters. Swifts, nighthawks, and flycatchers feed entirely on insects caught on the wing. Robins, wrens, chickadees, cuckoos, quail, and prairie chickens live almost entirely on insects when the latter are abundant. During insect outbreaks many birds of omnivorous food habits switch temporarily to an insect diet. Crows, blackbirds, gulls, owls, and small hawks are in this group and have been noted especially feeding on grasshoppers during periods of abundance.

All these animals are abundant and, being comparatively large individuals, eat proportionately large numbers of insects. In doing so they exert a steady ecological force against insect populations.

Predaceous Plants. A list of insect predators would not be complete without mention of those curious plants that trap animal prey and digest them. Bladderworts (*Utricularia*) are aquatic plants that trap small organisms in bladder-like pouches; sundews (*Drosera*) are bog plants having sticky tentacle hairs on their leaves that encompass prey; and pitcher plants (*Sarracenia*) have leaves in the shape of pitchers, partially filled with water, with stiff hairs pointing to the water; the hairs allow insects to get to the bottom of the pitcher but prevent their escape. None of these plants is sufficiently abundant to be of importance ecologically by reducing insect numbers.

PROTECTION AGAINST ENEMIES

Insects appear to have little or no protection against several groups of their enemies, notably fungi and bacteria. Against insect parasites and predators their only protection seems to be evasion; many insect stages are extremely limited in locomotion and obtain no protection by this means.

The principal group of enemies against which insects have achieved some measure of protection is the land-vertebrate group. Against this group such devices as protective resemblance, the building of protective structures, poison hairs, bites, stings, noxious secretions, and the mimicking of species possessing some of the foregoing offer protection to some extent.

Protective Resemblance. We have all been surprised at one time or another to discover a "stick" come to life in the net, or, in examining a tree trunk, to see what appeared to be a section of bark take wings and fly away. This protective resemblance is common in several insect groups. The walkingstick insects (Phasmidae), fig. 373, resemble sticks; in spring they are green; when they mature in autumn many become brown, resembling the foliage or twigs on which they feed and rest. Many moths at rest resemble bark. Some of the larger forms are the underwing or Catocala moths, perfect bark mimics with wings folded, but often conspicuous in flight owing to brightly colored underwings, fig. 373. Grasshoppers resemble lichens, various types of soil, dried leaves, or grass, depending on the species and its food. Psocids have similar protective color patterns, especially those that feed on algae or lichens on bark or rock bluffs. A few larvae at rest curl up and resemble a fresh bird dropping, notably the sawfly *Megaxyela aviingrata*. In general most of the leaf-feeding larvae are green or are mottled so that they blend into the foliage on which they feed.

Fig. 373. Protective mimicry and coloration. At left, a walkingstick insect on a twig; at right, an underwing (*Catocala*) with wings spread (*A*) and at rest on bark (*B*). (From Folsom and Wardle, *Entomology*, by permission of P. Blakiston's Son & Co.)

Building Protective Structures. Certain larvae build cases, houses, or canopies that give the occupant some physical protection and in addition may resemble the host or surroundings and result in protective resemblance. Bagworms (Lepidoptera), fig. 331, are a common example, and also caddisfly larvae of many kinds, fig. 320. Cases of the latter may be very difficult to see unless the insect is in motion. The larvae of several leaf-feeding beetles construct an urnlike case that is difficult to see in its natural surroundings.

Poisons, Bites, and Stings. Certain insects gain protection by inflicting pain on their assailants. Several caterpillars have sharp hairs, containing a poisonous fluid that causes a rash and extreme pain. So delicate are these hairs that one has only to brush against them lightly to feel the excruciating nettling sensation that they produce. Tussock moth larvae and eucleid moth or saddleback larvae, fig. 374, are protected by these. Other insects bite the aggressor, as, for instance, ants. Still others, such

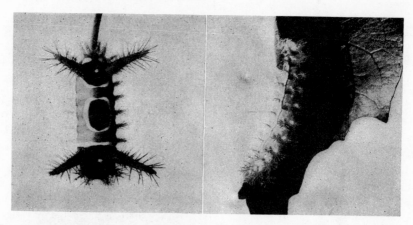

Fig. 374. Caterpillars having nettling hairs annoying to man.   Left, the saddleback *Sibine stimulea;* right, *Automerus io.* (From U.S.D.A., E.R.B.)

as bees and wasps, are provided with painful stings; bees use these only for protection; some wasps use these both for protection and for paralyzing prey.

Noxious Secretions.   A large number of insects have mechanisms for producing and ejecting noxious smelly substances.   Swallowtail butter-fly larvae have an eversible pair of horns, the osmeterium, on the pronotum, that give off an odor thought to be repellent to some animals.

Fig. 375.   A brightly colored mal-tasting butterfly *Papilio ajax.*   (From Folsom and Wardle, *Entomology,* by permission of P. Blakiston's Son & Co.)

Nymphal stages of most Heteroptera have stink glands on the dorsum of the abdomen. Many beetles, cockroaches, and others have less conspicuous repellent-producing glands (see p. 161). Other insects, without such definite glands, apparently have a disagreeable taste, because birds especially refuse them as food. Swallowtail butterfly and milkweed butterfly adults, and both larvae and adults of many brightly colored leaf-feeding beetles are in this category.

It is pertinent to note here that a large number of species possessing protective devices discussed in the two preceding paragraphs are strikingly marked or gaudily colored, fig. 375. This striking ornamentation may be a display of warning colors, to aid the memory of an assailant who has attacked a protected species and become aware of its defense. It is certain that birds and other vertebrates have no instincts to avoid protected species, so that each individual must learn for itself.

Protective Mimicry.   Among insects having no known protective mechanism, there are some that share the protection of those that do by looking almost exactly like them. Thus there are harmless plant bugs that look like ants, harmless flies that resemble bees or wasps, fig. 376, and edible butterflies that have the appearance of distasteful species. Among our fauna the best example of the latter type, fig. 377, is the viceroy *Basilarchia archippus,* an edible species that resembles in pattern and general color the distasteful milkweed butterfly *Anosia plexippus.* In tropical faunas similar cases are common.

The advantage gained by a species through these types of protection is average, for none of these methods are absolute. Insects inedible to birds may be delectable to other animals, or if refused by one kind of bird may be eaten readily by another. But if even a small advantage is gained by a species, that species has a tremendous advantage over a long period of time.

Fig. 376.   Mimicry in bees and wasps.   *A*, drone bee *Apis mellifera; B,* dronefly *Eristalis tenax.* (From Folsom and Wardle, *Entomology,* by permission of P. Blakiston's Son & Co.)

Fig. 377.   The viceroy, below, and the milkweed butterfly, above.   (From Folsom and Wardle, *Entomology,* by permission of P. Blakiston's Son & Co.)

COMPETITION

If we suppose an individual of a species in a situation having suitable climate and conditions of medium and food, and, further, that its enemies are not a critical factor, that individual may discover that another individual having similar wants is there also.   If there is only sufficient food for one, then the two individuals are in vital competition from which only one can emerge as the survivor.

Among insects, competition is chiefly for food.   This competition may be between either individuals of the same species or individuals of different species.

Frequently there is no reaction to critical competition, and all individuals may starve.   If, for instance, sawfly larvae overpopulate a host, they feed quietly until the entire host is stripped; then all of them wander until exhaustion and death if additional food is not found. In the case of critical competition involving two or more insect species, their different requirements may mitigate in favor of one of them. An interesting example is cited by Willard and Mason (1937) regarding two hymenopterous genera *Opius* and *Tetrastichus*, that parasitize the Mediterranean fruit fly larvae in Hawaii. Within a single fruit fly larva there can develop to maturity only a single larva of a species of the braconid *Opius*, but as many as ten to thirty individuals of the minute chalcid *Tetrastichus*, fig. 378. If both oviposit in the same fruit fly larva, the *Opius* larva kills most of the *Tetrastichus* larvae, but a few of the latter escape destruction. These develop more rapidly than the *Opius* larva and reach maturity, but leave too little food for the larger braconid larva, so that the *Opius* larva invariably dies.

Competition for food is frequently active and aggressive. Pemberton and Willard (1918) give an account of such an example occurring in wasps of the genus *Opius* previously referred to. In Hawaii three species, *tryoni, fullawayi,* and *humilis,* parasitize fruit fly larvae. The female wasps lay their eggs in the fly larvae, and several individual wasps of all three species may oviposit in the same larva. Only one survives, and this one is the result of a battle among the newly hatched larvae. The first instar

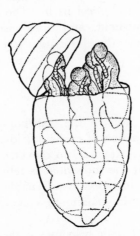

Fig. 378. Parasitic wasps *Tetrastichus giffardianus,* in puparium of fruit fly. (After Pemberton and Willard)

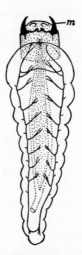

Fig. 379. First-instar larva of *Opius humilis.* Note the sharp heavily sclerotized mandibles *m.* (Redrawn from Pemberton and Willard)

of each *Opius* larva has a relatively large hard head bearing a pair of long sharp mandibles that can be opened and shut with great force and speed, fig. 379. These larvae are pugnacious and attack any other parasite larva within the fly larva, using these sharp mandibles to pierce and lacerate their antagonist's body. Whether the struggle is between individuals of the same species or of different species, only one *Opius* larva remains after the struggle is over. It was discovered that *O. tryoni* was almost invariably the victor over the other two species, owing to its greater agility, reaction time, force in use of mandibles, and other combative advantages.

Cannibalistic tendencies occur in many insect groups, and are invariably accentuated by crowding. The confused flour beetle lives and feeds on a variety of stored-grain products; generations are continuous, and adults, eggs, and all stages of larvae occur together in the food. Large larvae and adults may feed on eggs and small larvae of their own species but apparently make no effort to hunt them out. When the infestation of these beetles is small in relation to the volume of their food medium, the older individuals encounter the younger stages less frequently. As the infestation increases per unit volume of food medium, these encounters are more frequent, and cannibalism increases accordingly. By this mechanism a population point is reached where the losses due to cannibalism are equal to the reproductivity of the adults, and overcrowding beyond this point is prevented.

It is pertinent to note in connection with these phenomena, that there exists no conscious sense of competition by the insect itself. The competing individuals react instinctively throughout, and these instincts under certain conditions of crowding produce the elimination of excess numbers.

Tropisms

Instinctive behavior plays an important role in the distribution of members of an insect population. The reaction of each individual to stimuli or to a pattern of stimuli causes the individual to remain in an environment compatible with its needs. If the individual is removed from such an environment, the reactions to the stimuli will enable it to return or find a new environment with the maximum compatible components.

The basis of instinctive behavior is in automatic responses to definite stimuli, and each such response is called a tropism. Each insect species exhibits a wide range of tropisms, a great number relating to sexual behavior and mating, fig. 380, and others relating to ecological factors of the environment. It is this latter group of tropisms that we shall consider here.

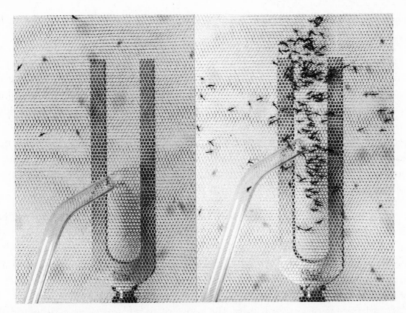

Fig. 380. A response to sound associated with sexual behavior. Area in front of tuning fork in cage of male mosquitoes. Left, before sounding fork; right, after sounding fork. (Courtesy of L. J. Roth)

Phototropism, Reaction to Light. Most insects have an extremely well-developed response to light, moving toward the light source or away from it. Cockroaches move away from the light and are termed negatively phototropic; bees and wasps move toward it, and are thus positively phototropic. But the reaction often is different in various stages of an insect's life cycle. Housefly maggots are negatively phototropic and move away from light; adult houseflies are positively phototropic and move toward light.

Some aquatic insects, such as mayfly nymphs, maintain their dorso-ventral position, that is, stay right side up, by orientation to light from above. This response is most pronounced when the insect is swimming, and is inhibited if the insect is at rest.

There is a definite response by some insects not merely to light in general but also to certain wavelengths of light. In most cases this is an aid to finding food or, in the case of ovipositing females, placing their eggs on the correct type of foliage. Thus, butterflies in search of food are guided by their perception of color in distinguishing yellow, red, and blue from green and approach the flowers of the former colors in preference to the green foliage. But some of these same butterflies will lay their eggs only on a green surface, which under natural conditions would be a healthy leaf suitable for larval food.

Geotropism, Reaction to Gravity. Many insects if placed in a vertical tube will go steadily to the top or bottom, rather than wandering haphazardly around the tube. Leafhoppers always go up; if the tube is inverted so that the insects are again at the bottom, they will start their upward climb again. This is a negative response to gravity, or negative geotropism. Other insects have a positive geotropism, normally going down or toward the earth. Many soil-inhabiting larvae such as wireworms have this reaction; thus, if they hatch from eggs laid on or near the soil surface, they burrow down into the soil.

Thigmotropism, Response to Contact. Many insects that normally live under bark, in soil, or in curled leaves have a well-developed touch or tactile reaction that causes them to remain in contact with some object, fig. 381. This is known as positive thigmotropism. Observation of the behavior involved indicates that the touch sensation acts as a sort of hypnosis, temporarily immobilizing the insect.

In all insects of active habits, the sense of touch serves as a detector of enemies. Frequently some area or structure at the apex of the abdomen, such as the cerci, has tactile hairs of extreme sensitivity to aid in these "escape" reactions.

Chemotropism, Response to Odors. The number of responses that insects make to various odors is legion. In relation to the environment these are mostly correlated with food, as in the case of an individual locating food for its immediate use, or of a female finding a suitable place for laying eggs in relation to the food of the resultant immature stages, fig. 382. In general, each insect is responsive to only the particular food odors that immediately concern the species. For instance, butterfly females of the genus *Macroglossa* will oviposit only on a surface having the odor of the plant *Galium,* on which the larvae feed; other odors cause no oviposition response.

Most insects follow odor-laden air currents, orienting their line of approach either by direction of air current or by increase or decrease in odor intensity. There are some insects, however, that follow the trail of scent left by their prey as a dog follows a rabbit. The braconid wasp *Microbracon* follows the scent trail of its host, the larva of the mealworm *Ephestia,* the hunting wasp running along with its antennae held close to the ground. Ants use this method to follow trails to and from the nest, locating it by the routes marked with chemical substances secreted and dropped by the ants.

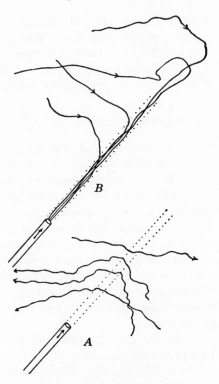

Fig. 381. An example of positive thigmotropism. Position taken by the earwig *Forficula* in a circular glass container. (From Wigglesworth, after Weyrauch)

Fig. 382. An illustration of chemotropism. Tracks followed by *Drosophila* flies (deprived of wings) when exposed to (*A*) an odorless stream of air, and (*B*) air carrying odor of pears. (From Wigglesworth after Flügge)

Thermotropism and Hygrotropism. Insects respond to various degrees of heat and humidity, moving towards the condition closer to their optimum. Insects that feed on warm-blooded animals use temperature as a guide to their hosts. Thus, mosquitoes and bedbugs are positively thermotropic to temperatures near 98°F., that is, near mammal blood heat.

Coordinated Tropisms. Many activities of insects are dependent on responses involving two or more tropisms at the same time. It has been demonstrated, for instance, that ovipositing *Macroglossa* butterflies require both a green color and the odor of *Galium* to induce egg laying. Certain newly hatched caterpillars that feed in trees have both a negative geotropism and a positive phototropism, insuring that the larvae travel upwards to the natural food.   In many other activities there is a fixed chain of responses, following each other in definite order.   With the stable fly *Stomoxys*, different reactions to smell, taste, warmth, and moisture control the fly's approach to its animal host, the extension of its proboscis, probing of the host tissues, and, finally, feeding.

Population Dynamics

The distribution and abundance of an insect species are measures of its success under the effect of the sum total of its environmental conditions.   In usual years a number of conditions will be favorable and others unfavorable to the increase of the species.   Furthermore the combination will be different in various parts of the species range.   The two sets of factors, favorable and unfavorable, tend to offset each other and "balance out."   As a result the abundance of a species in the wild varies from year to year but normally within moderately narrow limits.   Occasionally, however, the favorable factors predominate and the insects multiply to unusual or outbreak numbers.   Frequently, but not always by any means, these favorable factors are the result of man's changing of the environment.

Outbreaks may expand from a local start and follow an amoeboid type of distribution, such as the tent caterpillar outbreak in Minnesota in 1949–1954, fig. 383.   Other outbreaks, especially of grasshoppers and

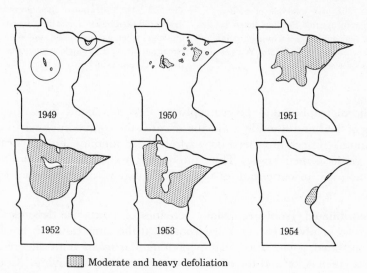

░░ Moderate and heavy defoliation

Fig. 383.   History of an outbreak of the forest tent caterpillar *Malacosoma disstria* in Minnesota during 1949–1954.   (After Butcher)

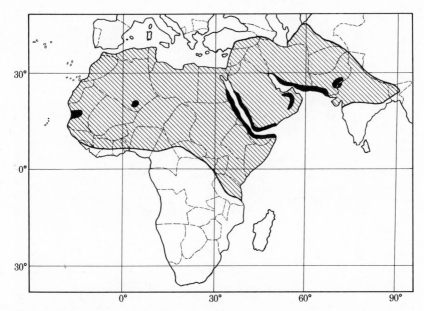

Fig. 384.   Outbreak areas of the desert locust *Schistocerca gregaria*.   Maximum invasion area shaded; known and suspected outbreak areas of 1950, black.   (After Uvarov)

locusts, may result in migratory swarms which leave their point of origin and travel from area to area.   The most persistent, destructive, and well-known example is the Old World desert locust *Schistocerca gregaria*, which periodically migrates through and devastates large areas in Africa and southwestern Asia, fig. 384.   In North America two of the most persistent migratory species are the Mormon cricket *Anabrus simplex* and the Rocky Mountain locust *Melanoplus mexicanus*.   The latter is common over much of the continent every year in a nonmigratory stage but on occasion builds up huge local populations which become migratory swarms.   The last spectacular swarming occurred in the summers of 1938 and 1939, fig. 385. The following flights in 1940 were fewer and shorter, and the swarming phenomenon disappeared in 1941.

Many outbreaks are not spectacular but nevertheless inflict large monetary losses if they affect something valuable to man.   The causes of outbreaks are therefore a matter of great interest.   Detailed studies have been made on many species, and complex statistical methods are frequently employed in attempting to verify correlations between insect abundance and different factors of the environment.   The ultimate aims of these studies are better prediction of outbreaks and better methods of control.

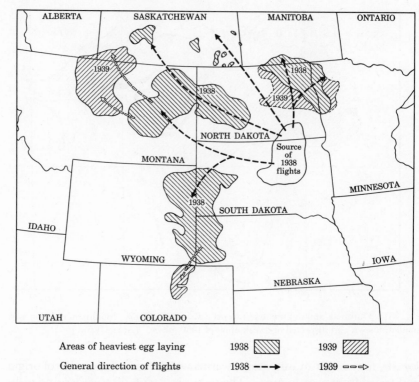

Areas of heaviest egg laying   1938 ▨   1939 ▨

General direction of flights   1938 ---▸   1939 ⇨

Fig. 385.   Main migration routes and areas of heaviest egg laying by the migratory grass-hopper *Melanoplus mexicanus* in 1938 and 1939.   (After Parker, Newton, and Shotwell)

Complexity of Habitat.   Some investigators believe that the chances of insect outbreaks are inversely proportional to the complexity of the eco-logical community to which the insect belongs.   According to this view the abundance of a species is the result of a balance between all the factors in the community, both climatic and biotic.   In a complex ecological community such as a mixed forest there are a vast number of species of different organisms, hence a very large total number of factors (some favorable, some unfavorable) governing the population of any one species. Should only one factor change favorably for a species, it would have a relatively small influence on the abundance of the species.   On the other hand, in a simple ecological community such as a wheat field there are few species in the total biota, and hence few factors balancing the popu-lation of any one species.   In this situation a single factor changing in favor of a particular species would theoretically have a much greater influence toward increasing the population of that species.

This viewpoint agrees with the observation that insect outbreaks are much more frequent in forests composed of extensive pure stands of single tree species as compared with mixed forests, and more frequent in single-crop areas as compared with areas of highly diversified farming.

Outbreaks and Damage. It should be pointed out that insects affecting man and his commodities need not occur in outbreak numbers to be of economic importance. A relatively small population (ecologically speaking) of a disease vector might be sufficient to cause tremendous losses to animals and plants because of the disease involved. In the case of fruit, a small infestation of certain insects results in lowering the market grade of the product, with greatly reduced cash profit to the grower. At the same time there are other species which do little noticeable damage at low levels of abundance, such as aphids on grain crops, but which cause tremendous losses when occurring in outbreak proportions.

REFERENCES

Allee, W. C., A. E. Emerson, O. Park, T. Park, and K. P. Schmidt, 1949. *Principles of animal ecology.* Philadelphia: W. B. Saunders Co. 837 pp.

Allee, W. C., and K. P. Schmidt, 1951. *Ecological animal geography,* 2nd ed. New York: John Wiley & Sons. 597 pp.

Andrewartha, H. G., and L. C. Birch, 1954. *The distribution and abundance of animals.* Chicago: University of Chicago Press. 782 pp.

Chapman, R. N., 1931. *Animal ecology.* New York: McGraw-Hill Book Co. 464 pp.

Clements, F. E., and V. E. Shelford, 1939. *Bio-Ecology.* New York: John Wiley & Sons. 425 pp.

Elton, Charles, 1927. *Animal ecology.* New York: The Macmillan Co. 207 pp.

Folsom, J. W., and R. H. Wardle, 1934. *Entomology with special reference to its ecological aspects,* 4th ed. Philadelphia: P. Blakiston's Son & Co. 605 pp.

Glen, Robert, 1954. Factors that affect insect abundance. *J. Econ. Entomol.,* **47:**398–405.

Glick, P. A., 1939. The distribution of insects, spiders, and mites in the air. *USDA Tech. Bull.* **673:**1–151.

Parker, J. R., R. C. Newton, and R. L. Shotwell, 1955. Observations on mass flights and other activities of the migratory grasshopper. *USDA Tech. Bull.,* **1109:**1–46.

Pemberton, C. E., and H. F. Willard, 1918. A contribution to the biology of fruit-fly parasites in Hawaii. *J. Agr. Research,* **15:**419–465.

Richards, O. W., and N. Waloff, 1954. Studies on the biology and population dynamics of British grasshoppers. *Anti-Locust Bull.,* **17:**1–182.

Tauber, O. E., C. J. Drake, and G. C. Decker, 1945. Effects of different food plants on egg production and adult survival of the grasshopper, *Melanoplus bivittatus* (Say). *Iowa State Coll. J. Sci.,* **19:**343–359.

Uvarov, B. P., 1928. Insect nutrition and metabolism. *Roy. Entomol. Soc. London Trans.* f. **1928:**255–343.

——— 1931. Insects and climate. *Roy. Entomol. Soc. London Trans.,* **79:**1–247.

——— 1951. Locust research and control, 1929–1950. *Colonial Research Publication No. 10.* 67 pp.

Willard, H. F., and A. C. Mason, 1937. Parasitization of the Mediterranean fruitfly in Hawaii, 1914–1933. *USDA Circ.,* **439:**1–18.

Chapter Ten

Control
Considerations

THE over-all picture of insect control in the United States and Canada has been changing steadily. In the first place, data show that insects do more damage to all types of food and fiber than was formerly thought. For many years 10 per cent was used widely as the estimated insect damage to food and fiber. Better surveys and better controls have demonstrated that this loss is at least 20 per cent, and in the case of crops such as clover seed there may be a perennial potential of 50 to 75 per cent. The monetary values of both the marketed crop and the cost of production have increased. These facts together mean that it is now necessary for operators of many types (individual, corporate, and governmental) to follow practices giving insurance against excessive losses from insects.

Another control factor has changed which is of especial importance to the crop entomologist. Not many decades ago wormy apples and caterpillars in the flour were no novelty and were tolerated in this country. Now the consumer demands insect-free products, and this is reflected in higher legal and market standards which must be met by the producer. Whenever this challenge has been met with more thorough chemical control efforts, there have arisen questions concerning residual insecticides on possible human food. This has led to the necessity of determining the permissible residual tolerances of insecticides and their breakdown chemicals. In practice, therefore, the entomologist is faced with the problem of achieving a more nearly perfect control of insects attacking edibles, but at the same time having fewer chemicals left on the marketed product.

During the late 1950's, another problem arose concerning insecticides. Programs using chlorinated hydrocarbons such as DDT for the control of

Fig. 386. Potatoes showing virtually complete destruction by the potato leafhopper; insert is a stand of healthy, sprayed potatoes taken on the same day to show contrast and condition. (From Illinois Natural History Survey)

forest insects over large areas, for the attempted extermination of the fire ant, and heavy local applications in quarantine and shade tree treatments drew attention to losses of fish and wild life caused by these highly residual insecticides. This occurred at a time when agricultural, urban, and industrial expansion following World War II was making serious inroads into natural recreational areas, yet burgeoning populations were making greatly increased demands on these same recreational outlets. There has thus developed a potential and sometimes actual conflict of human values between the economic benefits from certain types of insect control and the recreational desirability of preventing losses to fish and wildlife. In response to this situation, an increased emphasis has developed on the use of insect control methods and practices that are not harmful to wildlife.

HOW INSECTS CAUSE DAMAGE

Insects cause injury or damage to man and domestic animals, to a wide variety of crops, to stored products, and to buildings and many of their contents. Each item may be attacked by several or by many insect species, each causing a different type of damage. Sometimes the damage is direct, in that it is the result of the insect's own activities, such as feeding or oviposition. In other cases the principal damage is caused by a disease organism introduced into a plant or animal by the insect.

Direct Damage

Feeding-Chewing Type. Insects with chewing mouthparts, such as grasshoppers and caterpillars, as a rule cause the most conspicuous damage, because they remove a noticeable portion of the host.

Most readily noticed on plants is the work of forms feeding externally above the ground on leaves, fruits, buds, or twigs. Familiar examples are caterpillars on cabbage, fig. 387; grasshoppers feeding on corn, wheat, soy beans, and other crops; and adult Japanese beetles feeding on foliage and fruits of many trees. Another group includes a variety of small beetles, flies, and moths, whose larvae feed within the leaf tissues, and are called leaf miners. The spinach leaf miner *Pegomyia hyoscyanii* makes an

Fig. 387. Two heads of cabbage from adjoining plats. *A*, sprayed for protection from insects; *B*, not sprayed and badly injured by chewing insects. (From Metcalf and Flint, after Wilson and Gentner)

Fig. 388.   Mine of the aquilegia leaf miner.   (After Frost)

irregular blotch in the leaf; the aquilegia leaf miner, a minute fly, makes a winding serpentine mine, fig. 388.

Roots and underground tubers and bulbs are eaten by larvae of many beetles, flies, and a few moths.   Root feeders may be inconspicuous, but, if they destroy sufficient root material, the plants either fall over or suffer from lack of moisture and nourishment.   Excessive feeding by rootworms (larvae) of *Diabrotica* and related beetle genera causes serious lodging or stunting and death to corn stands, fig. 389.   White grubs, Japanese beetle larvae, and wireworms are other larvae that feed on roots.   A variety of dipterous larvae, or maggots, attack roots; these include species feeding on onions, cabbage, corn, and soy beans.   The maggots tend to feed on and in the larger roots and, when numerous, inflict serious damage or death to the plants.   Larvae of a few species of the fly family Syrphidae attack underground bulbs of tulips and other bulb crops, and a wide variety of beetle larvae attack tubers of Irish potatoes, sweet potatoes, and root crops.

Fig. 389. Serious root damage to corn caused by rootworms of the genus *Diabrotica*. Normal roots at left, damaged ones at right. (After Tate and Bare)

Boring insects are seldom seen, but there are many such species, and they cause considerable damage. Those attacking plants may bore into leaf petioles, branches, trunks, crown, fruits, or roots. Trunk and stem borers of living plants include larvae of clear-wing moths (Aegeriidae), longhorn beetles (Cerambycidae), bark beetles (Scolytidae), and many miller moths. Twig borers and borers in herbaceous plants may remove so much tissue that the plant dies or is greatly weakened, becoming a prey to wind. Borers in trees may girdle the cambium layer, as do bark beetles, fig. 280, causing the death of the host. Others may tunnel through the heartwood and cause injury to the living host, and their tunnels may greatly reduce the value of the wood for lumber, fig. 390. All stages of powder post beetles (Lyctidae) and termites (Isoptera) bore into and eat dead wood. Borers in fruits include some of our worst agricultural pests such as codling moth larvae in apples; plum curculio larvae in plums, peaches, and cherries; and oriental fruit moth larvae in both twigs and fruits of peach and related fruits.

Man and other animals are also attacked by chewing insects. The entire suborder Mallophaga, the chewing lice, live externally on vertebrates, where they feed on skin, feathers, and surface debris. Several fly

Fig. 390.   Injury by the locust borer.   (From the Connecticut Agricultural Experiment Station)

families have species whose maggots live in dead animal carcasses and a few that attack living animals.   The screw-worm fly larva *Callitroga hominivorax* gains entrance into the body of mammals by way of a wound and continues to eat through live tissue under the skin.   Maggots of the fly families, Oestridae and Gasterophilidae, live internally in the bodies of animals—in the stomach, nasal passages, or under the skin.   There is some question as to how these maggots feed, but at least in the full-grown stages they simply suck in gastric contents or secretions produced by the host animal, without any cutting or chewing by the larval mouthparts.

Feeding-Sucking Type.   Insects having piercing-sucking type of mouthparts leave no gaping wounds but sap the vitality of the host.   As these insects feed, they pump saliva into the feeding puncture or wound, and frequently the physiological reaction of the host to this saliva is worse than the effect of the withdrawal of blood or sap.   For instance, horses may be killed by the bites of blackflies, death resulting from a pathologic reaction to the blackfly saliva rather than from loss of blood.

Fig. 391. Hollyhock leaves showing effect of feeding by plant bugs. At left, little feeding; at right, excessive feeding that has caused complete etiolation. (From Illinois Nat. Hist. Survey)

On living plant leaves sucking insects empty the plant cells, removing the green color and causing a whitening or etiolation followed by production of scar tissue. Each feeding puncture results in a tiny white spot, and, when they are extremely numerous, the entire leaf may appear blanched, fig. 391. Frequently curling of the leaf follows heavy feeding, as in fig. 392. On fruits the feeding punctures cause the formation of scar tissue called catfacing, fig. 393. Sucking insects attacking roots or stems rarely produce feeding symptoms other than the reduced vitality or wilting of the host.

Fig. 392. Leaf curl on snowball caused by aphids. Normal foliage at left, infested foliage at right. (From Illinois Nat. Hist. Survey)

Fig. 393. Catfacing of peaches caused by feeding of plant bugs. (From Illinois Nat. Hist. Survey)

Vertebrates suffer from many more species of sucking insects than from chewing insects. Sucking lice, mosquitoes, horseflies, and fleas are examples of large groups that attack vertebrates almost exclusively. Feeding punctures, or "bites," of these insects usually cause a local irritation accompanied by swelling. Individuals react differently, however, so that no general diagnosis of the effect of bites can be given. The actual damage inflicted by these insects is twofold: (1) irritation, loss of blood, pathologic reaction, and loss of condition of the victim, and (2) the possible transmission of certain disease-producing organisms by some of the insect species. This latter is discussed in a later section of this chapter.

Toxins. When feeding, certain phytophagous sucking insects apparently inject a material which is poisonous to its host. These materials, called toxins, usually produce many of the same symptoms as do virus infections and are extremely destructive. Two well-known North American examples, both occurring on potatoes, are the potato or bean leafhopper *Empoasca fabae*, fig. 386, and the potato psyllid *Paratrioza cockerelli*. In this latter species only the nymphs produce the toxin, and a small number of nymphs can destroy a healthy potato plant.

Injury by Oviposition. A few groups of insects damage plants by laying eggs in them, fig. 394. Tree crickets of the genus *Oecanthus* drill rows of egg cavities in raspberry and blackberry canes and in twigs of fruit trees, causing a later splitting or decay of the injured stems. Cicadas (Cicadidae) and treehoppers (Membracidae) cause the same type of

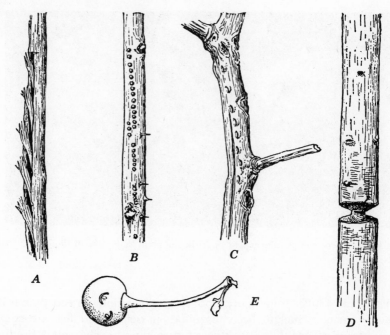

Fig. 394. Injury to plants caused by egg laying. *A*, twig split by periodical cicada; *B*, holes in raspberry cane made by tree cricket; *C*, slits in bark of apple twig made by treehopper; *D*, twig of pecan nearly cut in two by twig girdler; *E*, cherry showing two egg punctures of plum curculio. (After Metcalf and Flint, *Destructive and useful insects,* by permission of McGraw-Hill Book Co.)

injury to many kinds of fruit and shade trees. The feeding of these insects causes little damage, or it may occur entirely on noneconomic herbaceous plants in which the insects do not oviposit. Fruits may be injured and buds stunted by egg punctures.

Spoilage. The damage some insects cause is due to spoiling a product rather than feeding on it. The hop aphid *Phorodon humuli* usually causes little injury to the development of the hops; aphids feeding in the hop cones, however, produce honeydew (feces) which provides a growth medium for molds. This discolors the hops and greatly reduces their market value. Cockroaches in houses and stores drop feces on various merchandise, causing discoloration and sometimes an offensive odor, which reduces the value of the merchandise. On greenhouse and truck crops, insect webbing, aphid honeydew, or frass will often result in drastic reductions in sale value of the crop.

Stings and Other Irritants. There are some protective devices of insects that cause injury or extreme irritation, such as bee and wasp stings, ant bites, and nettling or poison hairs of certain caterpillars. Although very unpleasant and painful, these are only a negligible part of insect injury as a whole.

Transmission of Plant Pathogens

Insects affect certain plants seriously by disseminating disease-producing organisms. In many instances the diseases are much more destructive than the insect injury by feeding. Under these circumstances control of the disease may resolve itself into a problem of very thorough control of the insect, because even a few insects would be able to inflict, indirectly, staggering losses.

A number of plant pathogens, not actually carried by insects, gain entrance to the plant through insect feeding or oviposition punctures. Brown rot of peach commonly enters through feeding punctures of plum curculio adults, and bacterial rot of cotton through feeding and oviposition punctures of various insects.

Insects assist in the dissemination of some plant diseases by transporting them on the body or in the digestive tract. Fire blight bacteria are carried on the legs and body of bees, beetles, and some other insects, as well as by birds and other animals. Spores of certain fungus diseases, such as apple canker, are eaten by insects and pass through the digestive tract in healthy condition. In these cases insects are only one of many ways by which the disease is spread.

More important are cases in which insects are the principal or sole transmitters or vectors of a disease organism from one plant to another. The insects become infected with the pathogen, usually either bacterial or virus, by feeding on an infected plant; some of the disease organisms are injected either mechanically or with the saliva into the tissues of the next plant on which the insects feed. Various species of leafhopper transmit aster yellows, and the beet leafhopper transmits curly top of sugar beets, both virus diseases. Many other virus diseases are transmitted by other insects. Bacterial diseases such as cucurbit wilt disease are carried by insects. In cucurbit wilt the bacteria pass the winter in the digestive tract of the hibernating vectors, the cucumber beetles, which start the next year's infections. Many of these virus and bacterial diseases are exceedingly destructive.

Transmission of Animal Pathogens

Insects are vectors of some of the most important diseases of man and other vertebrates. As with plant diseases, in some cases insect transmission is only one of the several ways by which the disease organism is spread, and in other cases the insect vector is the only known agent by which the disease organism is disseminated from one host individual to another.

In the first category are typhoid fever, summer diarrhea, and some kinds of dysentery, all caused by species of the bacterial genus *Bacillus*. Houseflies get the disease organisms on feet or mouthparts through contact with sewage, saliva, or other infected material and then contaminate food or other items on which they alight later. These disease organisms are transmitted in a variety of mechanical ways, but under some particular conditions flies may be the principal effective method of dispersal.

Bubonic plague (the black death), caused by *Bacillus pestis,* is another contagious disease belonging in this first category. Rats and small mammals serve as the reservoir of the disease, and rat fleas carry the bacteria from rat to rat or from rat to human.

Insects are the sole vectors of several important human diseases. Malaria is caused by species of the protozoan genus *Plasmodium,* transmitted from one person to another by some species of mosquitoes belonging to the genus *Anopheles;* yellow fever and dengue (breakbone fever) are caused by virus organisms carried by several species of mosquitoes of which *Aedes aegypti* is the chief vector in North America; African sleeping sickness is caused by protozoans of the genus *Trypanosoma,* carried by flies of the genus *Glossina;* elephantiasis (filariasis) is caused by nematode worms of the genus *Filaria,* transmitted by several species of mosquitoes. In all these instances the mosquito or fly, when feeding on a diseased person, draws up into its buccal cavity or digestive system some of the disease organisms; some of these are discharged during feeding at a later date into the tissues of another person. In this manner healthy persons are inoculated with the disease.

Typhus is caused by an almost ultramicroscopic organism called *Rickettsia,* which is carried by body lice or cooties. These take up disease organisms when feeding and then later expel them in the feces. Scratching on the part of the bitten person works the disease organism into the skin and effects inoculation.

Ticks and mites are the only known vectors of several important diseases, of which three are of especial interest. Texas fever, lethal disease of cattle, is caused by a species of Sporozoa, *Babesia bigemina.* The disease organisms are transmitted by the cattle tick *Margaropus annulatus.* Rocky Mountain spotted fever, a highly fatal human disease of increasing incidence, caused by a *Rickettsia* organism, is maintained in some of the small wild rodents, and a few species of ticks of the genus *Dermacentor* effect the transfer of the disease organism by feeding on infected rodents during nymphal development and afterwards, when adult, biting man. A third disease is scrub typhus, an oriental disease caused by another *Rickettsia* organism. This is transmitted from wild rodents to man by chiggers (early instars of the mite family Trombidiidae) and was a serious hazard to humans in both the Burma and Pacific theaters during World War II.

Under most circumstances the practical control of these diseases is obtained by control of the vectors. This has been particularly effective in the case of Texas fever; control of the cattle tick has virtually eliminated the disease from the United States. Extremely satisfactory results have been obtained also in reducing outbreaks of typhus by controlling body lice. Mosquitoes, flies, and *Dermacentor* ticks are more difficult to control, and the species involved have a wide dispersal range so that measures aimed at control of these vectors have not always produced such remarkable results as those obtained with Texas fever and typhus.

A SURVEY OF PEST INSECTS AND ARACHNOIDS

In North America some ten thousand different species of insects are of economic importance in varying degrees. Of these about one thousand species are the persistent pests that cause the greater proportion of our insect damage. It is not proposed to give here a detailed discussion of these, but instead to present a brief survey of the most destructive species in relation to the crops or commodities they attack, or the damage they do.

Agricultural Crop Pests

By far the greatest number of important insect pests attack farm crops and animals. Injury is of many types, and insect species of widely different habits are involved.

Field-Crop Insects. All the major field crops, with the possible exception of soy beans, suffer high losses from insect attack. Serious enemies of cotton are the boll weevil, which feeds inside the boll, destroying the developing cotton fiber; the cotton leafworm, whose larvae eat the foliage; and the cotton aphid, which sucks juices from the leaves and stems. Corn may be almost completely destroyed by grasshoppers feeding on the foliage or by chinch bugs sucking the plant juices. The corn yield is annually reduced by several species of borers in cobs and stalks, including the European corn borer and the southwestern corn borer. Wheat and other small grains are injured extensively by various species of cutworms, wireworms, aphids, and grasshoppers, depending on climatic conditions and region. Larvae of the hessian fly attack the stems and crowns of grains; and this species is the most destructive single pest attacking wheat. Field enemies of tobacco, a high cash-value crop, are chiefly leaf feeders, such as hornworms and flea beetle adults; cutworms and tobacco budworms also cause serious damage. The potato beetle feeds on potato foliage, and various leafhoppers suck the plant juices. Potato tubers are injured by soil-inhabiting larvae such as wireworms and flea beetle larvae.

Other field crops are attacked by insects of general feeding habits, such as grasshoppers. Each crop has in addition certain pests more specific in their host preference. The most notable exception was the soy bean crop which for many years had no serious insect enemies in North America. About 1954, however, it was apparent that certain species of grasshoppers, root maggots, and white grubs were becoming pests of importance on this crop.

Truck Crop and Garden Insects. Each plant species grown in truck farm or garden is subject to ravages from one or more insects specific in their food preference. These include such insects as cabbage loopers, cabbage butterflies, and cabbage aphids, that feed on cabbage, cauliflower, and other cruciferous crops; the carrot rust fly, specific on carrots; melonworms, asparagus beetles, and the Mexican bean beetle. In addition to pests specific to each crop, there are many general feeding insects that may attack almost any of these crops. Garden webworms, grasshoppers, blister beetle adults, cutworms, and fall armyworms are among the group most likely to occur occasionally in destructive numbers.

Greenhouse Insects. In the greenhouse, warm humid conditions are maintained throughout the winter months. As a result we find in them many insect species that are normally tropical and subtropical in distribution. Most troublesome of these are several species of thrips, mealybugs, and scale insects. In addition, several species that are outdoor in habit during the summer invade greenhouses and continue active all winter, instead of becoming dormant. The melon aphid, green peach aphid, and the greenhouse leaf tier are examples of this type.

The different kinds of plants grown under glass are legion, and few are not attacked either by general feeding insects like larvae of leaf tiers or by specific pests such as the chrysanthemum midge, whose larvae make galls on leaves and stems, fig. 395. Normally forty to fifty species of potentially destructive insect species are found during the winter season in greenhouses. When one considers the variety of hosts involved and the fact that these may all occur in a range of glass of only a few thousand square feet, it poses a serious control problem and demands constant alertness on the part of the operator.

Fig. 395. Injury caused by the chrysanthemum midge *Diarthronomyia hypogaea.*
About natural size. (From Metcalf and
Flint, *Destructive and useful insects,* by
permission of McGraw-Hill Book Co.)

Fruit Insects. All classes of fruit—citrus, deciduous, and small—
suffer heavily from insect damage, and in each group the major pests are
different.

Citrus-fruit trees are injured mostly by scale insects, mealybugs, whiteflies, thrips, and mites. The purple scale, California red scale, and black
scale are especially important, damaging fruit and trees or producing
honeydew on which grows a black sooty fungus that discolors the fruit.
Many of the scales on citrus have a wide host range, but, being subtropical
in distribution, they are pests of other fruits only in the citrus belt, in
Florida, southern Texas, and southern California.

Deciduous fruits, including apple, pear, cherry, peach, plum, and their
allies, have many destructive pests. Apple fruit is attacked chiefly by
larvae of the codling moth. This insect is the most important species on
the apple control calendar. Peaches, cherries, and other soft fruits are
entered by larvae of the plum curculio, which also attacks apples. The
branches and foliage of the entire group suffer from San Jose scale,
oriental fruit moth, aphids, red spiders, a host of leaf-feeding insects, and
many that bore in the tree or deform the fruit.

Small fruits are a group of wide taxonomic composition and have
more specific insect pests. Grapes are attacked by the grape berry moth,
many aphids and leafhoppers, and leaf-eating beetles that eat roots and
leaves. Currants, raspberries, and strawberries are attacked by a variety
of aphids, leaf-feeding larvae, and stem or crown borers.

Insects Affecting Man and Domestic Animals

Both man and domestic animals suffer annoyance and exposure to disease from the activities of insects. Certain of these insects, such as the Anoplura, confine their attacks to one or two closely related species of animals. Others, such as mosquitoes, are general feeders on a wide variety of warm-blooded vertebrates.

Domestic fowl are attacked chiefly by Mallophaga (chewing lice) and mites, several of which live on hens, ducks, turkeys, and geese. On young fowl infestations of lice often cause death; on older birds the lice cause lack of condition and lower egg production. Mites sometimes become very injurious by reducing the general health of the flock. Blackflies are vectors of at least one duck disease similar in many respects to malaria. Hens suffer also from attacks of specific fleas, of which the southern stick-tight flea is the most persistent.

Domestic animals and man have a variety of specific parasites, including Anoplura (sucking lice), fleas, bedbugs, a few Mallophaga, and several kinds of mites. These latter include such annoying forms as itch mites, chiggers, and ticks. Sheep have in addition "sheepticks"; these are odd wingless flies of the family Hippoboscidae. Attacks by ectoparasites result in irritation and loss of condition, but seldom in death unless disease transmission is involved.

Several vertebrates are attacked internally by larvae of bot flies and warble flies. In the horse the larvae attach to various regions of the digestive tract and cause severe loss of weight and condition. Certain warble flies develop in the sinuses of sheep, and other species along the back of cattle where they form a pocket just beneath the hide. Larvae of the screw-worm fly enter wounds, feed beneath the skin, and annually cause large losses to all kinds of livestock.

In addition to these and other specific pests, all warm-blooded vertebrates are attacked by a great number of bloodsucking flies—mosquitoes, horseflies, blackflies, *Symphoromyia* flies, stable flies, and horn flies. Some of the fleas, ticks, mites, and bedbugs are also general feeders. The annoyance these cause is often severe. Blackflies especially may be destructive and occasionally cause the death of large numbers of horses and mules in local areas. Mosquitoes, blackflies, and horseflies are at times abundant enough to cause an exodus of tourists from an area, to reduce land values near suburban centers, or to retard settling of large tracts, as in the extreme northern part of Canada and in Alaska. The effect of these attacks on livestock in general may result in a loss of condition equal to or greater than that caused by specific parasites.

The greatest potential injury to man and animals by insects is through insect-borne diseases. As previously mentioned, insects transmit the pathogens of some of the most destructive diseases of vertebrates. During war the danger from insect-borne diseases is greatly increased, because men are concentrated under conditions in which sanitation and insect control may be difficult, and the crowding offers good opportunities for the rapid spread of disease organisms.

Stored-Food-Products Pests

Grain and meat, flour, grain meals, and other highly nutritious foodstuffs are eaten by many insects. When in storage, these commodities suffer a heavy loss from insect ravages and necessitate constant preventive and remedial measures to keep them to a minimum.

In North America the chief pests of stored grains and grain products are the adults and larvae of the sawtooth grain beetle, the confused flour beetle, cadelle, mealworms, and the granary and rice weevils; and the larvae of the Indian meal moth and the Mediterranean flour moth. Peas and beans in storage are eaten by various pea weevils (Bruchidae). Meats and cheeses are eaten by larder beetles and maggots of the cheese skipper.

Large quantities of stored foods are attacked first by the group of insects just listed. After a certain amount of damage is done molds enter, followed rapidly by a host of other insect species, and soon the entire mass of food may be reduced to a small percentage of the original.

Pests of Human Habitations

Some insect species have become almost "domesticated," especially north of the frost line, in that they are found almost entirely in human habitations. In the case of ectoparasitic species the relationship antedates civilization and is due to the parasites staying with the warm-blooded host. With other species, however, the relationship is more recent and is due to the relatively high temperatures at which houses and buildings are maintained even through severe winters. Thus some species, originally semitropical, are now found much farther north and are able to maintain themselves in human habitations.

Ectoparasites and pests of stored foods are of prime importance in human habitations. In addition, larvae of clothes moths and carpet beetles eat anything containing animal fibers, such as woolen garments, upholstery, and carpets. Silverfish and cockroaches are general feeders that eat starchy foods such as bookbindings and are an unsightly nuisance. Cockroaches drop excrement promiscuously and spot and taint food and quarters; when very abundant, they will give a house, store, or restaurant a disagreeable and penetrating odor. Ants frequently invade buildings and may become a serious nuisance in the kitchen and food-storage rooms.

Termites are the most destructive pests of buildings. They eat the wood in foundations, flooring, and walls, necessitating extensive repairs. Other insects live in wood in dwellings, such as Lyctidae beetles, and carpenter ants may eat out extensive galleries in wood of buildings to use for nests. But of all insects that attack the actual building, termites are by far the most formidable.

Shade Tree and Forest Insects

Trees in general support thousands of insect species, which may defoliate, girdle, or bore into the tree, or suck its juices. Many of these species have only a slight effect on the host tree, but some damage the tree severely or may even kill it. As a result there is a high annual loss in both shade and forest trees.

Shade trees in the northeastern states are attacked especially by the gypsy and brown-tail moths. Elms suffer most from the elm leaf beetle and from Dutch elm disease, carried from tree to tree by the small European elm bark beetle. Direct injury by bark beetles and wood borers weakens and kills trees of many species.

Fig. 396. Distribution of the European spruce sawfly in North America during the epidemic of 1938. (Modified after Balch)

Forest trees are visited periodically with insect outbreaks that kill huge tracts of timber. This is a loss of natural resources that in past years was given little attention, but, now that our forests are dwindling, increased efforts are being made to find means of checking losses. Larvae of forest tent caterpillars, gypsy and brown-tail moths, hemlock loopers, budworms, and tip moths are perennial defoliators of various deciduous and evergreen trees. Bark beetles are the greatest single enemy of conifers, especially in the West. Sawflies feeding on conifers occasionally appear in outbreak numbers and may cause tremendous damage. The most recent sawfly outbreak was of the introduced European spruce sawfly, which in 1938 defoliated about twelve thousand square miles of spruce timber, chiefly in the eastern provinces of Canada, fig. 396.

NATURAL vs. ARTIFICIAL CONTROL

From time to time insect pests of many kinds are reduced to insignificant numbers by inimical factors of the environment, such as drought, parasites, or disease, as discussed in Chapter 8. The European spruce sawfly, for instance, has been virtually exterminated in large areas by a virus disease. In 1935 the chinch bug was reduced to the status of a rarity in many corn-belt states by adverse winter conditions plus a fungus disease. But these phases of natural control are unpredictable and may be at a low ebb for long periods. Furthermore some of our worst pests, such as the codling moth, apparently have at most only partial or insignificant natural checks and so are a menace every year.

In order to protect his interests, it has therefore been necessary for man to devise means of combating insects by his own efforts. This type of control is called *artificial control,* in contrast to the natural control effected by the unaided environment.

CONTROL METHODS

In artificial control a great many different methods have been found to reduce the numbers of individual pests. These methods fall into a few general categories and are treated briefly in the following paragraphs.

Quarantine

The most obvious way to avoid damage by an insect is to prevent its becoming established in a country if it is not already there. There are

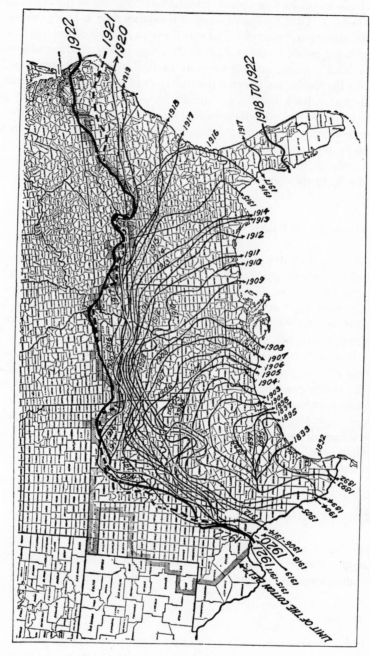

Fig. 397.   The spread of the cotton boll weevil in the United States, up to 1922.   (From U.S.D.A., E.R.B.)

hundreds of insects in other parts of the world, especially in temperate areas of Europe and Asia, which we believe might become pests of great economic importance if established in North America. To prevent their entrance, the United States Federal Government maintains an inspection of imports into the country, especially living plants or animals or packing material that is likely to harbor pests and serve as a carrier for them. Most or all of this material is fumigated before being allowed into the country. In addition, states may have restrictive regulations regarding the movement of critical materials within the state or into the state. The Canadian Government maintains a similar service.

It is admittedly impossible to prevent indefinitely the entrance of all potential new pests into the country, but quarantine records show that hundreds and sometimes thousands of new importations are prevented every year. It is impossible also to estimate how much we gain by this. Experience with such destructive importations as the cotton boll weevil, fig. 397, the European corn borer, and the Japanese beetle, however, emphasizes that we cannot afford to take the chance of allowing free entry to every insect species.

Biological Control

Present types of biological control fall into three categories—use of insect parasites or predators, use of pathogenic organisms, and reproductivity control.

Insect Parasites and Predators. The possibility of propagating and distributing natural enemies for the control of destructive insects has kindled the imagination of the entomologist for many decades. It has been found, however, that with destructive insects endemic to the United States we can do little to improve on existing natural control. Representing the evolutionary product of great geologic time, natural control has usually reached a peak that cannot be raised profitably by artificial means.

With introduced pests the situation is entirely different. The particular species may have an abundance of parasites or predators holding its numbers in check in its native land. When it is accidentally introduced into another country, usually only the pest without its parasites is transported. Freed from enemies, the pest in the new land is able to flourish at an unimpeded rate.

The ideal control for such an introduced pest would be to establish efficient enemies of it so that they would reduce the numbers of the pest to insignificant proportions. This might result in a permanent control that would obviate the necessity for an annual program of more expensive measures.

This ideal has been achieved only rarely. The most outstanding example has been the control of introduced cottony-cushion scale by the importation and establishment of the Australian vedalia ladybird beetle. So effective are the beetle and its larvae in controlling the scale in California that only occasionally and locally does the scale become important as a pest. Many parasites, especially of introduced pests such as the Japanese beetle, gypsy moth, and European corn borer, are imported by the U. S. Bureau of Entomology and Plant Quarantine, and released in the United States. Many imported parasites fail to maintain themselves in the United States under natural conditions, owing undoubtedly to their lack of adjustment to climate or the lack of availability of suitable hosts at the right time. Some species have become successfully established and aid in controlling the pest species. It is hoped that eventually sufficient parasite populations will be built up so that the populations of many pest species will drop well below their present destructive level. In some areas this result has already been achieved for the satin moth by introduced hymenopterous parasites, especially in Washington State and British Columbia. Propagation and dispersal of bacterial diseases of Japanese beetles have also given promise of being effective. Considerable success has been achieved in islands such as Hawaii by using a variety of parasites against insects attacking many crops.

Sufficient work has been done in biological control to show that a number of factors influence its success or failure. A few of these factors are the ecological requirements of the parasites, their effect on each other (see under COMPETITION, p. 472), their host specificity, their rate of increase, and the character of their dispersal. To be tried effectively, well-trained personnel and a great amount of specialized equipment are necessary, together with an organization for gathering parasite material in foreign countries and getting it into the United States alive and healthy.

Because of these conditions, the work on biological control in the United States is done chiefly by the federal government; a notable exception is much intensive work done by the state of California. The final distribution and liberation of parasites are often performed cooperatively by scientists of the federal government and interested state agencies. The Canadian Government is also extremely active in biological-control efforts.

Pathogenic Organisms. Spectacular natural outbreaks of fungus or bacterial diseases of chinch bugs and various other insects early led entomologists to hope for insect control through dissemination of pathogenic organisms. Early efforts failed, because years which did not naturally favor spread of the known diseases were ecologically unsuited also for their artificial propagation. Since about 1930 several organisms have been found that are more successful for artificial control. The milky disease of the Japanese beetle has proved useful in some parts of the American range of the beetle. Several kinds of polyhedrosis viruses are proving excellent controls for certain sawflies destructive to conifers. In California both a polyhedrosis virus and the Thuringian bacterium have proved effective as a field control for the alfalfa caterpillar. These successes point to a profitable field of investigation in insect control.

Reproductivity Control. During the 1950's an entirely new kind of biological control was invented by entomologists in the U. S. Department of Agriculture. The target species was the destructive screw-worm fly. The method consisted of two steps: (1) rendering males sterile but still active and sexually aggressive through exposure of fly pupae to gamma radiation and (2) rearing and releasing such males in much greater numbers than existed in the natural population. Females mated by these males received abnormal spermatozoa at copulation and the eggs laid were unfertilized and inviable. Release of sterile males continued until the species was annihilated locally. By this method the screw-worm was eliminated from the 170-square-mile Island of Curacao in 1955, and from Florida and the southeastern states in 1958. This method has one great disadvantage: it is extremely expensive and not adapted to use over large geographic areas.

A group of organic compounds called chemosterilants also produce varying degrees of sterilization in insects and, used as field sprays, may offer a possibility of producing sterilized males in sufficient numbers to swamp out normal males over large areas.

A different twist to the same problem involves efforts to manipulate the sex-determining apparatus of a species so that few females are produced. Genetic strains producing such results have been isolated in mosquitoes and described in certain moths. How these genetic strains can be used effectively and maintained in field control operations presents a real challenge.

Cultural and Management Control

Some insect pests of agricultural or forest crops may be kept below the damage level by various cultural or management practices.

An important general approach is keeping crops healthy by proper fertilizing, drainage, irrigation, and cultivation, and by planting crops that are well adjusted physiologically to the climate and soil.

Against certain pests specific cultural methods are of value, such as clean cultivation, crop rotation, certain times of harvesting or planting, and the use of insect-resistant or tolerant varieties.

Clean cultivation eliminates weeds that may serve as host to insects that attack the crop. The buffalo treehopper breeds on many herbaceous weeds; the adult hoppers fly into adjacent fruit trees, cut slits in the twigs, and in them lay their eggs. Clean cultivation of an orchard prevents this injury by eliminating the primary host. Weeds and soil debris also serve as hibernating or pupating quarters for a wide variety of harmful insects, and clean cultivation tends to discourage a build-up of population in that area.

Crop rotation has been found especially effective against some insects whose larvae feed on roots. Some species of *Diabrotica* rootworms can be controlled by crop rotation. These beetle larvae feed primarily on corn roots. If corn is grown continuously on the same ground for over three years in localities favoring these insects, they build up large populations and cause severe damage to corn. If, however, corn is eliminated and wheat or legumes substituted for a year, the rootworms starve. For this reason a rotation of corn with wheat or other crops arranged so that corn follows corn for no more than 2 or 3 years eliminates rootworm damage almost completely.

Choice of time of planting crops is useful as a control measure for certain insects. The hessian fly, a serious wheat pest whose larva feed in the leaf sheath of wheat, can be successfully controlled by regulating the time of planting winter wheat. The entire fall generation of adult midges normally emerges within a short period, following late summer rains. The adults live only 3 or 4 days, laying their eggs in grooves of wheat leaves. If winter wheat is planted after this generation is past, the plants will have no eggs laid on them and consequently will be entirely free from attack until spring. To take advantage of these conditions entomologists in wheat-growing areas annually establish dates for sowing winter wheat that will (1) allow the plants enough good weather to attain satisfactory growth before winter, and yet (2) be late enough to avoid all but a light infestation of hessian flies, fig. 398. The spring generation of hessian flies attacks only late shoots (tillers) of winter wheat and does little damage.

Early or late planting of corn and other crops is sometimes of assistance in reducing infestation and damage by such pests as rootworms and European corn borers.

In forestry, management practices that tend to keep down the abun-

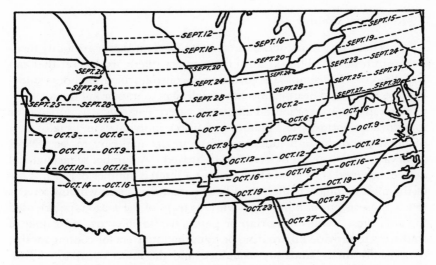

Fig. 398. Sample of a chart showing safe dates for sowing wheat in several north central and eastern states to escape injury by the hessian fly. (From U.S.D.A., E.R.B.)

dance of insect pests are essentially various ways of changing the ecological nature of the forest community. It has been found that some of the most destructive species of bark beetles build up outbreak populations in over-age stands of pine. By cutting the trees for timber before they reach old age, this beetle population increase is prevented, and younger trees in the stand are given a better chance of development. By selective cutting and logging, diversity of the forest can be achieved in regard to both species of trees and age groups represented by these trees. In general, the greater the diversity, the fewer insect outbreaks will occur.

Another important phase of cultural control is the selection of species or varieties of plants that are resistant to or tolerant of insect attack. This is as yet the most successful general method for preventing damage by the European corn borer. It has been found that certain varieties of corn have an unusually strong stalk that will withstand heavy infestations of corn borer larvae without becoming critically weakened. Less tolerant varieties having the same borer infestation break off and lodge, resulting in lower yield and greater difficulties in harvesting the crop.

This same crop-selection principle is used as a remedy for the gypsy moth. We have already mentioned that, for its successful early development, one of the species' favored hosts is necessary, but that later instars move over to pines and other nonfavored species, which they may defoliate. To decrease infestation by this pest, the favored host species are kept to a minimum and replanting is done with other species.

Mechanical Control

If it is necessary to use more direct methods than those previously mentioned in order to obtain control of an insect, there are several that can be followed. The simplest is mechancial control, which includes removing insects by hand or using mechanical devices to trap or kill them.

Hand picking is practiced on large caterpillars such as tobacco or tomato hornworms. The number of insect individuals is usually only moderate, and the individuals are large in size and easy to see. Nests of larvae can be cut out of trees and destroyed. A number of mechanical devices are used with good effect against a limited number of pests. One of the most common is the use of both screen doors and window screens to keep insects out of buildings. Various traps of the maze type are used to catch flies. Bands of burlap or paper are fastened around trunks of fruit trees to provide hibernating or pupating quarters for codling moths; periodically these bands are examined and the insect occupants killed. Bands of screen, gauze, or sticky substances are put around trees to prevent ascent of wingless female moths and larvae, fig. 399.

Fig. 399. Banding traps used to prevent ascent of larvae and wingless female moths. (From U.S.D.A., E.R.B.)

Against migrating wingless insects, such as Mormon crickets or chinch bugs, attacking field crops, various mechanical barriers were formerly used, especially furrows in the soil or wooden or paper barriers.

Physical Control

We have noted previously that insects can endure only limited extremes of heat, cold, and other physical phenomena. This limited endurance is utilized to kill insect pests. It is difficult to control such physical factors over a large space, so that with a few exceptions their use is restricted to buildings and tight enclosures.

Superheating is employed by many mills and elevators as a control measure. During hot weather in summer, the building heating plant is used to raise the temperature to about 140°F. for several hours, and this kills all the insects in the building. Heat transmitted by irradiation is being used to control insects in stored products.

Cooling is used extensively in storage for insect control. Furs, tapestries, and other valuable articles of animal origin are kept in lockers below 40°F. This does not kill all the insects, but at this temperature they are completely inactive and do no damage.

Electricity is used to some extent to kill insects. Screens and lights can be fitted with electrically charged grills that electrocute insects coming between the elements.

Limited control of some insect species has been obtained using light traps to attract and kill adults. Continued research may show this method to be practical for controlling certain pests, especially Lepidoptera.

Chemical Control

Various chemical compounds are toxic or repellent to insects and are used extensively for their control. Because it is usually more expensive, such chemical control is applied when control by other methods is too slow or too ineffective.

Types of Chemicals. Insecticides may be divided into seven categories: stomach poisons, contact poisons, general purpose, systemics, fumigants, repellents, and attractants.

STOMACH POISONS are substances that kill the insect when they are eaten and taken into the digestive tract. The most widely used are the arsenicals, especially Paris green, lead arsenate, and calcium arsenate. Hellebore and fluorine compounds such as sodium fluoride and sodium fluosilicate are also used.

Stomach poisons are used primarily against insects that have chewing mouthparts and bite off and swallow portions of the food. The poison is applied to the surface of the food as a spray, dip, or dust, and the insect is sure to take some of it with its meal. Against some insects such as ants and cockroaches, the poison is mixed with an attractive bait put where the insect can find and eat it. For ants the baits are usually liquid and are set near nests or runways.

CONTACT POISONS kill the insects by contact without being swallowed. Often the lethal agent is a gas that enters the spiracles and causes suffocation, as in the case of nicotine; in other instances the lethal compound may affect the nervous system.

Until about 1940, the principal contact poisons in use against insects were nicotine alkaloid (volatile), extracted from tobacco; pyrethrum, extracted from the dried flower heads of certain species of the aster genus *Chrysanthemum;* sulphur and several sulphur compounds; and several lubricating oils, miscible oils, and oil emulsions. Since 1940, there have been developed a number of synthetic organic compounds (discussed in general purpose compounds, below) that are far more toxic to many insects than the contact poisons of older vintage.

Contact poisons are of especial use against insects having sucking mouthparts, such as aphids, which do not take up poisons applied to the food surface; insects that cannot be reached when feeding, such as bottom-feeding mosquito larvae which are killed by contact poisons when they come to the surface for oxygen; and insects such as adult mosquitoes that are scattered generally throughout an area. Contact poisons toxic to a variety of insects have the advantage of killing both the sucking and chewing insects at the same time. These poisons are applied as a dip, dust, spray, or colloidal mist (aerosol).

GENERAL PURPOSE COMPOUNDS include an array of substances which act either as a stomach poison or contact poison, depending on the manner in which the insect encounters them. These include chlorinated hydrocarbons, organic phosphates, carbamates, thiocyanates, and sulphur compounds. Commonly used are DDT, DDVP, chlordane, heptachlor, aldrin, dieldrin, endrin, methoxychlor, toxaphene, parathion, malathion, lindane, naled, and carbaryl. Each of these has advantages under specific conditions; information is readily available from local entomological centers.

SYSTEMICS are poisons which are absorbed by the host and translocated to various parts of the organism, from which they may be taken up by the feeding insect. This phenomenon was recognized early with regard to selenium. Ornamental plants grown on soil containing selenium absorb the latter, which kills mites and insects feeding on the plants; this selenium method cannot be used on edible crops, however, because of the extreme toxicity of selenium to humans and livestock. Potatoes sprayed with Bordeaux mixture (a copper compound) and bean plants sprayed with derris show this phenomenon. More recently special insecticides have been developed using this principle. Systox, disyston, phorate, and others being developed are used on various plants. Systemics such as coumaphos and ronnel are being used on livestock. Systemics may have a great future in insect control because they afford predators of pest species a better chance of survival.

FUMIGANTS are toxic gases, usually applied in an inclosure such as a box, box car, drum, bin, building, or tent. Compounds in general use are hydrocyanic acid, paradichlorbenzene (PDB), methyl bromide, ethylene dichloride, carbon tetrachloride, carbon disulfide, and chloropicrin. The use of fumigants is aimed at killing all the insects in the inclosure, and fumigants are employed commonly to rid houses, greenhouses, warehouses, stores, mills, and grain elevators of insects. Box fumigation is used for small quantities of material, such as clothing. In World War II fumigation stations were used extensively to rid clothing of lice and other vermin. Prior to fumigation buildings need to be checked for gas leaks and the leaks closed in order to maintain the desired gas concentration as long as needed.

These fumigants are dangerous to humans, and fumigation jobs must be handled with care during the application and ventilated properly when the treatment is completed. Hydrocyanic acid is deadly to humans and especially dangerous because it is nearly odorless. Chloropicrin, even in minute quantities, is distressing and virtually incapacitating to humans because of its penetrating odor and irritating properties.

REPELLENTS are used to keep insects away from something and are not necessarily toxic. Naphthalene and camphor have been used for decades in homes for keeping insects out of stored clothing. Creosote is used to keep termites out of wood.

Of many compounds that have been tested as repellents of mosquitoes, other biting flies, ticks, and chiggers, the most effective are dimethyl phthalate, Indalone, DET and an organic compound known as Rutgers 612. Each of these compounds is effective against only a certain number of species, but various mixtures of them applied to clothing and skin give fair protection to the user for a short period. Dimethyl phthalate and other repellents act primarily as a killing agent on mites rather than as true repellents.   Benzyl benzoate is proving especially effective for giving protection against mites.

ATTRACTANTS are used in traps of various types to catch insects.   Chief compounds include sex attractants extracted from the insects themselves, sugars, and a wide variety of aromatic or volatile compounds.   Attractants are not used as a direct control measure but as a tool in sampling field populations.   The samples are then used in forecasting the time and extent of insect occurrences. In the future, attractants may provide a basis for actual control programs.

Injury and Residues.   Insecticides must be used with caution, because they may damage the host as well as the insects, or leave a residue that is toxic to man or other animals.   Some plants are highly sensitive to one insecticide but not to others.   Lists outlining these "do's" and "don't's" can be obtained from local extension offices.   For each insecticide there are legal limits to the amount of residue that is permitted on the marketed product.   These limits must be taken into consideration in planning the control program for each crop.

An important characteristic of insecticides is the length of time they remain toxic after being applied.   Most stomach poisons remain toxic for long periods and are spoken of as having a high residual action.   Nicotine alkaloid and pyrethrum have practically no residual action, losing their potency almost immediately after they are applied, owing to chemical deterioration.   Some of the synthetic contact poisons, such as DDT, have a high residual action; others, such as some of the organic phosphates, have a short residual life.   The length of residual action governs to a large extent the frequency with which the insecticide must be reapplied.

Plant Conditioning.   We have mentioned that certain sprays such as Bordeaux mixture induce a poisoning of plant sap that is toxic to insects. There are other cases of somewhat similar results that are thought to be due to induced physiological change of the plant, such as plant acidity or alkalinity.   An example is the reaction of certain arsenicals on cotton. After being treated with acid calcium arsenate, cotton is unusually susceptible to the cotton aphid, which multiplies in great numbers.   On the other hand, cotton treated with basic copper arsenate normally suffers no more than ordinary attack by the cotton aphid.   The nature of this plant conditioning is not fully understood, but it promises to be an interesting and profitable field for investigation.

Application.   Putting or getting an insecticide where it will do the most good presents problems of many kinds.   First, the insecticide must be applied at the correct season and in some cases (as with mosquito

Fig. 400. Application of insecticides; the use of airplane equipment has been of great aid in many situations. (From Ohio Agricultural Experiment Station)

larvae) within a period of only a few days. In the second place, weather conditions must be considered, because sprays and dusts cannot be applied during rains or in high winds, and some crops are more susceptible to insecticide burning during periods of high temperature and humidity.

A specialized set of machinery is available for applying sprays, dusts, and aerosols, for dipping cattle, or administering fumigation materials. A careful choice of these must be made for each control project, the area and local condition being taken into consideration, such as topography, height and spacing of the crop, and labor conditions.

Insecticide Resistance. Continued exposure to chemical insecticides has led to the selection of insecticide-resistant strains or populations in about 140 species of insects, mites, and their allies. The resistance is primarily to organic insecticides such as DDT and organophosphates. In each group of insects the physiological and genetic mechanisms of resistance may differ; in the same species these mechanisms may differ in populations occurring in different areas. Typical species in which resistance has been observed include mosquitoes, cooties, foliage mites, aphids, weevils, flies, bugs, and others. The appearance of resistant strains in pest species necessitates considerable ingenuity on the part of the entomologist in devising new control schedules using insecticides to which the target species is susceptible.

Integrated Control. Organic chemical insecticides are toxic to most of the arthropod members of an ecosystem. When applied for the control of a particular pest, these chemicals may kill biological control agents of the target pest or of other potential pests, resulting in either a rapid resurgence of the target pest or excessive population expansion of other pests previously below economic levels. In attempts to remedy these unsatisfactory features, control programs have been devised whose aim is to combine the maximum benefits through chemical control with the minimum disturbance of the ecosystem. Devices utilized in these programs include (1) timing chemical treatments to maximize the reduction of pest species and minimize losses to parasites and predators, (2) treating restricted areas in which pest populations reach high proportions, (3) choosing chemicals with short residual action, (4) reducing dosages of high toxicity chemicals, and (5) treating alternate strips of the crop, thus leaving islands or strips of untreated sections from which parasites and predators can spread readily into adjacent treated areas. Few rules of general application can be given. In any situation, detailed knowledge must be secured of the biologies of the pest and its enemies, and other potential pests in the ecosystem. A program must then be devised taking all factors into account.

CONTROL PROBLEMS

In reviewing the question of insect control there are some pertinent general considerations that must be borne in mind.

Cost. We have already mentioned that biological control programs are carried on and financed by state or federal agencies. There are other control projects of such great magnitude and such significance nationally that they are planned and financed by these same agencies. For example, about 1929 the Mediterranean fruit fly became established in Florida and was recognized as a pest that might ruin the American citrus fruit industry. Immediately a federal government project was initiated to attempt the eradication of the pest. Quarantines were set up, cleanup measures enforced, thousands of tons of suspected fruit destroyed, and exhaustive surveys made. The fly colony apparently was completely extirpated, and the nation bore the cost. Periodic grasshopper outbreaks threaten to consume all growing crops in entire states; here the federal government assists the farmers by supplying materials and machinery to fight a menace national in scope.

But to the householder with moths in the closet, the farmer with his usual array of insect enemies, the mill operator with bugs in his products, in short, to everyone faced with the necessity of controlling insects by his own efforts, cost is a paramount consideration. The control cost must be low enough to allow the control application to be profitable. If, for instance, an insect threatened to reduce the yield of corn 10 bushels per acre, and a control program would avert 80 per cent of this loss, control cost per acre would have to be less than the price of 8 bushels of corn. Otherwise control would not be attempted, because, if it were, either the farmer would break even on the deal and be out the extra work involved, or he would lose money. The same principle holds with all control done by private means.

In devising control methods, therefore, the entomologist must always strive for practical ones from the cost standpoint. With low-priced crops such as field crops, which seldom have a value of more than $300 per acre, the premium is definitely on low-cost control even at a sacrifice of some efficiency in control obtained. In the case of greenhouse crops, the cash value of the product may be $10,000 or more per acre of glass, and the market price may drop disastrously with only a small insect infestation. Here the demand is for perfect control even at a high price.

The use for a crop may influence control considerations. For example, a small infestation of borers is negligible in field corn that will be harvested as ripe grain. Such an infestation would be disastrous in corn, pears, tomatoes, and other crops that are to be canned. Canned goods must be free of insects in order to satisfy government regulations for interstate shipment. Because poor insect control could result in the condemnation of an entire pack from a cannery if insect parts are found in the product, the most rigid insect control is an economic necessity.

The Weakest Link. In order to achieve most in both efficiency and economy, it is necessary to apply control measures at that point in the life history when the insect is most vulnerable or control is most practical. In the life history of many insects there is a point at which the insect may be reached easily by control applications. The cabbageworm, for example, is vulnerable at any time during its larval stage to poisons applied to its host. In the codling moth this vulnerable period is much shorter, being the interval of larval life between hatching and entrance of the young larva into an apple. Here the control is applied. An insecticide with a high residual action, such as lead arsenate or DDT, is applied that will cover the late blossoms or fruit. The young larvae will be caught in their attempt to enter an apple, either by eating a little arsenate as they bite through the apple skin, or by contact with the DDT.

During chinch bug outbreaks, it is impractical, because of expense, to apply an insecticide to all the acreage of grass or grain that harbors the bugs. When, however, the bugs migrate from these crops to corn, it is practical to put a repellent barrier or a strip of insecticidal material along a line that the bugs must cross to reach a new food supply. In this way a small strip of applied insecticide is an effective control of millions of bugs on the march.

Community Projects. In the case of some insects it does one person little good to effect control on his premises if the neighbors for some distance around fail to do the same. The ox-warble or cattle grub is easily controlled by squeezing the full-grown larvae out of the pockets they make beneath the hide of the cow and killing them, even though this method of control does little for the immediate season, since the damage is already done. But, if the neighbors don't do it, warble flies from their cattle will fly over and reinfest those of a person who has attempted control. If done thoroughly over a large section of country, the destruction of this year's warbles will prevent their recurrence next year. The same is true of the control of many species of pest mosquitoes; coordinated control over a large area is usually necessary for relief. In this case, however, the control need is annual.

Dispersal of Information. To be successful, insect control must be based on a detailed knowledge of the insects, local conditions, and possible control methods; and any one of these can change in many details from year to year. Most of the control, moreover, must be done by individuals who are not specially trained in the work and who do not have this information readily available. To put the known information into the hands of these people is the field of *extension*.

Extension translates the findings of the research entomologist into specific control recommendations for the farmer, forester, home-owner, and others. Extension in entomology is most highly developed in relation to farm operations. Successful extension operations are based on a detailed knowledge of the potential insect pests of the area, their life histories and habits, and up-to-date control measures to be used against them. Armed with this knowledge, regular surveys (preferably weekly) are made along transects across the various crop plantings of the area, and the local stage and abundance of each pest is mapped. From this information the extension entomologist can formulate immediate control instructions or warnings of probable future insect outbreaks, fig. 401. To reach the maximum number of interested farmers, all possible communication media are used—bulletins and circulars by direct mail, newspaper releases, and radio and television announcements. Greater detail in control instruction is accomplished by special schools, lectures, and demonstrations timed for periods when activities on the farms are at low ebb. County agents and other farm advisory groups work closely with the entomologist and are extremely helpful in all phases of agricultural extension.

Surveying and forecasting are being developed rapidly in other economic areas also, especially for forest insects. Forecasts of the abundance

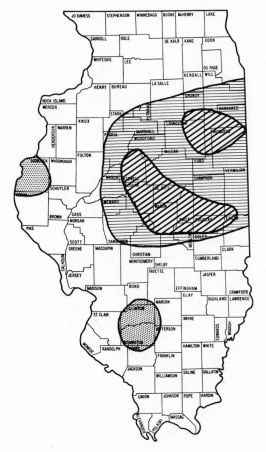

Fig. 401. Forecast of danger areas for chinch bug outbreaks in Illinois in 1963, based on survey of hibernating insects. *Unmarked areas,* below economic levels; *dotted areas,* light losses probable; *horizontally lined areas,* light to medium losses probable; *diagonally lined areas,* heavy losses probable. Maps of this type are the basis for much extension information. (From *The Illinois Natural History Survey Reports.*)

of insects biting humans, such as mosquitoes and deer flies, are used by mosquito abatement districts and by organizations interested in attracting tourists.

Perhaps the most difficult job of extension is to place needed information in the hands of the private gardener and householder who may have aphids on the roses, ants in the kitchen, or termites in the house. Efforts in this direction include chiefly a scatter-shot distribution by mail, press or over the air, and demonstration clinics for personnel of various types of pest control businesses that have become widespread in the United States.

Extension is a vital phase of insect control. State and federal agencies, private research centers, and industrial corporations cooperate in furthering the work. By these efforts it is hoped ultimately that in the United States anyone with an insect-control problem will be able to find the best known way to handle it.

REFERENCES

Ann. Rev. Entomol., 1–9(1956–1964). Articles about many topics discussed in this chapter.

Armitage, H. M., 1954. Insect eradication procedures in incipient infestations. *J. Econ. Entomol.,* **47**:6–12.

Balch, R. E., 1952. The spruce budworm and aerial forest spraying. *Can. Geog. J.,* **45**:201–209.

Baumhover, A. H., et al., 1955. Screw-worm control through release of sterilized flies. *J. Econ. Entomol.,* **48**:462–466.

Bigger, J. H., G. C. Decker, J. M. Wright, and H. B. Petty, 1947. Insecticides to control the European corn borer in field corn. *J. Econ. Entomol.,* **40**:401–407.

Black, L. M., 1953. Transmission of plant viruses by cicadellids. *Advances in Virus Research,* **1**:69–89.

Bořkovec, A. B., 1962. Sexual sterilization of insects by chemicals. *Science,* **137**:1034–1037.

Brown, A. W. A., 1951. *Insect control by chemicals.* New York: John Wiley & Sons. 817 pp.

Brown, A. W. A., 1961. The challenge of insecticide resistance. *Bull. Ent. Soc. Amer.,* **7**(1):6–19.

Craighead, F. C., 1950. Insect enemies of eastern forests. *USDA Misc. Publ.,* **657**:1–679 + suppl.

Graham, S. A., 1951. Developing forests resistant to insect injury. *Sci. Monthly,* **73**:235–244.

Gunther, Francis A., and Roger C. Blinn, 1955. *Analysis of insecticides and acaricides.* New York, Interscience Pub. 696 pp.

Herms, William B., 1950. *Medical entomology,* 4th ed. New York, The Macmillan Co. 643 pp.

Horsfall, W. R. 1962. *Medical Entomology.* New York: The Ronald Press Co. 467 pp.

Isley, Dwight, 1937. *Methods of insect control.* Minneapolis: Burgess Publishing Co. Pt. 1, 121 pp.; Pt. 2, 135 pp.

Jones, T. H., R. T. Webber, and P. B. Dowden, 1938. Effectiveness of imported insect enemies of the satin moth. *USDA Circ.,* **459**:1–24.

Knipling, E. F., 1955. Possibilities of insect control or eradication through the use of sexually sterile males. *J. Econ. Entomol.,* **48**:459–462.

Lindquist, A. W., 1955. The use of gamma radiation for control or eradication of the screw-worm. *J. Econ. Entomol.,* **48**:467–469.

Mallis, A., 1954. *Handbook of pest control.* New York: MacNair-Dorland. 1068 pp.

Metcalf, C. L., W. P. Flint, and R. L. Metcalf, 1951. *Destructive and useful insects, their habits and control,* 3rd ed. New York: McGraw-Hill Book Co. 1071 pp.

Painter, R. H., 1952. *Insect resistance in crop plants.* New York: The Macmillan Co. 520 pp.

Peairs, L. M., and R. H. Davidson, 1956. *Insect pests of farm, garden, and orchard,* 5th ed. New York, John Wiley & Sons. 661 pp.

Riley, W. A., and O. Johannsen, 1938. *Medical entomology,* 2nd ed. New York, McGraw-Hill Book Co. 483 pp.

Shotwell, R. L., 1953. The use of sprays to control grasshoppers in fall-seeded wheat in western Kansas. *USDA, BEPQ,* **E-868**:1–14.

Steinhaus, E. A., 1948. Polyhedrosis ("Wilt Disease") of the alfalfa caterpillar. *J. Econ. Entomol.*, **41**:859–865.

1951. Possible use of Bacillus thuringiensis Berliner as an aid in the biological control of the alfalfa caterpillar. *Hilgardia,* **20**:359–381.

Sweetman, H. L., 1936. *The biological control of insects.* Ithaca: Comstock Publishing Co. 461 pp.

Thompson, C. G., and E. A. Steinhaus, 1950. Further tests using a polyhedrosis virus to control the alfalfa caterpillar. *Hilgardia,* **19**:411–445.

van den Bosch, R., and V. M. Stern, 1962. The integration of chemical and biological control of arthropod pests. *Ann. Rev. Ent.,* **7**:367–386.

Wardle, Robert A., 1929. *The problems of applied entomology.* New York, McGraw-Hill Book Co. 587 pp.

Index

(In a series of entries, important references are indicated by **boldface** type)